AF389229

Advances in Direct Injection Reciprocating Internal Combustion Engines

Advances in Direct Injection Reciprocating Internal Combustion Engines

Selected Articles Published by MDPI

MDPI • Basel • Beijing • Wuhan • Barcelona • Belgrade • Manchester • Tokyo • Cluj • Tianjin

This is a reprint of articles published online by the open access publisher MDPI from 2017 to 2019 (available at: http://www.mdpi.com). The responsibility for the book's title and preface lies with José Ramón Serrano, Ricardo Novella and Pedro Piqueras, who compiled this selection.

For citation purposes, cite each article independently as indicated on the article page online and as indicated below:

LastName, A.A.; LastName, B.B.; LastName, C.C. Article Title. *Journal Name* **Year**, *Article Number*, Page Range.

ISBN 978-3-03936-634-7 (Pbk)
ISBN 978-3-03936-635-4 (PDF)

Cover image courtesy of José Ramón Serrano.

Contents

Editorial

Why the Development of Internal Combustion Engines Is Still Necessary to Fight against Global Climate Change from the Perspective of Transportation

José Ramón Serrano *, Ricardo Novella and Pedro Piqueras

CMT—Motores Térmicos, Universitat Politècnica de València, 46022 València, Spain; rinoro@mot.upv.es (R.N.); pedpicab@mot.upv.es (P.P.)
* Correspondence: jrserran@mot.upv.es

Received: 26 September 2019; Accepted: 4 October 2019; Published: 29 October 2019

Internal combustion engines (ICE) are the main propulsion systems in road transport. In mid-2017, Serrano [1] referred to the impossibility of replacing them as the power plant in most vehicles. Nowadays, this statement is true even when considering the best growth scenario for all-electric and hybrid vehicles. The arguments supporting this position consider the growing demand for transport, the strong development of cleaner and more efficient ICEs [2,3], the availability of fossil fuels, and the high energy density of said conventional fuels. Overall, there seems to be strong arguments to support the medium-long-term viability of ICEs as the predominant power plant for road transport applications. However, the situation has changed dramatically in the last few years. The media and other market players are claiming the death of ICEs in the mid-term [4]. Politicians from several G7 countries, such as France, Spain, and the United Kingdom, have announced the prohibition of ICEs in their markets [5], in some cases, as early as 2040. Large cities, such London, Paris, Madrid, and Berlin, are also considering severe limits to ICE-powered vehicles. What is the analysis that can be made from this new situation?

1. What Is the Problem with ICE (Internal Combustion Engines)?

The media's arguments against ICEs range from the need to reduce CO_2 emissions (global warming) to the need to improve the air quality in cities (NOx and particulate matter emissions).

Much of this debate about the future of ICEs has arisen from the Dieselgate scandal [6,7]. A horrible wrong decision from a management and engineering point of view at a specific time and place has generated a worldwide butterfly effect in the automotive industry. However, making the problem a virtue, Dieselgate has forced new regulations to obtain much more efficient and cleaner ICEs [8–11].

As commonly takes place, old and lax pollutant regulations have now resulted in a pendular effect toward radically contrary positions, delighting the media and generating excessive political reactions without a clear scientific basis. All this is reflected in the look to publish a sufficiently popular or good news novelty. We could define the situation as energy populism. Although new regulations that force ICE technology to be more environmentally friendly must always be welcome, prohibitions motivated by a poor diagnosis of the situation will not help at all, neither to improve air quality nor to mitigate global warming.

2. What Is the Problem with Electric Vehicles?

What should be the alternative to the current ICE in the mid-to-long term? Combining the pendulum effect of public opinion with the excellent marketing of new actors in the passenger car sector, a confusion cocktail is served for the media. After all, one might ask if the use of the conventional propulsive systems over 120 years was the right path. How can such an old concept be innovative?

How can the ICE be great and technologically advanced at burning fossil fuels? An easy but wrong conclusion comes without the need of reflection: Let us welcome "new electric motors and batteries" in zero-emission cars!

The bad news is that energy is neither created nor destroyed, only transformed. Electric motors and batteries are not new, nor are they clean and, in general, are not free from problems. One can directly identify two relevant problems.

The first problem is that vehicle propulsion involves energy transformations and the electric motor does not use a primary energy source but an energy vector. Although public opinion has a clear idea of how some processes like friction can negatively affect transport applications, the understanding of the impact of the second law of thermodynamics is limited. The problem is that electricity must be produced, most usually from non-renewable energy sources, which equals around 60% in energy losses, and then transported, which adds 20% of additional losses. Unfortunately, renewable sources are barely 10% of the global energy mix, as observed in Figure 1 [12] without a medium-term forecast of significant increase.

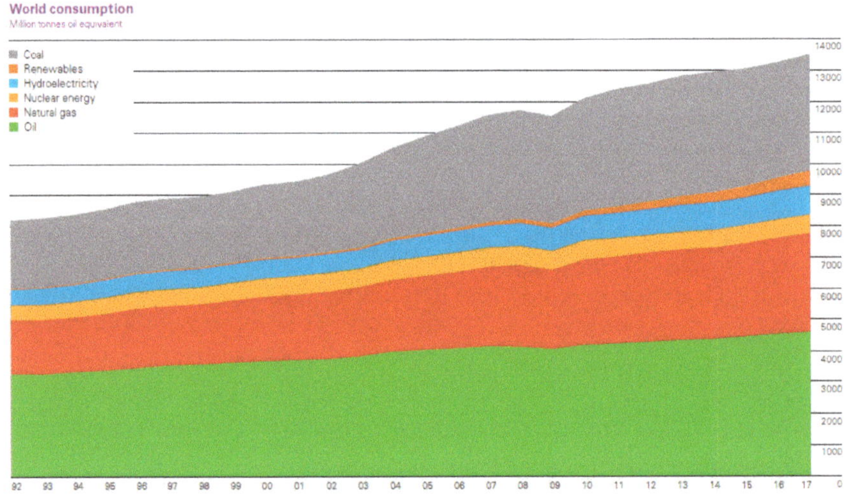

Figure 1. Evolution of world energy consumption by origin during the last 25 years [12].

In some countries like the USA, China, Russia, Poland, South Korea, or Germany, fossil fuels, including a good percentage of coal, remain the largest source of energy as a raw material for electricity production. In a first approach, the only G8 country with real alternatives to CO_2-emitting technologies is France due to its continued commitment with nuclear energy. Therefore, with the current energy mix and with an analysis of the complete life cycle, the so-called analysis from the cradle to the grave, the alternative to electric motors will not eliminate global CO_2 emissions.

On this concern, Figure 2 [13] which takes the data from the cradle to the grave analysis elaborated by the JEC—Joint Research Centre-EUCAR-CONCAWE collaboration [13] effectively shows how with the European electricity production mix the shift to battery electric vehicles (BEVs) would reduce but not remove CO_2 emissions. The reduction of the EU electricity mix is estimated as 40 gCO_2/km (from 210 to 170 gCO_2/km) in a total shift from ICEs to BEVs. However, the European Union reaches 35% of the mix between renewable and hydraulic energy sources [12], while worldwide, it is just 10% (Figure 1). If one thinks that CO_2 emission is a global problem, energy policies on this regard cannot be acceptable being regional.

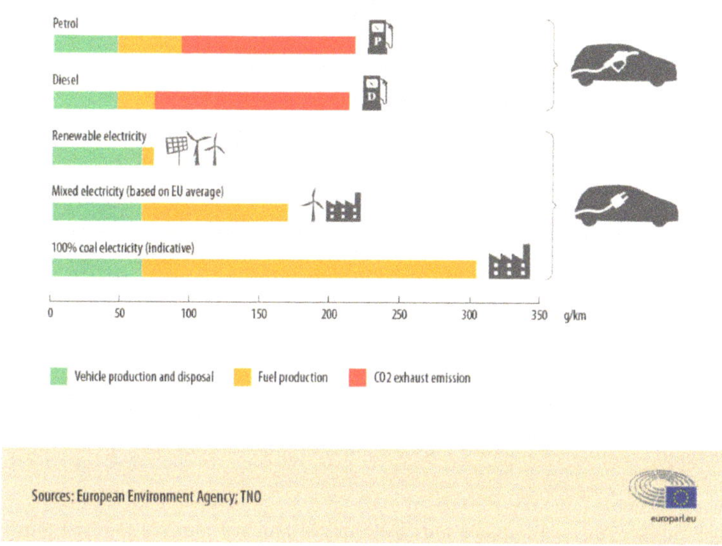

Figure 2. Life-cycle CO_2 emissions as a function of the energy source [13] with data from [14].

More recently, in April 2019, the international media echoed a recent study performed by the German IFO (Institute Center for Economic Studies, CESifo GmbH) conducted by Sinn et al. [15], who calculated that a Tesla Class 3 emits from 156 to 180 gCO_2/km during its lifetime with the German energy mix. This result in CO_2 emission ranges from 11% to 28% more than the modern Diesel E6d Temp engines. In addition, a life cycle analysis of the full electrification of road transport shows that the gaseous emissions would only be relocated from cities to the surroundings of large thermal power plants and manufacturing centers, as pointed out by Messagie [16]. Unfortunately, global warming cannot be relocated and atmospheric phenomena do not know the boundaries, as acid rain and clouds of particulate material (PM 2.5) have repeatedly demonstrated, as shown in Figure 3 [17]. In summary, for the combination of a massive electrification of road transport and the current global energy mix, the maximum benefit is a relocation of the emitted CO_2. As no substantial changes are anticipated in the current electric mix until 2030, the electrification of transport as a clear solution to the problem of climate change should be postponed [17].

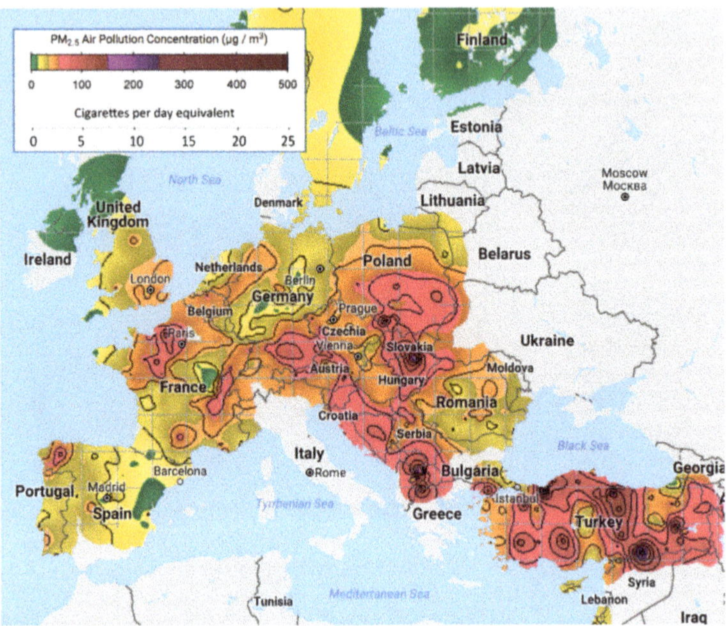

Figure 3. European PM2.5 levels. From [17].

The second problem with electric vehicles comes from the need of electricity storage. In a simple basic way, electricity must be generated as it is consumed. Of course, one can resort to batteries as the electricity storage solution although not in a significant amount for road transport applications. Like the ICE, batteries are an old well-known concept. In addition, batteries also involve harmful chemical compounds. Despite progressive improvements, batteries are a totally immature technology in the range of power and energy required for most road transport applications, where there is no competitor against successfully flexible liquid fuels [18]. There are four challenges that the development of batteries must deal with:

- The charging time of the battery is unacceptably long for many users [19].
- The energy density is unacceptably low with real autonomies below 250 km in compact vehicles [20] and around 300 km in sport urban vehicles (SUVs) [21].
- The lifetime of the batteries is limited and less than the vehicle life. Several studies [22,23] show this fact and discuss the risks and costs associated with their recycling or disposal in a proper way.
- The supply of raw materials for manufacturing such as nickel, lithium, cobalt, copper, and manganese among others is an emergent obstacle as they are reaching high prices quickly and gaining in importance on geopolitical strategies. According to Sarah Maryssael, global manager of metal supplies for Tesla [24], the main problem is currently the supply of cobalt, which is necessary for the cathode of lithium-ion batteries; a Tesla Model X needs 7 kg per vehicle and a Tesla Model 3 about 4.5 kg [25]. This mineral is extracted mainly from the Democratic Republic of Congo, where human rights are violated through child labor and mines stand out by their poor safety conditions, among others [26]. Then, cobalt reaches the international markets and its origin is diluted due to the low traceability of the production chain. Finally, it is fundamentally processed in China, what exemplifies the potential of this technology for further economic stress and uncertainties. What would be the cost of these materials refined in western countries with EU security, environmental, and health standards? It would probably be exorbitant.

3. What Can the New Generation of ICE Provide?

Limitations to greenhouse gases (CO_2), gaseous pollutants, and noise emissions will be increasingly severe, forcing the automotive industry to invest in more innovative technologies for their reduction [10,27–30]. Real driving emissions tests are being adopted in the major global economic zones as this strategy expands the ICE operational range in which the pollutant emissions must be kept below the approval limits [31–33]. A revolution is approaching with respect to traditional gasoline and diesel engines making the boundaries between them disappear as deeper knowledge and greater control of the combustion process is acquired [34]. Advanced injection systems [35,36], turbochargers [37,38], organic Rankine cycles (ORC) [39], hybridization [40,41], multifuel solutions [42–44], or advanced combustion concepts are becoming a part of the ICE context. All these strategies are dedicated to extract every Joule of energy from the fuel. The research on aftertreatment systems based on monolithic reactors offers interesting possibilities to effectively clean the exhaust gases to incredible limits [45]. Nowadays, the automotive industry does not find anything too innovative or risky to meet the expected, medium-term demand for cleaner and more efficient ICE. Finally, fossil fuels are cheap and available. Oil depletion is no longer a topic of discussion as fracking technology has offered a new paradigm, leading the USA to the largest producer of fossil fuels in the world [12].

4. What Is Improving the Expectations in the New-Generation ICE?

ICE emits particulate matter, gaseous pollutants, and CO_2 locally. This is accepted as it is accepted that electric cars do not. Assuming both particulars are great arguments for the replacement of ICEs, what would happen if the situation were somehow the opposite? In a life cycle analysis, neither the production of the batteries nor the production of electricity is free of CO_2 emissions and pollutants [15]. The generation of electricity causes CO_2 emissions much greater than the synthesis of liquid fossil fuels, as shown in Figure 2, as it is an energy vector more difficult to obtain and transport. It can be similarly stated that the manufacturing of ICE generates CO_2 emissions, although less than in the case of batteries and electric motors [46,47], as also shown in Figure 2.

What can ICE do to increase air quality? We can affirm that modern Euro 6d Temp Diesel engines can clean the air from particulate and smog in heavily polluted areas, such as the situations referred in China [48]. The particulate filters of modern internal combustion engines reduce the level of PM10 below the mean atmospheric value, as shown in Figure 4 [49]. If one combines data in Figures 3 and 4, the advantages of ICE with particulate filters in countries such as Poland, where almost 50% of their energy mix depends exclusively on coal [12], are evident.

The technology is available, and the research is driven to allow the next generation of ICEs to act as air pollutant cleaners in large cities, whose source of pollution is not only road traffic of old ICEs. This is something that electric motors with batteries cannot do. The new Diesel Euro 6d Temp is emitting 80% less NOx than stipulated by the standard. This means that they are cleaning the air of emissions coming from other sources [50]. Effective energy policies are needed to renew transport fleets around the world, as justified in [51] for the Europe case, and promote in all countries the same emission standards for ICE as in the United States, Japan, or Europe. The discussion should not focus on the type of technology but on having in the streets the most modern versions of it.

Another important fact concerning ICEs is that the contribution of transport to global emissions of GWPs (Global Warming Potential) has historically remained at 11%. As shown in Figure 5 elaborated from the United Nations Food and Agriculture Organization (FAO) data [52], industry, agriculture, resource extraction, waste processing, and residential and commercial consumption do the rest. Therefore, a worldwide massive change to electric vehicles would mean a potential worldwide reduction of 11% of the equivalent tons of CO_2 that is emitted under the assumption of using fully CO_2-free energy sources for the BEVs batteries charge.

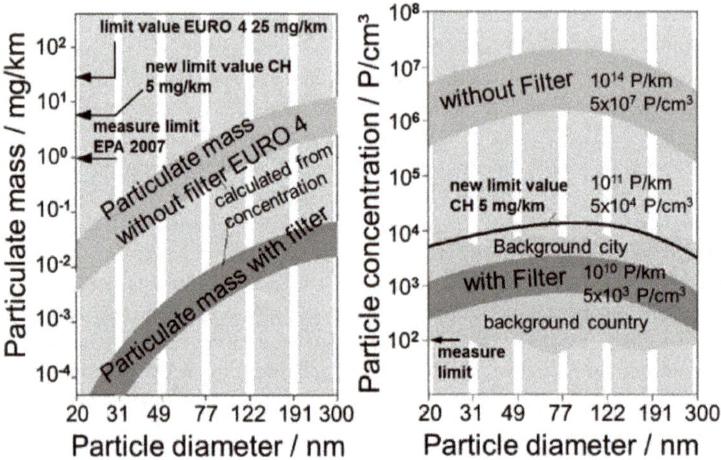

Figure 4. Diesel internal combustion engine (ICE) equipped with particulate filters as air cleaners in urban areas [49].

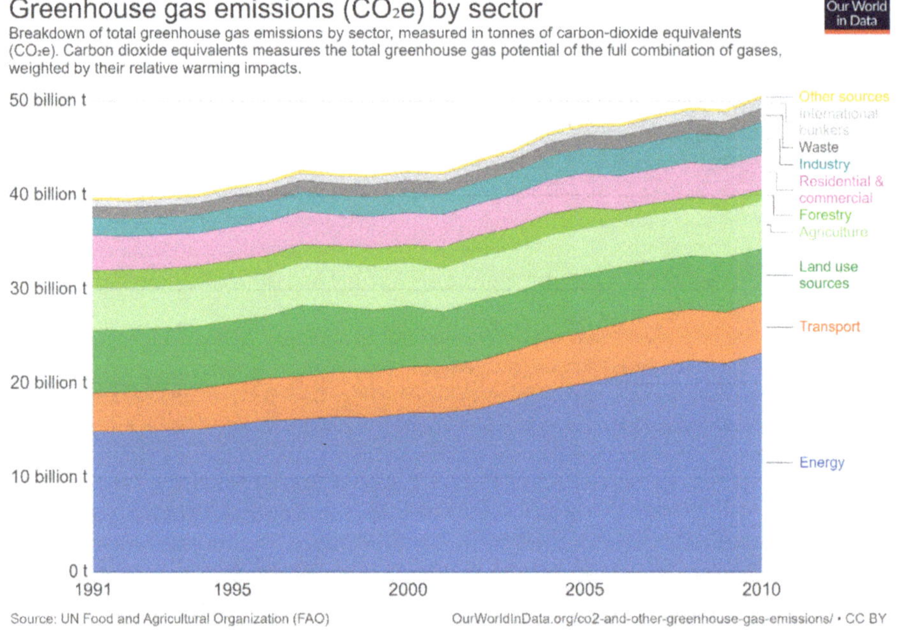

Figure 5. Breakdown of total greenhouse gas emissions by sector, measured in tons of carbon-dioxide equivalents [52].

However, only 10% of world energy consumption is free of CO_2 [12], which means that in best of cases, the reduction would be 10% of 11%. Even the previously calculated 1.1% is not fully reachable in a life cycle analysis, as shown in Figure 2. In the very long term, it can be argued that electric cars will substantially lower their CO_2 emissions if electricity comes exclusively from renewable or nuclear sources. As can be seen in Figure 6, despite the large dispersion of data in the sources [13,15,16,18], this scenario is far from being fulfilled today in most of the European Union countries. Considering countries like Germany or Spain with around 35% of renewable sources in the mix, the average equivalent CO_2 emissions are slightly better than 2019 Diesels E6d Temp. If we make an extrapolation to the future, in the central scenario, we would need to increase the mix of renewables by over 60% to have the same competitive advantage over combustion technologies based on compression ignition (CI) in CO_2 emissions [15,18]. Even if we reach 100% renewable, electric vehicles would never have zero CO_2-equivalent emissions if one considers the life cycle and not only local use.

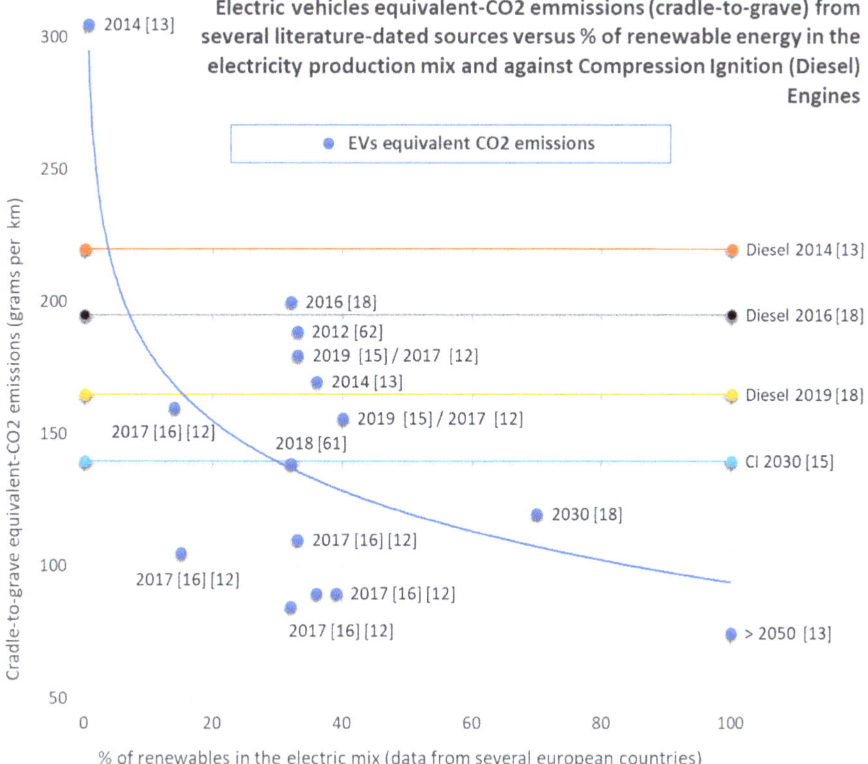

Figure 6. Equivalent life-cycle CO_2 emissions of electric vehicles as a function of the percentage of renewables in the electric energy production mix. Comparison with compression ignition (CI) or diesel engines. Elaborated from the references of this publication.

If one imagines that 60% of renewable sources in the energy mix were the medium-long term standard, could the ICE do so well? The answer is yes and even better if we use synthetic fuels from the capture and use of atmospheric CO_2 (CCU) [53]. There are already several R&D projects in Switzerland, Germany, and Canada focused on CCU. These are systems capable of transforming CO_2 taken directly from the air into liquid fuels called 'PtX fuels' (e-fuels, including e-Diesel). This is done by hydrogenation of CO_2 using H2 produced by electrolysis from renewable sources [54]. There are also projects to pump the captured CO_2 from the power plants to the oil wells and subsequently convert it into neutral oil from the CO_2 point of view. Other studies approach CO_2-capturing vehicles, both its own CO_2 emission and atmospheric CO_2, to generate CO_2 neutral fuel on board [55]. That way, the self-CCU could even contribute to a reduction in atmospheric CO_2. If the fuels used in these CO_2 capture cars were mostly biofuels [56], as is the case in Brazil, this would represent an efficient way to remove CO_2 from the atmosphere. However, the development of this technology is subject to the efficient recovery of exhaust gas energy from ICEs [57]. In conclusion, if a paradigm change is required, then vehicles acting as CO_2 captors to create a CO_2 circular economy may arise as the most efficient solution. They would eliminate the other 90% of CO_2 that transport is not producing [57]. This is an opportunity that BEVs cannot offer.

Public funding and government efforts should promote research to reduce polluting emissions, rather than to choose winners for an uncertain future. Direct subsidies to any industry or technology and the banning of others, without enough scientifically proved arguments, is the type of risk exercise that has never been successful. It seems that the European authorities have finally begun to listen to scientists and engineers, which are claiming for the cleaning potential of cities air contamination by the last-generation ICEs depollution systems [58,59], and that explain the facts of the situation [60–62]. The German Bundestag in May 2019 proposed that Euro 6d Temp diesel engines cannot be banned in German cities, not even the Euro 4 and Euro 5 when they emit less than 270 mgNOx/km (retrofit) that is pending being certified by the German Supreme Court [63]. In France, it is being studied to give the maximum environmental rating to Euro 6d Temp Diesel engines after finding that they are as much or cleaner than gasoline engines [64]. In general, promoting research activities of any technology, regardless of the field of research, has always provided great benefits for future generations, and has usually been the cheapest path for the society to progress.

Conflicts of Interest: The authors declare no conflict of interest.

References

1. Serrano, J.R. Imagining the Future of the Internal Combustion Engine for Ground Transport in the Current Context. *Appl. Sci.* **2017**, *7*, 1001. [CrossRef]
2. Ding, Y.; Sui, C.; Li, J. An Experimental Investigation into Combustion Fitting in a Direct Injection Marine Diesel Engine. *Appl. Sci.* **2018**, *8*, 2489. [CrossRef]
3. Nguyen, D.V.; Duy, V.N. Numerical Analysis of the Forces on the Components of a Direct Diesel Engine. *Appl. Sci.* **2018**, *8*, 761. [CrossRef]
4. EL MUNDO. España pretende prohibir las matriculaciones de coches diésel, gasolina e híbridos a partir de 2040. *El Mundo*. 2018. Available online: https://www.elmundo.es/motor/2018/11/13/5beab545e2704eb15b8b45ec.html (accessed on 15 October 2019).
5. Financial Times. Dyson Presses UK Government for Earlier Petrol Car Ban. *Financial Times*. 2019. Available online: https://www.ft.com/content/9b078162-7195-11e9-bf5c-6eeb837566c5 (accessed on 15 October 2019).
6. Brand, C. Beyond 'Dieselgate': Implications of unaccounted and future air pollutant emissions and energy use for cars in the United Kingdom. *Energy Policy* **2016**, *97*, 1–12. [CrossRef]
7. Dey, S.; Caulfield, B.; Ghosh, B. The potential health, financial and environmental impacts of Dieselgate in Ireland. *Transp. Plan. Technol.* **2018**, *41*, 17–36. [CrossRef]
8. Normativas de Emisiones Contaminantes en Europa (Versión Completa). Available online: https://www.dieselnet.com/standards/eu/ld.php#stds (accessed on 15 September 2019).
9. Zhang, M.; Zhong, J.; Capelli, S.; Lubrano, L. Particulate Matter Emission Suppression Strategies in a Turbocharged Gasoline Direct-Injection Engine. *J. Eng. Gas Turbines Power Trans. ASME* **2017**, *139*, 102801. [CrossRef]
10. Payri, R.; De La Morena, J.; Monsalve-Serrano, J.; Pesce, F.C. Alberto VassalloImpact of counter-bore nozzle on the combustion process and exhaust emissions for light-duty Diesel engine application. *Int. J. Engine Res.* **2019**, *20*, 46–57. [CrossRef]
11. Lapuerta, M.; Ramos, Á.; Fernández-Rodríguez, D.; González-García, I. High-pressure versus low-pressure exhaust gas recirculation in a Euro 6 Diesel engine with lean-NOx trap: Effectiveness to reduce NOx emissions. *Int. J. Engine Res.* **2019**, *20*, 155–163. [CrossRef]
12. BP Statistical Review of World Energy. June 2018. Available online: https://www.bp.com/en/global/corporate/energy-economics/statistical-review-of-world-energy.html (accessed on 15 September 2019).

13. European Environment Agency. 2019. Available online: http://www.europarl.europa.eu/news/es/headlines/society/20190313STO31218/emisiones-de-co2-de-los-coches-hechos-y-cifras-infografia (accessed on 15 September 2019).

14. Dwards, R.; Hass, H.; Larivé, J.-F.; Lonza, L.; Maas, H.; Rickeard, D. *Well-to-Wheels Analysis of Future Automotive Fuels and Powertrains in the European Context*; Report EUR 26236 EN. JRC Technical Reports; European Commission. Publications Office of the European Union: Brussels, Belgium, 2014. [CrossRef]

15. Christoph, B.; Hans-Dieter, K.; Sinn, H.W. Kohlemotoren, Windmotoren und Dieselmotoren: Was zeigt die CO_2-Bilanz? *IFO Schnelld.* **2019**, *72*, 40–54.

16. Messagie, M. Life Cycle Analysis of the Climate Impact of Electric Vehicles. *Transp. Environ.* **2017**. Available online: https://www.transportenvironment.org/publications/electric-vehicle-life-cycle-analysis-and-raw-material-availability (accessed on 15 September 2019).

17. Available online: http://berkeleyearth.org/wp-content/uploads/2017/01/Europe-air-pollution.png (accessed on 15 September 2019).

18. Blaich, M. El Motor Diésel y la Conflictiva Tendencia a la Electrificación. *Interempresas.net. Automoción.* **2019**. Available online: http://www.interempresas.net/Sector-Automocion/Articulos/232843-El-motor-diesel-y-la-conflictiva-tendencia-de-la-electrificacion.html (accessed on 15 September 2019).

19. Neaimeh, M.; Salisbury, S.D.; Hill, G.A.; Blythe, P.T.; Scoffield, D.R.; Francfort, J.E. Analysing the usage and evidencing the importance of fast chargers for the adoption of battery electric vehicles. *Energy Policy* **2017**, *108*, 474–486. [CrossRef]

20. Los Coches Eléctricos y su Autonomía Limitada. Available online: https://www.ocu.org/coches/coches/noticias/autonomia-coches-electricos# (accessed on 15 October 2019).

21. Autobahn Test: Tesla Model X Beats Audi e-tron & Jaguar I-Pace; Nextmove GmbH. Available online: https://nextmove.de/autobahn-test-tesla-model-x-beats-audi-e-tron-jaguar-i-pace/ (accessed on 15 September 2019).

22. Tang, L.; Rizzoni, G.; Cordoba-Arenas, A. Battery Life Extending Charging Strategy for Plug-in Hybrid Electric Vehicles and Battery Electric Vehicles. *IFAC Pap. Online* **2016**, *49*, 70–76. [CrossRef]

23. Blooma, I.; Colea, B.W.; Sohna, J.J.; Jonesa, S.A.; Polzina, E.G.; Battagliaa, V.S.; Henriksena, G.L.; Motlochb, C.; Richardsonb, R.; Unkelhaeuserc, T.; et al. An accelerated calendar and cycle life study of Li-ion cells. *J. Power Sources* **2002**, *101*, 238–247. [CrossRef]

24. García, F. *Alarma ante la Posible Escasez de Baterías Para Vehículos Eléctricos*; El Mundo: Madrid, Spain, 2019; Available online: https://www.elmundo.es/motor/2019/05/08/5cd2ea68fdddff87418b4576.html (accessed on 15 October 2019).

25. Fuentes, V. Tesla Prevé una Escasez Mundial de Minerales que son Clave Para Fabricar las Baterías de los Coches Eléctricos. Available online: https://www.motorpasion.com/tesla/tesla-preve-escasez-mundial-minerales-que-clave-para-fabricar-baterias-coches-electricos (accessed on 15 October 2019).

26. Broom, D. *The Dirty Secret of Electric Vehicles; Formative Content*; World Economic Forum: Geneva, Switzerland, 2019; Available online: https://www.weforum.org/agenda/2019/03/the-dirty-secret-of-electric-vehicles/ (accessed on 15 September 2019).

27. Boccardo, G.; Millo, F.; Piano, A.; Arnone, L.; Manelli, S.; Fagg, S.; Gatti, P.; Herrmann, O.E.; Queck, D.; Weber, J. Experimental investigation on a 3000 bar fuel injection system for a SCR-free non-road Diesel engine. *Fuel* **2019**, *243*, 342–351. [CrossRef]

28. Puskar, M.; Kopas, M. System based on thermal control of the HCCI technology developed for reduction of the vehicle NOX emissions in order to fulfil the future standard Euro 7. *Sci. Total Environ.* **2018**, *643*, 674–680. [CrossRef]

29. Marcin, N. Selected Issues of the Indicating Measurements in a Spark Ignition Engine with an Additional Expansion Process. *Appl. Sci.* **2017**, *7*, 295.

30. Benajes, J.; Novella, R.; De Lima, D.; Tribotté, P. Analysis of combustion concepts in a newly designed two-stroke high-speed direct injection compression ignition engine. *Int. J. Engine Res.* **2015**, *16*, 52–67. [CrossRef]

31. Luján, J.M.; Bermúdez, V.; Dolz, V.; Monsalve-Serrano, J. An assessment of the real-world driving gaseous emissions from a Euro 6 light-duty Diesel vehicle using a portable emissions measurement system (PEMS). *Atmos. Environ.* **2018**, *174*, 112–121. [CrossRef]

32. Grigoratos, T.; Fontaras, G.; Giechaskiel, B.; Zacharof, N. Real world emissions performance of heavy-duty Euro VI Diesel vehicles. *Atmos. Environ.* **2019**, *201*, 348–359. [CrossRef]

33. Green Car Congress. ADAC Testing Finds New Diesel Cars Cleaner than Required; Euro 6c and 6d-Temp Vehicles Well below the Permissible NOx Limits. Available online: https://www.greencarcongress.com/2019/02/201902-22-adac.html (accessed on 15 September 2019).

34. Serrano, J.R.; Novella, R.; Gomez-Soriano, J.; Martinez-Hernándiz, P.J. Computational methodology for knocking combustion analysis in compression ignited advanced concepts. *Appl. Sci.* **2018**, *8*, 1707. [CrossRef]

35. Chiatti, G.; Chiavola, O.; Frezzolini, P.; Palmieri, F. On the link between diesel spray asymmetry and off-axis needle displacement. *Appl. Sci.* **2017**, *7*, 375. [CrossRef]

36. Sangik, H.; Juhwan, H.; Jinwook, L. A Study on the Optimal Actuation Structure Design of a Direct Needle-Driven Piezo Injector for a CRDi Engine. *Appl. Sci.* **2017**, *7*, 320.

37. Dimitriou, P.; Burke, R.; Zhang, Q.; Copeland, C.; Stoffels, H. Electric turbocharging for energy regeneration and increased efficiency at real driving conditions. *Appl. Sci.* **2017**, *7*, 350. [CrossRef]

38. Serrano, J.R.; Arnau, F.J.; Dolz, V.; Tiseira, A.; Lejeune, M.; Auffret, N. Analysis of the capabilities of a two-stage turbocharging system to fulfil the US2007 anti-pollution directive for heavy duty diesel engines. *Int. J. Automot. Technol.* **2008**, *9*, 277–288. [CrossRef]

39. Fernández-Yáñez, P.; Armas, O.; Gómez, A.; Gil, A. Developing computational fluid dynamics (CFD) models to evaluate available energy in exhaust systems of diesel light-duty vehicles. *Appl. Sci.* **2017**, *7*, 590.

40. Huang, Y.; Surawski, N.C.; Organ, B.; Zhou, J.L. Fuel consumption and emissions performance under real driving: Comparison between hybrid and conventional vehicles. *Sci. Total Environ.* **2019**, *659*, 275–282. [CrossRef]

41. Andwari, A.M.; Pesiridis, A.; Karvountzis-Kontakiotis, A.; Esfahanian, V. Hybrid electric vehicle performance with Organic Rankine Cycle Waste Heat Recovery system. *Appl. Sci.* **2017**, *7*, 437. [CrossRef]

42. Benajes, J.; García, A.; Monsalve-Serrano, J.; Boronat, V. Dual-fuel combustion for future clean and efficient compression ignition engines. *Appl. Sci.* **2016**, *7*, 36. [CrossRef]

43. Mustafa, A.; Ahmet, I.; Çelik, M. The Impact of Diesel/LPG Dual Fuel on Performance and Emissions in a Single Cylinder Diesel Generator. *Appl. Sci.* **2018**, *8*, 825.

44. Torregrosa, A.J.; Broatch, A.; Novella, R.; Gomez-Soriano, J.; Mónico, L.F. Impact of gasoline and diesel blends on combustion noise and pollutant emissions in premixed charge compression ignition engines. *Energy* **2017**, *137*, 58–68. [CrossRef]

45. Bermúdez, V.; Serrano, J.R.; Piqueras, P.; Sanchis, E. On the impact of particulate matter distribution on pressure drop of wall-flow particulate filters. *Appl. Sci.* **2017**, *7*, 234. [CrossRef]

46. Qiao, Q.; Zhao, F.; Liu, Z.; Jiang, S.; Hao, H. Comparative Study on Life Cycle CO_2 Emissions from the Production of Electric and Conventional Vehicles in China. *Energy Procedia* **2017**, *105*, 3584–3595. [CrossRef]

47. ACEA—The Automobile Industry Pocket Guide 2018–2019. Available online: https://www.acea.be/publications/article/acea-pocket-guide (accessed on 15 September 2019).

48. Kan, H.; Chen, R.; Tong, S. Ambient air pollution, climate change, and population health in China. *Environ. Int.* **2012**, *42*, 10–19. [CrossRef]

49. Mayer, A.; Wyser, M.; Czerwinski, J. Erfahrungen mit Partikelfilter-Nachrüstungen bei Baumaschinen in der Schweiz. In Proceedings of the FAD-Konferenz, Dresden, Germany, 12–13 November 2003.

50. Has the Government Got It Wrong on 'dirty Diesel' Cars? Tests Show Some BMW, Mercedes and Vauxhall Models Produce almost ZERO Harmful NOx Emissions. Available online: https://www.thisismoney.co.uk/money/cars/article-6733271/Are-diesel-cars-really-dirty-Tests-reveal-models-produce-zero-NOx-emissions.html (accessed on 26 September 2019).

51. Serrano, J.R.; Piqueras, P.; Abbad, A.; Tabet, R.; Bender, S.; Gomez, J. Impact on Reduction of Pollutant Emissions from Passenger Cars when Replacing Euro 4 with Euro 6d Diesel Engines Considering the Altitude. *Energies.* **2019**, *12*, 1278. [CrossRef]

52. Ritchie, H.; Roser, M. CO_2 and Greenhouse Gas Emissions. Available online: https://ourworldindata.org/co2-and-other-greenhouse-gas-emissions (accessed on 15 September 2019).

53. Cormos, A.M.; Cormos, C.C. Techno-economic evaluations of post-combustion CO_2 capture from sub- and super-critical circulated fluidized bed combustion (CFBC) power plants. *Appl. Therm. Eng.* **2017**, *127*, 106–115. [CrossRef]

54. Kramer, U. Defossilizing the Transportation Sector. Options and requirements for Germany. Fvv Prime Movers. Technologies. 2018. Forschungsvereinigung Verbrennungskraftmaschinen e.V. Lyoner Strasse 18 60528 Frankfurt/ M., Germany. www.fvv-net.de/en. Available online:

https://www.fvv-net.de/fileadmin/user_upload/medien/materialien/FVV_Future_Fuels_Study_report_Defossilizing_the_transportation_sector_R586_final_v.3_2019-06-14__EN.pdf (accessed on 15 October 2019).

55. Hamad, E.Z.; Al-Sadat, W.I. Apparatus and Method for Oxy-Combustion or Fuels in Internal Combustion Engines. U.S. Patent 2013/0247886 A1, 26 September 2013.

56. Sun, H.; Wang, W.; Koo, K.P. The practical implementation of methanol as a clean and efficient alternative fuel for automotive vehicles. *Int. J. Engine Res.* **2019**, *20*, 350–358. [CrossRef]

57. Desantes, J.M.; Benajes, J.; Serrano, J.R.; Arnau, F.; Garcia-Cuevas, L.M.; Serra, J.M.; Catalán, D. *Motor de Combustión Interna*; OEPM Madrid: Madrid, Spain, 2019; 201930285.

58. Johnson, T.; Joshi, A. Review of Vehicle Engine Efficiency and Emissions. *Sae Tech. Pap.* **2018**. [CrossRef]

59. Nieuwste Diesels Reinigen de Lucht. Available online: https://autonieuws.be/uitlaat/4756-nieuwste-diesels-reinigen-de-lucht (accessed on 15 September 2019).

60. Wissenschaftliche Gesellschaft für Kraftfahrzeug-und Motorentechnik e.V. *The Future of the Combustion Engine/Assessment of the Diesel Engine Situation*; WKM: Berlin, Germany, 2017.

61. Aparicio, F.; Casanova, J.; López, J.M.; Payri, F.; Tinaut, F.V.; Wolff, G. *El Automóvil en la Movilidad Sostenible*; Informe ASEPA; ASEPA: Madrid, Spain, 2018.

62. Troy, R.; Singh, H.B.; Majeau-Bettez, G.; Strømman, A.H. Comparative Environmental Life Cycle Assessment of Conventional and Electric Vehicles. *J. Ind. Ecol.* **2012**, *17*, 53–64. [CrossRef]

63. Diesel-PKW Dürfen Nach Erfolgreicher Hardware-Nachrüstung Weiter Einfahren. Available online: https://www.bmu.de/pressemitteilung/bundestag-beschliesst-einheitliche-regeln-fuer-umgang-mit-verkehrsverboten/ (accessed on 15 October 2019).

64. Euro 6D-Temp Diesel Like Petrol. France Tries to Adapt the Anti-Pollution Stamps. Available online: https://www.diesel-international.com/automotive/france-euro-6d-temp-diesel/ (accessed on 15 October 2019).

Article

Numerical Analysis of the Forces on the Components of a Direct Diesel Engine

Dung Viet Nguyen and Vinh Nguyen Duy *

Modelling and Simulation Institute—Viettel Research and Development Institute, 100000 Hanoi, Vietnam; dungnv98@viettel.com.vn
* Correspondence: vinhnd12@viettel.com.vn; Tel.: +84-985-814-118

Received: 16 April 2018; Accepted: 8 May 2018; Published: 11 May 2018

Abstract: This research introduces a method to model the operation of internal combustion engines in order to analyze the forces on the rod, crankshaft, and piston of the test engine. To complete this research, an experiment was conducted to measure the in-cylinder pressure profile. In addition, this research also modelled the friction forces caused by the piston and piston-ring movements inside the cylinder for calculating the net forces experienced by the test engine. The results showed that the net forces change according to the crank angle and reach a maximum value near the top dead center. Consequently, we needed to concentrate on analyzing the stress of the crankshaft, rod, and piston at these positions. The research results are the foundation for optimizing the design of these components and provide a method for extending the operating lifetime of internal combustion engines in real operating experiments.

Keywords: internal combustion engine; mechanical stress; fine element method; friction force; in-cylinder pressure

1. Introduction

In the operation of internal combustion engines, the rod, crankshaft, and piston play crucial roles as they are considered the heart of engines. However, these components always operate in critical operating conditions, such as under very high thermal and mechanical stresses [1–5]. However, defining these loads is difficult due to complications in the constructive and operating characteristics of these above-mentioned parts. Consequently, classical methods have been used. In most cases, a highly simplified representation of geometric shape and load are used for analysis. As a result, the operating characteristics of the internal combustion engines cannot be correctly obtained. The finite element method (FEA) is a good solution to solve these problems. The biggest advantage of the FEA is the simplicity of its basic concepts. Users must learn and correctly understand these concepts, since they include certain hypotheses, simplifications, and generalizations [6,7]. To calculate objects, users have to model objects so that they align with real subjects. A model includes lines, planes, or curved surfaces and volumes created in a three-dimensional (3D) computer-aided design (CAD) environment. In this stage of development, the model is continuous with an infinite number of points, as the real pieces are analyzed. The main goal of the finite element method is to obtain the finite element mesh, transforming the continuous structure into a discrete model with a finite number of points. Using the finite element method (FEM), users can combine design, simulation, element mesh, and structure analysis. Therefore, users can easily and quickly apply this tool [8,9].

For example, Webster et al. [10] conducted a 3D analysis based on the FEA method to calculate the mechanical stress of a diesel engine connecting rod. They conducted an experiment to measure the maximum compressive load and the load distribution on the piston and crankshaft. However, the connecting rod caps and bolt pretension were modelled based on the FEM method and multi-point constraint equations. In another study [11], the 3D model of a passenger car's crankshaft was built using CATIA V5 software (Dassault Systèmes, Vélizy-Villacoublay, France) and transferred to ANSYS software (ANSYS, Inc., Cecil Township, PA, USA) for stress analysis to provide a conceptual method for optimizing the design using weight reduction. However, this research did not fully mention the net force on the piston, such as the friction forces or the role of gravity. In addition, this study only analyzed a single crankpin and ignored the interaction between other components. Lucjan et al. also performed a failure analysis of the crankshaft of a diesel engine in the experiment by observing the fatigue failure and determined the operational stress of the crankshaft at maximum engine power. They stated that the maximum tensile stress value reached 112.44 MPa around the fillet area of the crankpin at a crankshaft position of $10°$ at top dead center (TDC). However, this critical region stress only accounts for about 25% of the yield stress of the crankshaft material [12].

Normally, the maximum stresses on the components are designed to be significantly smaller than the yield stress; however, in real operation, some critical conditions occur, such as the overload of the engine or the failure of the lubricant system resulting in the destruction of engine parts. Notably, some engines may be modified from the original design to adapt to operation demands. For instance, in a previous study [13], an old model diesel engine was modified by retrofitting a turbocharger to enhance its performance. However, as mentioned in prior studies [13,14], the thermal and mechanical stresses of engines after turbocharging dramatically increased. Thus, evaluating the stress increase is necessary to find a solution for the stable and longtime operation of these engines.

In this research, we present a method to calculate the mechanical stress on some major components of the test engine based on the modeling of the test engine combining the FEM method with experimental data. The main objective of this study was to determine the stress state in the crankshaft, rod, and piston during operation. Furthermore, we aimed to explain the reasons for the failure of these components to determine a solution to extend their operating lifetime.

2. Experimental Procedure to Measure In-Cylinder Pressure

The test engine was a four-cylinder diesel engine with a maximum power of 110 kW after turbocharging. The engine characteristics are described in our previous paper [13]. The testing equipment used for measuring the in-cylinder pressure included a chassis dynamometer, a fuel-consumption measurement device, and a pressure measurement system. All these devices were connected to a server computer using dedicated software. In this research, the in-cylinder pressure was measured at the maximum power conditions. This means that the test engine was controlled to reach maximum power, and thus the in-cylinder was recorded by a pressure sensor installed in the head cylinder. The chassis dynamometer contained an absorption unit to measure the torque and rotational speed. The engine power and engine speed could be varied and measured in the range of 0 to 220 kW and 0 to 4500 rpm, respectively. The dynamometer was controlled using an installed PUMA computer (AVL, Graz, Austria) that receives signals from sensors equipped on the dynamometer and the test engine. In addition, to ensure the accuracy of the experimental test, we used other equipment, such as an external cooling system, named AVL 533 (AVL, Graz, Austria) that permits continuous temperature control. The quantity of fuel consumed by the engine was measured by fuel-consumption measurement equipment, called Fuel Balance AVL 733S (AVL, Graz, Austria), using the gravimetrical method. The AVL Fuel Balance system allows for high-precision fuel-consumption measurement even at low consumption and for a short measurement period. The recommended measuring range of the equipment is up to 150 kg/h with an accuracy of 0.12%. Figure 1 shows the test engine installed in the test room for the experimental process.

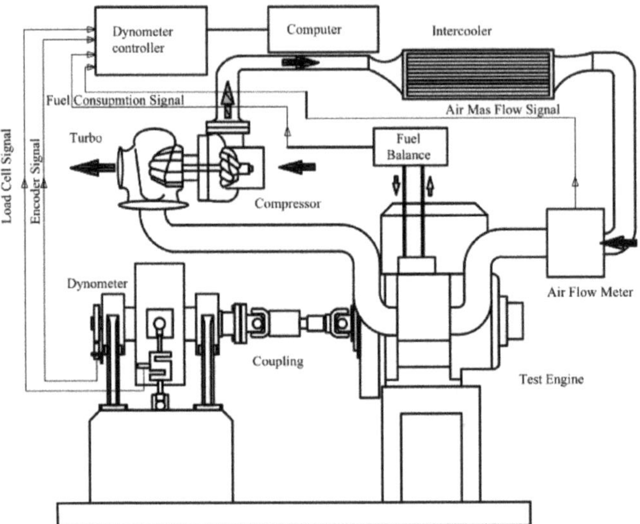

Figure 1. Schematic of the experimental procedure.

3. Kinetics and Dynamics of the Test Engine

A simple model of an internal combustion engine consisting of a piston located in a cylinder that is connected to the crankshaft via the rod and simplified to move in a plane is shown in Figure 2, with symbols $AB = l$ (rod length), $AR = jl$ (centre gravity of connecting rod from crankpin center), $OA = r$ (crankshaft length), and $OC = h$ (crankshaft mass centre from main bearing centre).

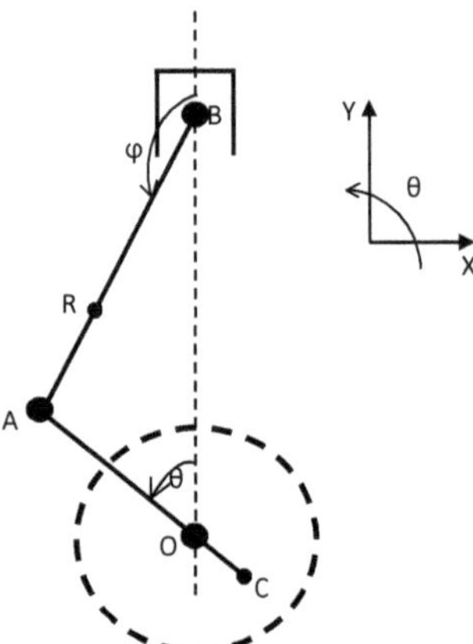

Figure 2. The simple model of piston, crankshaft, and transmission rod.

The pressure p exerts a force on the piston m_1 to create movement along the axis of the transmission cylinder to the transfer bar m_2 of the parallel plane, causing the crankshaft m_3 to rotate around the center O. m_1 is the mass of the piston, m_2 is the mass of the rod, m_3 is the mass of the crankshaft, l_2 is the length of the rod, l_3 is the crankshaft length, p is the pressure on the piston surface, θ is the angle between the crankshaft and the vertical axis, and ϕ is the angle between the rod and the vertical axis (Figure 2).

3.1. In-Cylinder Pressure and Friction Force Calculation

During the operation of internal combustion engines, combustion is an irreversible process that uses the reactions of fuel with an oxidizer to convert chemical energy into sensible energy, which also works to form products. In practice, carbon dioxide (CO_2), nitrogen, and water are the main products of the combustion process. However, other species, including nitric oxide, nitrogen oxide, carbon monoxide, hydrocarbon, and particulate matter, are also common products. The liquid state determination at each point in the process requires a detailed understanding of the intermediate reactions that change the original mixture into the final combustion products. However, in all cases, we can use the first law of thermodynamics to determine the correlation between the beginning and end of a combustion process.

The application of this law does not require users to know the development of the intermediate stages of the process. The first law of thermodynamics expresses the relationship between the variation of the internal energy (or enthalpy) and that of heat and work. When applying this law to a system where the chemical composition changes, we need to determine the zero state of the internal energy or enthalpy of all the substances in the system.

In specific cases, the calculation of combustion in internal combustion engines is based on the equation of the first law of thermodynamics:

$$\frac{d(m_c u)}{d\alpha} = -p_c \frac{dV}{d\alpha} + \frac{dQ_F}{d\alpha} - \sum \frac{dQ_w}{d\alpha} - h_{BB} \frac{dm_{BB}}{d\alpha}, \tag{1}$$

where $\frac{d(m_c u)}{d\alpha}$ is the change in internal energy inside the cylinder, $-p_c \frac{dV}{d\alpha}$ is the work cycle, $\frac{dQ_F}{d\alpha}$ is the heat input, $\sum \frac{dQ_w}{d\alpha}$ is the heat loss through the wall, $h_{BB} \frac{dm_{BB}}{d\alpha}$ is the enthalpy loss due to gas leakage, m_c is the volume of the liquid inside the cylinder, u is the internal energy, p_c is the pressure inside the cylinder, V is the cylinder volume, Q_F is the heat of the fuel supply, Q_W is the heat loss to the wall, α is the angle of the crankshaft rotation, h_{BB} is the enthalpy value, and $\frac{dm_{BB}}{d\alpha}$ is the variation in volumetric flow rate.

Equation (1) applies to both internal and external composite engines. However, the changes in the composition of the two cases are different. In the case of the formation of a mixture inside a cylinder, if we assume that the fuel in the cylinder is burned immediately, the combustion mixture is immediately blended with the residual gas in the cylinder. The air-fuel ratio (A/F) decreases continuously from a high value at the starting point to a low value at the end of the combustion. After the transformation, Equation (1) becomes:

$$\frac{dT_c}{d\alpha} = \frac{1}{m_c \cdot \left(\frac{\partial u}{\partial T} + \frac{\partial u}{\partial p} \cdot \frac{p_c}{T_c} \right)} \left[\frac{dQ_F}{d\alpha} \left(1 - \frac{u_c + \frac{\partial u}{\partial p} p_c}{H_u} \right) - \frac{dQ_w}{d\alpha} - \frac{dm_{BB}}{d\alpha} \right.$$
$$\left. \cdot \left(h_{BB} - u_c - p_c \frac{\partial u}{\partial p} \right) - m_c \frac{\partial u}{\partial \lambda} \frac{\partial \lambda}{\partial \alpha} - p_c \frac{dV_c}{d\alpha} \left(1 - \frac{\partial u}{\partial p} \frac{m_c}{V_c} \right) \right] \tag{2}$$

where T_c is the cylinder temperature, m_c is the volume of the fluid in the cylinder, p_c is the pressure in the cylinder, u_c is the specific internal energy of the volume of liquid inside the cylinder, H_c is the low calorific value, λ is the air residue ratio ($1/\Phi$), α is the equivalent ratio, and V_c is the cylinder volume.

The solution of the above equation depends on the combustion model, the laws of exothermic reactions, and the heat transfer through the cylinder wall, as well as the pressure, temperature, and composition of the gas mixture together with the equation of the state:

$$p_c = \frac{1}{V} m_c R_c T_c \tag{3}$$

Establishing the relationship between pressure, temperature, and density from Equation (3), we used the Runge–Kutta method to determine the temperature of the cylinder. The pressure will then be determined by the state equation. Figure 3 shows the in-cylinder pressure profile using the experimental and numerical methods. The in-cylinder pressure always changes according to the crankshaft angle. The in-cylinder pressure is nominal at the intake and exhaust strokes; however, it increases dramatically during the compression and power strokes. It reaches a very high value at the position near the top dead center. From there, we determined the force $P_T = p_c A$ caused by the pressure applied to the top of the piston.

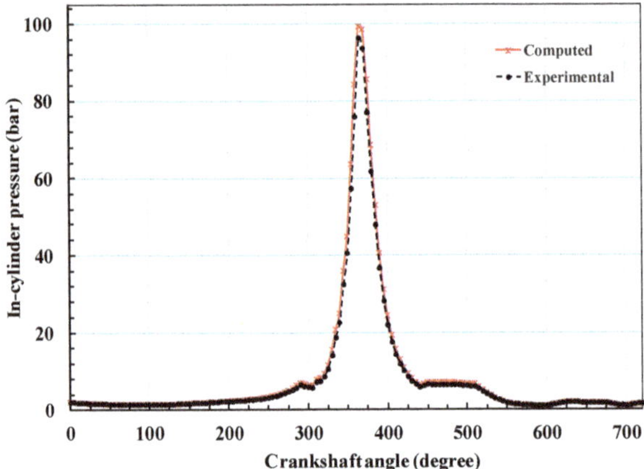

Figure 3. The in-cylinder pressure profile using the experimental and numerical methods.

In addition, the crankcase pressure generated by oil pressure on the underside is normally determined in the experiment. This force is very small compared to the total force exerted on the top of the piston. Consequently, it was ignored in this research. Furthermore, the friction force (F_{fr}) exerted on the piston needed to be determined in this calculated case. F_{fr} includes the piston-ring friction (F_r) and the friction from the piston surface and cylinder ($\mu_p S$). The total friction force can be calculated according to Equation (4) [15,16].

$$F_{fr} = \mu_p S + F_r = \mu_p S + \mu_r 2\pi D_3 r E \left(1 - \frac{D_3 - d_4}{4r}\right) \sqrt{1 - \frac{(D_3 - d_4)^2}{16r^2}} \tag{4}$$

where μ_r is the friction coefficient of the contact surface between the ring and cylinder, D_3 is the diameter of the cylinder, d_4 is the diameter of the ram, r is the width of the cross-section of the cylinder element, E is the elastic coefficient of the spring materials, S is the force resulting from the contact between the piston and the cylinder, and μ_p is the corresponding friction coefficient.

The sliding friction is dramatically reduced by adding a lubricant between the contact surfaces. In addition, in the operation of an internal combustion engine, the piston ring operates from the hydrodynamic lubrication to the boundary lubrication depending on the crankshaft angle, viscosity, and load. This phenomenon can be described by Stribeck's equation [17]. In this research, the friction coefficient μ_r was determined using Stribeck's equation as described in Equation (5):

$$\mu_r = c \left(\frac{\eta v}{pw}\right)^x \tag{5}$$

where c is constant, h is the dynamic viscosity, v is the piston speed, p is the pressure exerted on the piston ring, w is the width of the ring, and x is the exponent varying from 0.33 to 0.66. Therefore, we calculated the total friction force based on the number and parameter characteristics of the test engine. The inertia force caused by the movement of the piston and rod was also dramatically affected by the piston; it will be described in the following sections.

3.2. Kinetics and Dynamics of the Piston Mechanism

According to [18], assuming the displacement of the piston in the horizontal direction is trivial, applying the sine theorem in the OAB triangle (Figure 2), we have:

$$\frac{l_2}{\sin\theta} = \frac{r}{\sin\phi} \Rightarrow \sin\phi = \frac{r}{l}\sin\theta. \tag{6}$$

If we derive the above expression, we have:

$$-\dot{\phi}\cos\phi = -\frac{r}{l_2}\dot{\theta}\cos\theta \Rightarrow \dot{\phi} = \frac{r\dot{\theta}\cos\theta}{l_2\cos\phi}. \tag{7}$$

The z displacement of the piston in the vertical direction is defined by:

$$z = r\cos\theta + l_2\cos\phi. \tag{8}$$

We can infer that the velocity of the piston in the vertical direction is determined by Equation (8) and the acceleration of the piston in the vertical direction can be calculated with Equation (9).

$$\dot{z} = -r\dot{\theta}\sin\theta - l_2\dot{\phi}\sin\phi, \tag{9}$$

$$\ddot{z} = -\left(r\ddot{\theta}\sin\theta + r\dot{\theta}^2\cos\theta\right) - \left(l_2\ddot{\phi}\sin\phi + l_2\dot{\phi}^2\cos\phi\right), \tag{10}$$

The piston moves only along the cylinder, and thus the acceleration in the horizontal and vertical directions across the piston are determined in Equations (10) and (11), respectively.

$$a_{PX} = 0, \tag{11}$$

$$a_{PY} = \ddot{z} = \dot{\theta}^2\left(\frac{(r\cos\theta)^2}{l\cos^3\phi} - r\cos\theta - r\sin\theta\tan\phi\right) + \ddot{\theta}(r\cos\theta\tan\phi - r\sin\theta), \tag{12}$$

The cylinder has actually become elliptical due to the rotation of the piston about the wrist pin when the piston moves toward the bottom dead center. In addition, the high temperature during the operation process also affects the deformation of the piston and cylinder. Consequently, the forces exerted on the piston are not symmetric in the x direction. However, in this research, we assumed that the forces exerted on the piston were symmetric in the x direction. Therefore, the forces acting on the piston include $Q(t)$, the total external force acting on the piston, $m_P.g$, the force of gravity, and F_{BX}, F_{BY}, the link reaction forces in the x and y directions, respectively [17,19], as described in Figure 4.

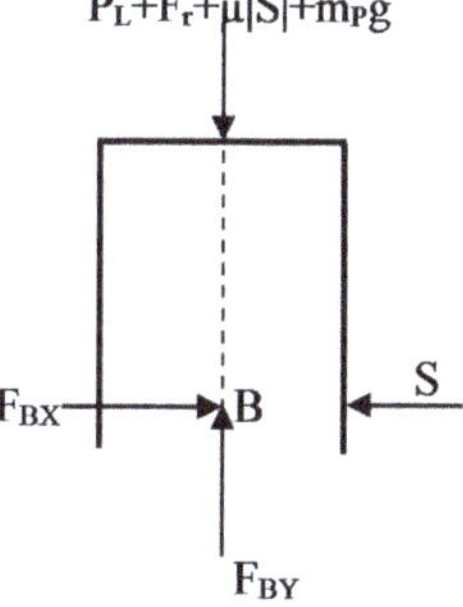

Figure 4. The modelled forces on the piston.

Since the piston does not move in the x direction, the total force acting on x is zero, inferring that:

$$\sum F_X = F_{BX} - S = 0 \Rightarrow F_{BX} = S. \tag{13}$$

The total force acting in the y direction is:

$$\sum F_Y = F_{BY} - m_P g - Q(t) = m_{pa} a_{PY}, \tag{14}$$

where $Q(t)$ is the total external force acting on the piston, including the force acting on the piston (P_T), the friction caused by the impacts on the cylinder wall (F_r), friction from the piston surface and cylinder $(\mu_p S)$, the force caused by pressure acting on the piston bottom (P_B is crankcase pressure), and the inertia force $m_{pa} a_{PY}$.

From Equation (13), we infer that the reaction force on the piston in the y direction is:

$$F_{BY} = m_P g + Q(t) + m_{pa} a_{PY}, \tag{15}$$

in which:

$$Q(t) = P_T + F_r \pm \mu_p S + P_B. \tag{16}$$

Note that the sign of $\mu_p S$ depends on the direction of piston movements. Thus, we obtained a graph illustrating the total force on the pistons of the test engine, as shown in Figure 5.

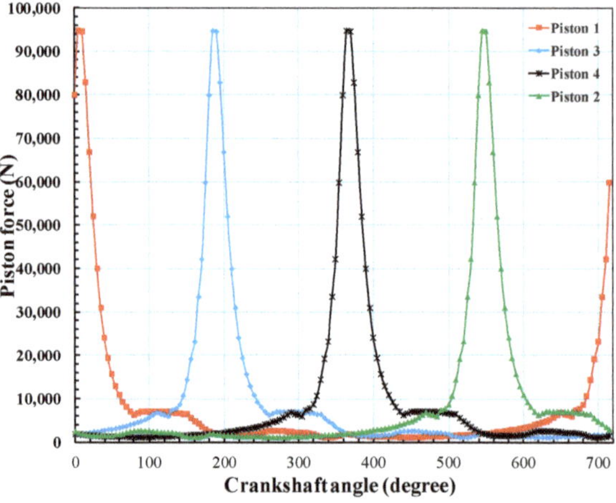

Figure 5. The total forces on the pistons in the test engine.

3.3. Kinetics and Dynamics of the Rod Mechanism

The acceleration of the transmission rod at the center is determined by the acceleration of the head of the rod connected to the piston. In the horizontal and vertical direction, we have:

$$a_{RX} = \dot{\theta}^2 (1 - j) r \sin\theta - \ddot{\theta}(1 - j) r \cos\theta, \tag{17}$$

$$a_{RY} = \dot{\theta}^2 \left[j \frac{(r\cos\theta)^2}{l_2 \cos^3\phi} - r\cos\theta - jr\sin\theta\tan\phi \right] + \ddot{\theta}(jr\cos\theta\tan\phi - r\sin\theta). \tag{18}$$

The total forces on the rod are as follows:

The forces acting on the rod include the gravity force at the center of the rod and the link reaction forces at the two ends of the rod as described in Figure 6. From this, we can identify the following.

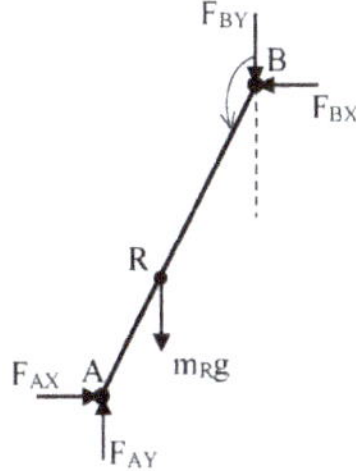

Figure 6. The model of forces on the rod.

(1)   Total force in the x direction:

$$\sum F_X = F_{AX} - F_{BX} = m_R a_{RX},\qquad(19)$$

(2)   Total force in the y direction:

$$\sum F_Y = F_{AY} - F_{BY} - m_R g = m_R a_{RY},\qquad(20)$$

(3)   Total torque on the axis passing through the center R:

$$\sum M_R = I_R \ddot{\phi}_R,\qquad(21)$$

$$-F_{BX}(1-j)l\cos\phi - F_{BY}(1-j)l\sin\phi - F_{AX}jl\cos\phi - F_{AY}jl\sin\phi = I_R\ddot{\phi}_R.\qquad(22)$$

3.4. Kinetics and Dynamics of the Crankshaft

The crankshaft converts the up and down movements of the pistons into rotary motion and drives the external drive. The crankshaft is connected to the pistons via the transmission rods.

Similar to the determination of the rod acceleration, we can determine the crankshaft acceleration at the center using the acceleration at the common point between the transmission rod and the crankshaft $\vec{a}_C = \vec{a}_A + \vec{a}_{C/A}$ and project it in the horizontal and vertical directions:

$$a_{CX} = -\dot{\theta}^2 hr\sin\theta + \ddot{\theta}hr\cos\theta,\qquad(23)$$

$$a_{CY} = \dot{\theta}^2 hr\cos\theta + \ddot{\theta}hr\sin\theta.\qquad(24)$$

The total force acting on the crankshaft is illustrated in Figure 7.

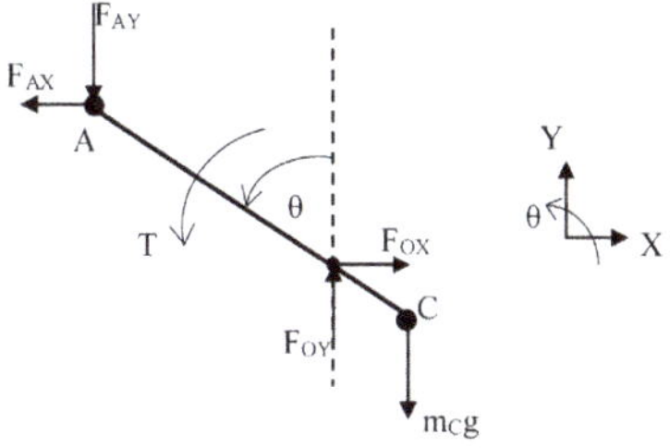

Figure 7. The forces acting on the crankshaft.

The total force in x direction is:

$$\sum F_X = F_{OX} - F_{AX} = m_C a_{CX},\tag{25}$$

The total force acting in the y direction is:

$$\sum F_Y = F_{OY} - F_{AY} - m_C g = m_C a_{CY},\tag{26}$$

The total torque around crankshaft is:

$$\sum M_C = I_C \ddot{\theta}_C,\tag{27}$$

$$T + F_{AX}(1+h)r\cos\theta + F_{AY}(1+h)r\sin\theta - F_{OX}hr\cos\theta - F_{OY}hr\sin\theta = I_C\ddot{\theta}_C.\tag{28}$$

Combining these above equations, we have:

$$
\begin{aligned}
S = &-r\ddot{\theta}\left[\frac{I_R\cos\theta}{(l\cos\phi)^2} + m_P\tan\phi(\cos\theta\tan\phi - \sin\theta) + jm_R\left(\frac{j\cos\theta}{\cos^2\phi} - \sin\theta\tan\phi - \cos\theta\right)\right] - \\
&-r\dot{\theta}^2\left[\frac{I_R}{(l\cos\phi)^2}\left(\frac{r\cos^2\theta\tan\phi}{l\cos\phi} - \sin\theta\right) + m_P\tan\phi\left(\frac{r\cos^2\theta}{l\cos^3\phi} - \sin\theta\tan\phi - \cos\theta\right) + \right. \\
&\left. + jm_R\left(\frac{jr\cos^2\theta\tan\phi}{l\cos^3\phi} - \cos\theta\tan\phi + \sin\theta - \frac{j\sin\theta}{\cos^2\phi}\right)\right] - g\tan\phi(m_P + jm_R) - Q(t)\tan\phi
\end{aligned}\tag{29}
$$

Substituting the known forces in Equation (8), we can obtain the torque acting on the crankshaft:

$$
\begin{aligned}
T = &I_C\ddot{\theta} + m_C a_{CX}hr\cos\theta - m_R a_{RX}r\cos\theta + m_C a_{CY}hr\sin\theta - m_R a_{RY}r\sin\theta - m_P a_{PY}r\sin\theta + \\
&+ m_C hgr\sin\theta - m_R gr\sin\theta - m_P gr\sin\theta - Q(t)r\sin\theta - Sr\cos\theta
\end{aligned}\tag{30}
$$

Substituting the known accelerations into the above equation, we have:

$$T = \ddot{\theta}I(\theta) + \frac{1}{2}\dot{\theta}^2 I'(\theta) + g(\theta) + Q(t,\theta),\tag{31}$$

where $I(\theta)$ is the inertia function, defined by:

$$
\begin{aligned}
I(\theta) = &I_C + m_C h^2 r^2 + I_R\left(\frac{r\cos\theta}{l\cos\phi}\right)^2 + m_P r^2(\cos\theta\tan\phi - \sin\theta)^2 + \\
&+ m_R r^2\left[(1-j)^2\cos^2\theta + (j\cos\theta\tan\phi - \sin\theta)^2\right]
\end{aligned}\tag{32}
$$

where $I'(\theta)$ is the rate of the inertial variable based on the rotational angle θ of the crankshaft determined by:

$$
\begin{aligned}
I'(\theta) = &2I_R\left(\frac{r\cos\theta}{l\cos\phi}\right)^2\left(\frac{r\cos\theta}{l\cos\phi}\tan\phi - \tan\theta\right) + \\
&+ 2m_P r^2(\cos\theta\tan\phi - \sin\theta)\left(\frac{r\cos^2\theta}{l\cos^3\phi} - \cos\theta - \sin\theta\tan\phi\right) - \\
&- 2m_R r^2(1-j)^2\sin\theta\cos\theta + 2m_R r^2(j\cos\theta\tan\phi - \sin\theta)\left(\frac{jr\cos^2\theta}{l\cos^3\phi} - \cos\theta - j\sin\theta\tan\phi\right)
\end{aligned}\tag{33}
$$

where $g(\theta)$ is the gravitational torque, determined by:

$$g(\theta) = gr[m_P(\cos\theta\tan\phi - \sin\theta) + m_R(j\cos\theta\tan\phi - \sin\theta) + m_C h\sin\theta],\tag{34}$$

and $Q(t,\theta)$ is the momentum or torque determined by the general load $Q(t)$ acting on the piston, determined by:

$$Q(t,\theta) = Q(t)r(\cos\theta\tan\phi - \sin\theta).\tag{35}$$

With the input pressure as mentioned in Figure 3 acting on the top of the piston, corresponding to the engine speed of 2200 rpm, and using a firing order of 1–3–4–2, we can determine the average load moment on the crankshaft over the entire cycle as follows:

$$T_{tb} = \int_{\theta=0}^{4\pi} T(\theta)d\theta = 308.5534201 \quad (Nm). \tag{36}$$

Figure 8a shows the torque acting on the four crankpins of the crankshaft. The maximum torque on each crankpin occurs when the crankshaft is positioned in the range of 360° to 370°, immediately after the top dead center. The main torque exerted on the crankpins is the in-cylinder pressure; however, the friction and inertia force also dramatically contribute to the total force varying according to the crankshaft angle. For instance, at the crankshaft angle of 363°, the total torque exerted on crankpin 1 reaches a maximum value of 6203 Nm. Meanwhile, the inertia force and the friction force contribute to approximately 8.9% and 4.6% of the total torque, respectively. Figure 8b shows the torque acting on the main journals of the crankshaft. They also vary depending on the crankshaft angle. However, the torque of each main journal is dramatically different from one other since it is calculated based on the different torque directions acting on the adjacent crankpins.

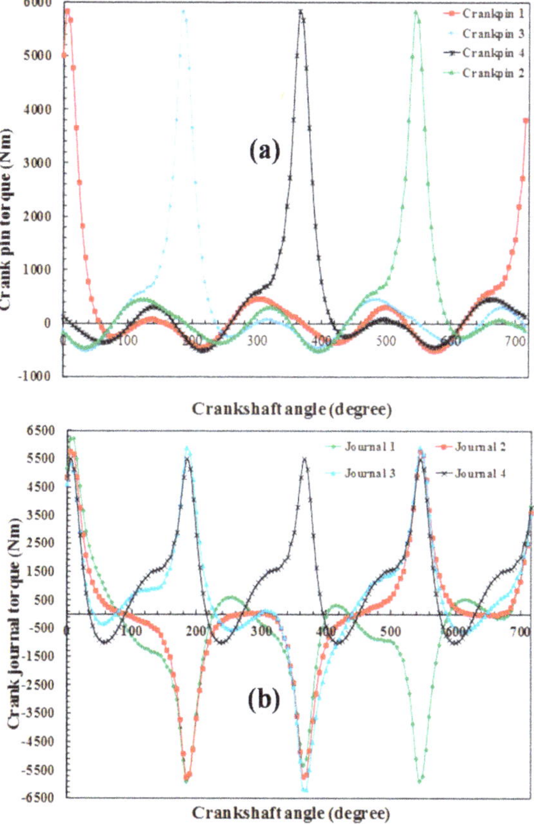

Figure 8. The forces acting on the crankpins and the crank journal of the crankshaft. (**a**) the torque acting on the crankpins ; (**b**) the torque acting on the crank journal.

4. Conclusions

This research shows an experimental and numerical method to calculate the net forces on the components of an internal combustion engine, including the piston, rod, and crankshaft. The forces reach a maximum value near the top dead center. The force caused by the in-cylinder pressure is the main force affecting the test engine components; however, the friction and inertia also significantly affect these components. These forces strictly depend on the engine speed and the geometrical parameters of these components.

Future research will focus on modeling the thermal dynamic and heat transfer phenomena of the engine to fully evaluate the effect of the deformation of the engine components on the friction forces. In addition, a stress analysis of the piston, rod, and crankshaft will also be conducted to extend the operational lifespan of the engine.

Author Contributions: Experimental design was performed by V.N.D. and D.V.N.; fieldwork was conducted by V.N.D.; statistical analyses were performed by D.V.N. and V.N.D. contributed to the writing of the paper.

Acknowledgments: This work was supported by the Modelling and Simulation Institute, Viettel Research and Development Institute.

References

1. Yu, Z.; Xu, X. Failure investigation of a truck diesel engine gear train consisting of crankshaft and camshaft gears. *Eng. Fail. Anal.* **2010**, *17*, 537–545. [CrossRef]
2. Sroka, Z.J.; Dziedzioch, D. ScienceDirect Mechanical load of piston applied in downsized engine. *Arch. Civ. Mech. Eng.* **2015**, *15*, 663–667. [CrossRef]
3. Satyanarayana, K.; Rao, S.V.U.M.; Viswanath, A.K.; Rao, T.V.H. ScienceDirect Quasi-dynamic and thermal analysis of a diesel engine piston under variable compression. *Mater. Today Proc.* **2018**, *5*, 5103–5109. [CrossRef]
4. Rao, S.V.U.M.; Rao, T.V.H.; Satyanarayana, K.; Nagaraju, B. ScienceDirect Fatigue Analysis of Sundry I C Engine Connecting Rods. *Mater. Today Proc.* **2018**, *5*, 4958–4964. [CrossRef]
5. Wang, M.; Pang, J.C.; Zhang, M.X.; Liu, H.Q.; Li, S.X.; Zhang, Z.F. Thermo-mechanical fatigue behavior and life prediction of the Al-Si piston alloy. *Mater. Sci. Eng. A* **2018**, *715*, 62–72. [CrossRef]
6. Romanov, V.A.; Lazarev, E.A.; Khozeniuk, N.A. The evaluation of the stress-strain state for the cylinder heads of high-powered diesel engines using the multiphysics ansys technology. *Procedia Eng.* **2015**, *129*, 549–556. [CrossRef]
7. James Prasad Rao, B.; Srikanth, D.V.; Suresh Kumar, T.; Sreenivasa Rao, L. Design and analysis of automotive composite propeller shaft using fea. *Mater. Today Proc.* **2016**, *3*, 3673–3679. [CrossRef]
8. Gunasegaram, D.R.; Molenaar, D. Towards improved energy efficiency in the electrical connections of Hall-Héroult cells through Finite Element Analysis (FEA) modeling. *J. Clean. Prod.* **2015**, *93*, 174–192. [CrossRef]
9. Reza, M.; Asad, A.; Ranjbarkohan, M.; Dardashti, B.N. Dynamic Load Analysis and Optimization of Connecting Rod of Samand Engine. *Aust. J. Basic Appl. Sci.* **2011**, *5*, 1830–1838.
10. Webster, W.D.; Coffell, R.; Alfaro, D. A three dimensional finite element analysis of a high speed diesel engine connecting rod. *SAE Tech. Paper.* **1983**, *9*, 83–96.
11. Thejasree, P.; Dileep Kumar, G.; Leela Prasanna Lakshmi, S. Modelling and Analysis of Crankshaft for passenger car using ANSYS. *Mater. Today Proc.* **2017**, *4*, 11292–11299. [CrossRef]
12. Witek, L.; Sikora, M.; Stachowicz, F.; Trzepiecinski, T. Stress and failure analysis of the crankshaft of diesel engine. *Eng. Fail. Anal.* **2017**, *82*, 703–712. [CrossRef]
13. Duc, K.N.; Tien, H.N.; Duy, V.N. A Study of Operating Characteristics of Old-Generation Diesel Engines Retrofitted with Turbochargers. *Arab. J. Sci. Eng.* **2017**. [CrossRef]
14. Mamat, A.M.I.; Romagnoli, A.; Martinez-Botas, R.F. Characterisation of a low pressure turbine for turbocompounding applications in a heavily downsized mild-hybrid gasoline engine. *Energy* **2014**, *64*, 3–16. [CrossRef]

15. Faisal, A.-G.; Faisal, M.M.T.; Muafag, S. Friction Forces in O-ring Sealing. *Am. J. Appl. Sci.* **2005**, *2*, 626–632. [CrossRef]

16. Guzzomi, A.L.; Hesterman, D.C.; Stone, B.J. Variable inertia effects of an engine including piston friction and a crank or gudgeon pin offset. *Proc. Inst. Mech. Eng. Part D J. Automob. Eng.* **2008**, *222*, 397–414. [CrossRef]

17. Richardson, D.E. Review of Power Cylinder Friction for Diesel Engines. *J. Eng. Gas Turbines Power* **2000**, *122*, 506–519. [CrossRef]

18. Guzzomi, A.L.; Hesterman, D.C.; Stone, B.J. The effect of piston friction on engine block dynamics. *Proc. Inst. Mech. Eng. Part K J. Multi-Body Dyn.* **2007**, *221*, 277–289. [CrossRef]

19. Mourelatos, Z.P. A crankshaft system model for structural dynamic analysis of internal combustion engines. *Comput. Struct.* **2001**, *79*, 2009–2027. [CrossRef]

 *applied sciences*

Article

A Study of the Transient Response of Duct Junctions: Measurements and Gas-Dynamic Modeling with a Staggered Mesh Finite Volume Approach

Antonio J. Torregrosa *, Alberto Broatch, Luis M. García-Cuevas and Manuel Hernández

CMT–Motores Térmicos, Universitat Politècnica de València, 46022 Valencia, Spain; abroatch@mot.upv.es (A.B.); luiga12@mot.upv.es (L.M.G.-C.); maher12@mot.upv.es (M.H.)
* Correspondence: atorreg@mot.upv.es; Tel.: +34-963-877-658

Academic Editor: Kuang-Chao Fan
Received: 20 March 2017; Accepted: 2 May 2017; Published: 8 May 2017

Abstract: Duct junctions play a major role in the operation and design of most piping systems. The objective of this paper is to establish the potential of a staggered mesh finite volume model as a way to improve the description of the effect of simple duct junctions on an otherwise one-dimensional flow system, such as the intake or exhaust of an internal combustion engine. Specific experiments have been performed in which different junctions have been characterized as a multi-port, and that have provided precise and reliable results on the propagation of pressure pulses across junctions. The results obtained have been compared to simulations performed with a staggered mesh finite volume method with different flux limiters and different meshes and, as a reference, have also been compared with the results of a more conventional pressure loss-based model. The results indicate that the staggered mesh finite volume model provides a closer description of wave dynamics, even if further work is needed to establish the optimal calculation settings.

Keywords: duct junction; staggered mesh finite volume model; multi-port

---

1. Introduction

Duct junctions are essential elements of numerous piping systems, including the intake and exhaust systems of reciprocating internal combustion engines. The use of one-dimensional time domain gas-dynamic codes has become commonplace in the numerical study of unsteady flows in such systems, both in terms of their effect on engine performance and on intake and exhaust orifice noise [1]. While assuming one-dimensional wave action may be acceptable when duct diameters are relatively small, as is the case in the majority of the ducts present in engine intake and exhaust systems of passenger car engines, in certain elements, and most notably in duct junctions, complex three-dimensional flow structures may occur [2]. Consideration of the effects of such structures on the one-dimensional flow in the adjacent ducts requires the definition of suitable boundary conditions at the junction, usually involving empirical information.

The effects of a junction on the flow in the neighboring ducts arise in different ways. From the point of view of the passive propagation of small amplitude pressure waves (i.e., in the acoustic range) the effect can be characterized in terms of length corrections, which have been reported to depend on the type of side-branch and the branch width and length [3], and with a rapid increase in the duct length corrections being associated with the excitation of non-planar higher order modes, which also results in lower sound transmission. This sort of representation has been quite successfully applied to the prediction of the effect on intake noise of a multi-pipe junction in the intake manifold [4]. It has also been reported [5] that for low Strouhal numbers based on the duct diameter, the acoustic transmission properties of T-junctions can be acceptably described by using an incompressible quasi-steady model, the upper limit of the Strouhal number being defined by flow-acoustic interaction effects, which differ significantly between different flow configurations: waves incident on the junction at the downstream side are attenuated, whereas waves incident at the other branches may be either amplified or attenuated, depending on the Strouhal number [6]. Such flow-acoustic interactions due to the coupling of the flow and the geometry are common to all intake and exhaust system elements [7].

When the focus is on the effect of the junction on the propagation of finite amplitude pressure waves and the resulting influence on engine performance, different approaches are found in the literature, most of them inspired by the seminal work of Benson [8]. The simplest approach is given by constant pressure models, in which is it assumed that the pressure at the end of all branches of the junction is the same at any time, so that the pressure is assumed to be uniform across the junction. The most comprehensive description of these models is given in [9], where it was shown that, besides the assumption of uniform pressure, additional closing equations must be added. While the choice of those equations is arbitrary, it was also shown in [9] that assuming that the total enthalpy for all the outgoing flows is the same provides suitable results.

More elaborated approaches are based on the consideration of the pressure differences existing between the different branches, which are incorporated in a quasi-steady manner, i.e., steady pressure loss coefficients (or more properly, as discussed in detail in [10], energy change coefficients accounting partly for losses and partly to a mutual energy transfer between the partial flows) are applied at each time step. The solutions proposed differ mainly in the origin of the pressure loss coefficients, in the hypotheses underlying their determination, and in the precise implementation of the solution method.

Regarding the origin of the coefficients, while there have been some attempts to obtain them from computational fluid dynamics (CFD) simulations [11–13], it appears that the results are strongly dependent on the numerical method used, both in the details of the flow and in the overall values of the coefficients obtained [14]. Therefore, usually the coefficients are either obtained from simple and robust models, or specific measurements are performed in order to characterize the junction under consideration. The most successful example of the first option is probably that presented in [15], where a remarkable agreement with experimental results was obtained from a model that extended the previous work performed in [16] and neglected any effects of mixing losses, compressibility and wall friction. Regarding the experimental characterization, it is usual to consider steady incompressible flow, as in [17], but more recently, specific studies accounting for the flow compressibility have been reported [18,19] that suggest that the total pressure loss coefficient is mainly dependent on the Mach number, mass flow rate ratio, and area ratio, and is almost independent of the Reynolds number.

Numerous implementations of the pressure loss model for multi-pipe junctions have been proposed in the literature, comprising implicit time formulations [20] and different explicit solutions, such as the supplier–collector strategy [21], the branch superposition method [22] and the generalization of the classical approach of Benson presented in [23]. The limitations of these approaches lie mainly in the fact that, even if steady flow coefficients contain information on three-dimensional separation effects around the junction, the results will be significant only if quasi-steady flow can be assumed, which requires that mass and energy storage at the junction are very small, which may not be the case in real manifold flows. Additionally, any information regarding the wave refraction characteristics of the junction is lost in the quasi-steady approximation.

Overcoming these limitations requires accounting for the unsteady and multi-dimensional character of the flow at the junction, but without incurring in an excessive computational cost. A suitable solution is thus to include a local multi-dimensional region within an otherwise one-dimensional wave-action engine simulation, as first suggested in [24]. In this first approach, an inviscid two-dimensional model was applied to the simulation of shock-wave propagation through different junctions, and the observed evolution of the wave fronts through the junctions and the measured high frequency pressure oscillations induced by the transverse reflections were successfully predicted. However, even if the increase in the computational cost was reasonable, it did not appear to be justified when compared with a conventional quasi-steady pressure loss model [25].

It appears, thus, that a full three-dimensional description of the junction should be used in order to describe its unsteady behavior. Such a description was presented in [26,27], successfully reproducing the flow field and the associated non-plane-wave motion. However, even if coarse 3D grids were used in the first simulation cycles that were switched to more refined grids during the last simulation cycles, the computational cost and time may still be regarded as excessive for the practical design and evaluation of full intake and exhaust systems.

A possible alternative to 1D–3D coupling, which could provide some accountancy for the three-dimensional effects at the junction and to the authors' knowledge has not been explored in some detail, would be the use in this context of a staggered mesh finite volume method [28]. Such methods have become standard in commercial codes, either as the core solver [29,30], or used locally for elements exhibiting significant three-dimensional features, such as plenums and mufflers [31]. Typically, when these methods are applied to simple duct junctions, a single volume is used for the junction with appropriate effective areas and characteristic lengths at each connection with the adjacent ducts. As these connections contain information on vector quantities (including the orientation of the branch duct) the momentum equation can be solved, even in an approximate way, so that all the effects of the junction of the flow need not be included through the pressure loss coefficients. Additionally, it would be possible to use a refined mesh locally at the junction, so that a first-order estimate of any three-dimensional features could be obtained.

The objective of this paper is precisely to establish the potential of these ideas as a way to improve the description of the effect of simple duct junctions on an otherwise one-dimensional flow system, as the intake or exhaust of an internal combustion engine. Specific experiments have been performed in order to obtain precise and reliable results on the propagation of pressure pulses across junctions. The results obtained have been compared to simulations performed with different versions of a staggered mesh finite volume method and different meshes and, as a reference, also with the results of a more conventional pressure loss-based model.

2. Materials and Methods

Two junctions, shown schematically in Figure 1, were manufactured. A T-junction and a Y-junction were considered, in order to allow the analysis of the effect of the angle of the side branch. An internal diameter of 51 mm was used in all the branches of the junctions.

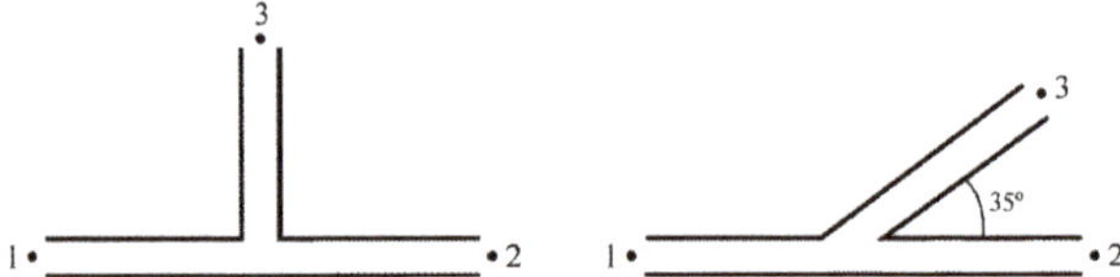

Figure 1. Junctions considered in the study.

While several formalisms may be used for the representation of the transient response of a system, the most intuitive one for the present case of a junction is that based on the consideration of wave components, so that the junction is actually regarded as a multi-port. In this framework, for a junction such as that represented in Figure 2, one has three excitations and three responses, and writing the relations between them directly in matrix form in the frequency domain, one has:

$$\begin{bmatrix} B_1 \\ B_2 \\ B_3 \end{bmatrix} = \begin{bmatrix} R_1 & T_{21} & T_{31} \\ T_{12} & R_2 & T_{32} \\ T_{13} & T_{23} & R_3 \end{bmatrix} \begin{bmatrix} A_1 \\ A_2 \\ A_3 \end{bmatrix}. \tag{1}$$

where, as indicated in Figure 2, A_i represents the wave component moving towards the junction in port i and B_j represents the wave component moving away from the junction in port j. Regarding the matrix elements, R_i denotes the reflection coefficient as seen from port i whereas T_{ij} denotes the transmission coefficient between ports i and j. All these magnitudes are functions of the frequency f.

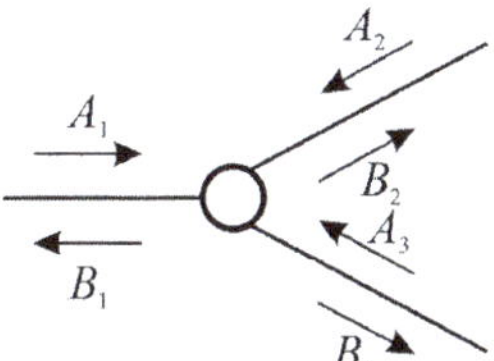

Figure 2. Wave components acting on a multi-port.

In this way, one has a reflection coefficient for each of the pipes arriving at the junction, and transmission coefficients for all the possible transmission paths, indicated by the corresponding subscripts. The experimental setup and the corresponding measurement procedure are described in detail in [32,33], and here only a brief overview is given in Appendix A.

Two different modeling approaches were evaluated: a staggered mesh finite volume method and, as a reference, a more conventional pressure loss-based model. The staggered mesh finite volume method used is described in detail in [34], where a flux-corrected transport (FCT) technique was used in order to suppress the spurious oscillations that these numerical methods exhibit in the vicinity of discontinuities in the flow variables. It was found that satisfactory results were obtained through the combination of dissipation via damping together with the phoenical form of the anti-diffusion term. As an alternative, the momentum diffusion term described in [35] was also used. A brief summary is given in Appendix B.

The pressure loss-based model uses a conventional one-dimensional finite volume model with a collocated mesh, derived from the code available in [36]. The junction is modeled as a small volume with three connections to which different pressure loss coefficients are assigned, and in which the mass and energy conservation equations are solved. For the connection to the duct where the incident pressure pulse propagates it has been assumed that the total pressure loss is zero, whereas for the other two connections their corresponding pressure loss coefficients are computed following the expressions given in [15,16]. The procedure is summarized in Appendix C.

3. Results and Discussion

In this section, first the experimental results obtained will be analyzed, both in the time (t) and the frequency (f) domains. Then, the performance of the different modeling approaches will be discussed, first in the case in which the junction itself is represented by a zero-dimensional element, and secondly in the case in which the staggered mesh method is used to provide a quasi-3D description of the junction.

3.1. Experimental Results

3.1.1. Time Domain Analysis

The results for the T-junction and shown in Figure 3. Ports are denoted as in Figure 1, and it is apparent that when the junction is excited at port 1 the pulse transmitted through port 2 (i.e., in the main propagation direction) has a higher amplitude than that transmitted through port 3 (the branched duct), as could be intuitively expected. It is also apparent, and equally expectable, that when the junction is excited at port 3, the pulses transmitted through ports 1 and 2 are very similar, the small differences observed being attributable to manufacturing issues.

Differences in the reflected pulses recorded at ports 1 and 3 are also apparent, even if the incident pulses do not have the same amplitude. The reflected pulse recorded at port 3 is noisier, and its amplitude is comparable to that of the reflected pulse recorded at port 1, while the corresponding incident pulse has a lower amplitude. This indicates that reflection is more intense when the junction is excited at the branch duct, as it is also intuitively reasonable in terms of the interaction of the incident pulse with the wall of the main duct. Of course, none of these effects, regarding both transmission and reflection, can be accounted for by a constant pressure model, and this is the reason why such a model will not be considered in the subsequent discussion.

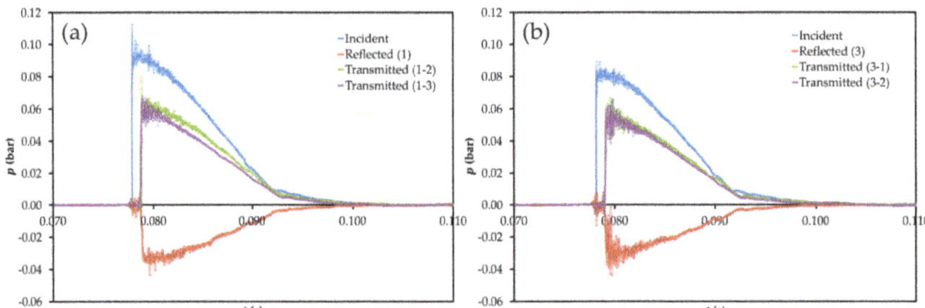

Figure 3. Experimental results for the T-junction in the time domain. (**a**) Excitation at port 1; (**b**) excitation at port 3. Ports are denoted as in Figure 1.

The results for the Y-junction are shown in Figure 4. Here the trends observed confirm those found for the T-junction regarding the difference between the main duct and the branch duct, but with additional issues related with the branch angle. Comparison of Figure 4a,b indicates that the difference in amplitude between the two pulses transmitted is more important when the branch direction is against that of the incident pulse (i.e., when the junction is excited at port 2), in which case the results are rather similar to those shown in Figure 3a for the T-junction. Regarding the reflected pulses recorded at ports 1 and 2, some differences may be observed mostly in the last part of the pulse, which suggests some difference in the dynamic behavior of the junction.

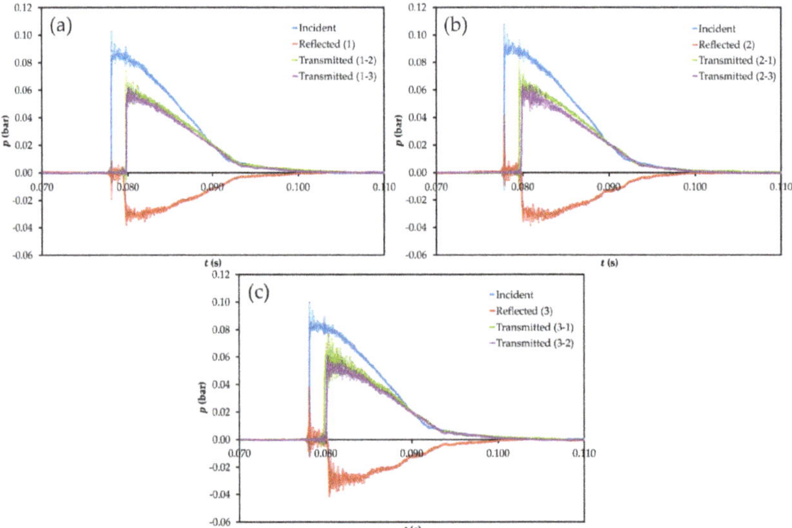

Figure 4. Experimental results for the Y-junction in the time domain. (**a**) Excitation at port 1; (**b**) excitation at port 2; (**c**) excitation at port 3. Ports are denoted as in Figure 1.

These statements are supported by the results obtained with the excitation at port 3, shown in Figure 4c. Here, it appears that again the amplitude of the transmitted pulse is higher when there is no significant change in direction along the transmission path (in this case, from port 3 to port 1). However, the differences are not as apparent as those seen in Figure 4b, which is reasonable considering that here there is some change in direction in the two transmission paths. It is also worth noticing the clear differences observed between the reflected pulse recorded at port 3 and those recorded at the other two ports. A much more complex time evolution can be observed in the case of port 3, which again suggests that wave dynamics inside the junction depend significantly on the port at which the junction is excited.

3.1.2. Frequency Domain Analysis

Here, the results obtained for the transmission and reflection coefficients defined in Equation (1) are analyzed. For brevity, only the modulus of these coefficients will be considered, as this contains significant information about the overall energetic behavior of the junction. The results for the T-junction are shown in Figure 5, where it can be observed that the values obtained in the very low frequency range (below 200 Hz) are fully consistent with the time domain results shown above in Figure 3: when exciting the junction at port 1, it is seen that $|T_{12}|$ is systematically larger than $|T_{13}|$ in this frequency range, whereas when the excitation is at port 3 the differences between $|T_{31}|$ and $|T_{32}|$ are significantly smaller.

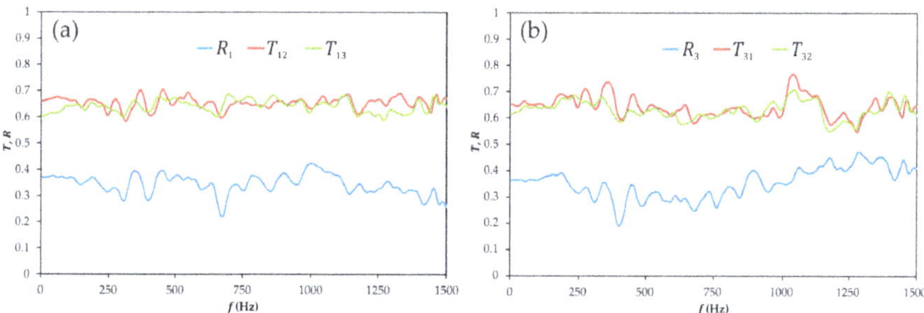

Figure 5. Experimental results for the T-junction in the frequency domain. (**a**) Excitation at port 1; (**b**) excitation at port 3. Ports are denoted as in Figure 1. R_i: reflection coefficient as seen from port i; T_{ij}: transmission coefficient between ports i and j.

At higher frequencies, above 200 Hz, it can be seen that the behavior of $|T_{12}|$ and $|T_{13}|$ is essentially flat around mean values of 0.65 and 0.64, respectively, with a maximum deviation from the mean of 0.065 in $|T_{12}|$ and of 0.05 in $|T_{13}|$. On the contrary, in the case of $|T_{31}|$ and $|T_{32}|$ their mean values are very similar to those of $|T_{12}|$ and $|T_{13}|$ (0.64 and 0.63, respectively) but some relevant acoustic features can be detected in both coefficients between 1000 and 1250 Hz, mostly in the case of $|T_{31}|$ where the deviation from the mean value reaches a maximum of 0.125, while $|T_{32}|$ follows the same trend but with a maximum deviation from the mean of 0.08 . This confirms, on one hand, that when the junction is excited at port 3 the two propagation paths are substantially equivalent and, on the other hand, that their behavior is different from that obtained when exciting the junction at port 1.

This second statement is fully supported by the spectra of the reflection coefficients $|R_1|$ and $|R_3|$: it is apparent that $|R_3|$ is overall larger than $|R_1|$ for frequencies below 200 Hz, as suggested by the results shown in Figure 4, but now without any uncertainty due to the difference in amplitude between the incident pulses used in each test. Additionally, the trend observed is rather different for frequencies above 200 Hz, and most notably above 1000 Hz, where $|R_1|$ shows a certain decreasing tendency whereas $|R_2|$ increases with frequency.

The corresponding results for the Y-junction are shown in Figure 6. Again, results below 200 Hz confirm the time domain tendencies observed in Figure 4. In this frequency range, it is seen that while $|T_{12}|$ is only slightly higher than $|T_{13}|$, when exciting at ports 2 or 3 one finds that the transmission coefficient corresponding to a smaller change in direction (that is, $|T_{21}|$ when exciting at port 1 and $|T_{31}|$ when exciting at port 3) is significantly larger than the other one, and that this effect is more noticeable the larger is the change in direction.

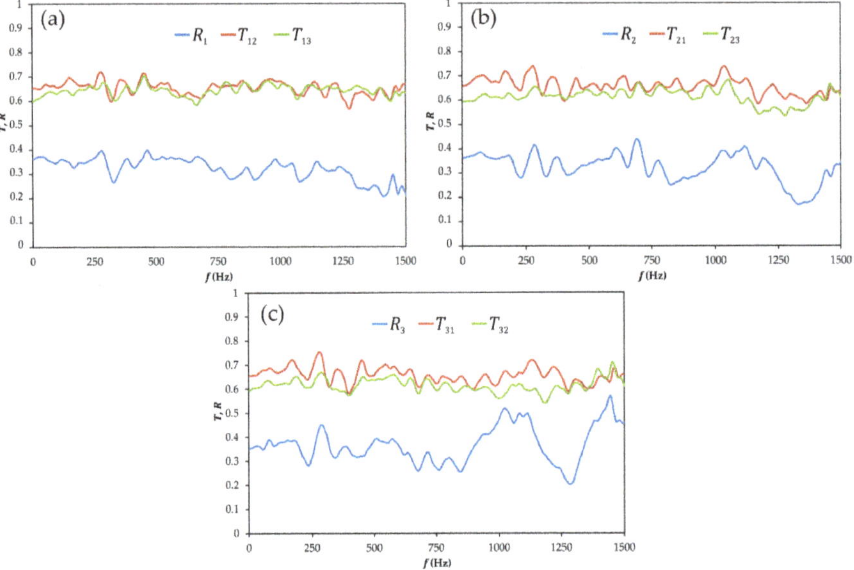

Figure 6. Experimental results for the Y-junction in the frequency domain. (**a**) Excitation at port 1; (**b**) excitation at port 2; (**c**) excitation at port 3. Ports are denoted as in Figure 1. R_i: reflection coefficient as seen from port i; T_{ij}: transmission coefficient between ports i and j.

When considering frequencies above 200 Hz, noticeable differences are also observed between the case with excitation at port 1, for which results very similar to those shown in Figure 5a are obtained, with small differences between $|T_{12}|$ and $|T_{13}|$ and an almost flat behavior with little dependency on frequency, and the other two cases, in which the transmission coefficients corresponding to the propagation path with the smallest change in direction ($|T_{21}|$ and $|T_{31}|$) are significantly and systematically higher than those implying an important change ($|T_{23}|$ and $|T_{32}|$, respectively) except at the highest frequencies represented.

However, it is in the reflection coefficients where the effect of the change in the excitation port is more apparent. In fact, the results for $|R_1|$ do not differ substantially from those obtained for the T-junction and shown in Figure 5a, neither in the low frequency values nor in the high frequency trend. On the contrary, the high frequency behavior seen in $|R_2|$ and $|R_3|$ is a clear indication of the change produced in the dynamic characteristics of the junction when the excitation port is changed, an effect that could be guessed from the time domain results of Figure 4 but now is fully confirmed. Actually, a well-defined trend of picks and troughs can be observed in both cases, with similar shape but a clear frequency shift, which provides a sort of acoustic signature of the dynamic behavior of the junction. The fact that such a behavior is not apparent in the spectrum of $|R_3|$ for the T-junction shown in Figure 5b indicates that such dynamic issues are suppressed by the symmetric nature of the excitation through a perpendicular branch.

3.2. Assessment of Modelling Approaches Considering a 0D Description of the Junction

Here, modeling approaches in which the junction itself is regarded as a 0-dimensional element, while the flow in the adjacent ducts is assumed to be one-dimensional, will be evaluated. In the context of the staggered mesh finite volume method, this corresponds to the case in which the junction is regarded as a single volume and the adjacent ducts are meshed only in the axial direction. The pressure loss-based model used here falls also within this category since, as commented in Appendix C, the junction branches are connected through an auxiliary 0D element.

Again, separate analyses for the time and the frequency domains are presented.

3.2.1. Time Domain Assessment

First, in Figure 7, direct comparison between the experiments and the method with the momentum diffusion term (MDT) as flux limiter is given for the case of the T-junction. The figures at the top provide a direct representation of the results obtained, whereas in the figures at the bottom the differences between the experimental and numerical results are represented (these are labeled ΔR and ΔT for reflected and transmitted pulses, respectively). In general, the model reproduces the experimental results within reasonable limits, but with a superimposed oscillation due to the development of the pulse from station 0 to station 1 (refer to Figure A1) and which is a consequence of the way in which the inlet boundary condition has been set. The scale of the vertical axis in the differences plots has thus been chosen so as to allow proper comparison for the times not affected by those oscillations.

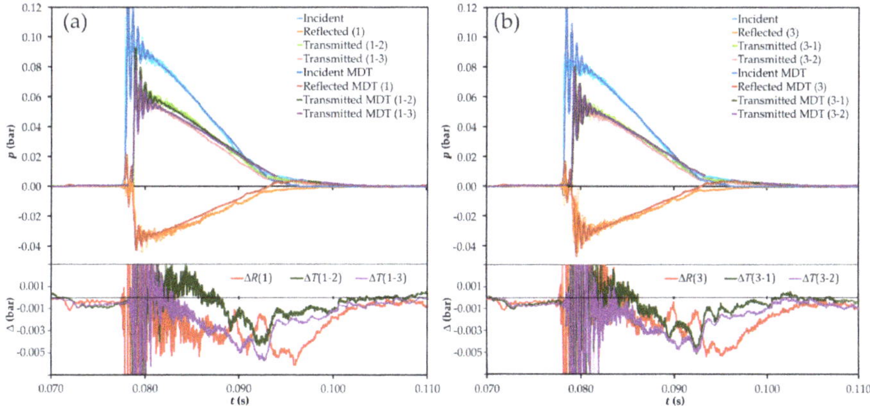

Figure 7. Experimental vs. modeled results for the T-junction: raw data (top) and differences (bottom) in the time domain, momentum diffusion term (MDT) method. (**a**) excitation at port 1; (**b**) excitation at port 3. Ports are denoted as in Figure 1. $\Delta R(i)$: difference in the pulse reflected at port i; $\Delta T(i-j)$: difference in the pulse transmitted between ports i and j.

From the differences plots, it is apparent that the numerical results tend to underestimate the actual measured values in the trailing part of the pulses, for $t > 0.85$ s, the differences being larger in general for the case of the reflected pulse. The situation is rather more complex for the previous instants, with different trends observed for the transmitted and reflected pulses, and with a noticeable influence of the port at which the junction is excited.

The results obtained for the rest of the modeling approaches considered are compared in Figure 8, where for clarity the experimental results are not shown in the top figures, but the bottom figures have been expanded to allow proper analysis of the behavior observed in each of the propagation paths. It is apparent that the conventional pressure loss method (labeled 1D in the figure) is much less dispersive than any of the staggered-grid methods, and thus better suited for this particular calculation setting.

This is particularly true in the case of the reflected pulses, where the conventional method approaches the measured values considerably earlier. It is also apparent that while no significant differences can be found between the MDT and the FCT methods in the reflection seen from port 1, this is not the case when the junction is excited at port 3: the FCT method exhibits larger differences, except in the last part of the reflected pulse.

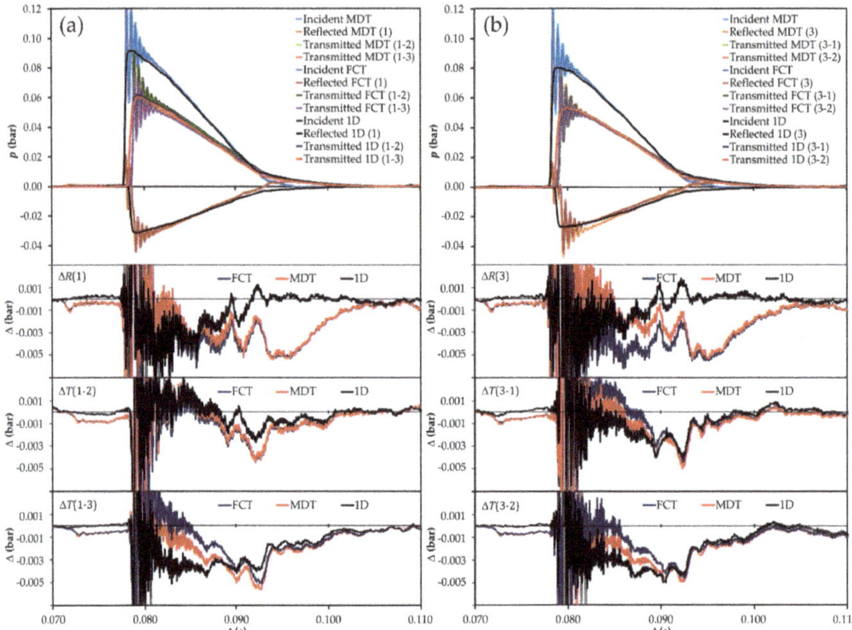

Figure 8. Comparison between the different models for the T-junction: raw data (top) and differences (bottom plots) in the time domain. (**a**) excitation at port 1; (**b**) excitation at port 3. Ports are denoted as in Figure 1. $\Delta R(i)$: difference in the pulse reflected at port i; $\Delta T(i - j)$: difference in the pulse transmitted between ports i and j. FCT: flux-corrected transport; MDT: momentum diffusion term; 1D: conventional pressure-loss model.

Regarding the different transmission paths, it can be observed that, while relatively small differences between all the methods are seen in the case of transmission from port 1 to port 2 (the performance of the conventional method being slightly better), significant differences appear at intermediate times in all the cases in which port 3 is involved. There is a clear trend in the results obtained in these cases, in the sense that the conventional method produces the lowest values, the FCT method the highest values, and those produced by the MDT method lie in between. However, the maximum differences are observed at time instants in which the amplitude of the transmitted pulses is relatively high.

As an additional criterion for the comparison of the performance of the different modeling approaches, the mean quadratic error corresponding to the differences shown in Figure 8 was computed. A time window with $0.82 < t < 0.95$ was chosen to avoid the large oscillations and to focus on those times for which the differences between the methods are more apparent.

The values obtained for the mean quadratic errors are summarized in Table 1, where it is confirmed that the best values for the reflection coefficients are those provided by the conventional method, while the FCT method gives the best approach to the transmission coefficients.

Table 1. Values of the mean quadratic error: T-junction.

Path	MDT	FCT	1D
$R(1)$	1.514×10^{-4}	1.842×10^{-4}	1.448×10^{-4}
$R(3)$	1.102×10^{-4}	2.046×10^{-4}	1.019×10^{-4}
$T(1\text{--}2)$	1.005×10^{-4}	9.932×10^{-5}	6.926×10^{-5}
$T(1\text{--}3)$	1.609×10^{-4}	1.298×10^{-4}	1.774×10^{-4}
$T(3\text{--}1)$	1.121×10^{-4}	1.094×10^{-4}	1.114×10^{-4}
$T(3\text{--}2)$	1.554×10^{-4}	1.245×10^{-4}	1.833×10^{-4}

Abbreviations: FCT: flux-corrected transport; MDT: momentum diffusion term; 1D: conventional pressure-loss model.

Similar comments can be made about the comparison shown in Figure 9 for the case of the Y-junction. Again, the conventional method reproduces better the behavior of the reflected pulses, regardless of the excitation port, and the MDT and the FCT methods exhibit significant differences only when the junction is excited at port 3, following the same trend as for the T-junction.

The trend is also very similar for the different transmission paths. Transmission between ports 1 and 2 is acceptably reproduced by all the modeling approaches, regardless of the exciting port, again with a slightly better performance of the conventional model. In those cases in which port 3 is included in the transmission path, the tendency observed is again the same when the junction is excited at port 1 or 2, with a small difference with respect to the T-junction when the excitation comes from port 3: in this case, the lowest values are those provided by the MDT method, most notably in the transmission from port 3 to port 2.

Again, the mean quadratic errors were calculated, and the corresponding results shown in Table 2 confirm the previous comments.

Table 2. Values of the mean quadratic error: Y-junction.

Path	MDT	FCT	1D
$R(1)$	1.575×10^{-4}	1.915×10^{-4}	1.273×10^{-4}
$R(2)$	1.992×10^{-4}	2.172×10^{-4}	1.681×10^{-4}
$R(3)$	1.648×10^{-4}	2.945×10^{-4}	1.394×10^{-4}
$T(1\text{--}2)$	1.171×10^{-4}	1.369×10^{-4}	9.186×10^{-5}
$T(1\text{--}3)$	1.514×10^{-4}	1.322×10^{-4}	1.742×10^{-4}
$T(2\text{--}1)$	1.992×10^{-4}	2.172×10^{-4}	1.681×10^{-4}
$T(2\text{--}3)$	1.336×10^{-4}	1.296×10^{-4}	1.141×10^{-4}
$T(3\text{--}1)$	1.669×10^{-4}	1.949×10^{-4}	1.355×10^{-4}
$T(3\text{--}2)$	2.117×10^{-4}	1.516×10^{-4}	1.751×10^{-4}

Abbreviations: FCT: flux-corrected transport; MDT: momentum diffusion term; 1D: conventional pressure-loss model.

From these results, it is apparent that the conventional pressure loss model, while is not able to account for all the differences observed between the two transmission paths studied in each test, could still provide a sufficient approximation to the real situation if the focus of the problem is on the reflection properties of the junction and only time domain issues are relevant for the problem under study (for instance, the eventual influence of a reflection at an intake junction on the volumetric efficiency on the engine). The staggered mesh finite volume method appears to be more sensitive to the relative directions of the different branches, mostly when the excitation comes from the side branch (port 3), as should be expected since the momentum equation is actually solved, albeit in an approximate way, at the junction, whereas in the conventional model such effects are included only through their influence on the pressure loss coefficients.

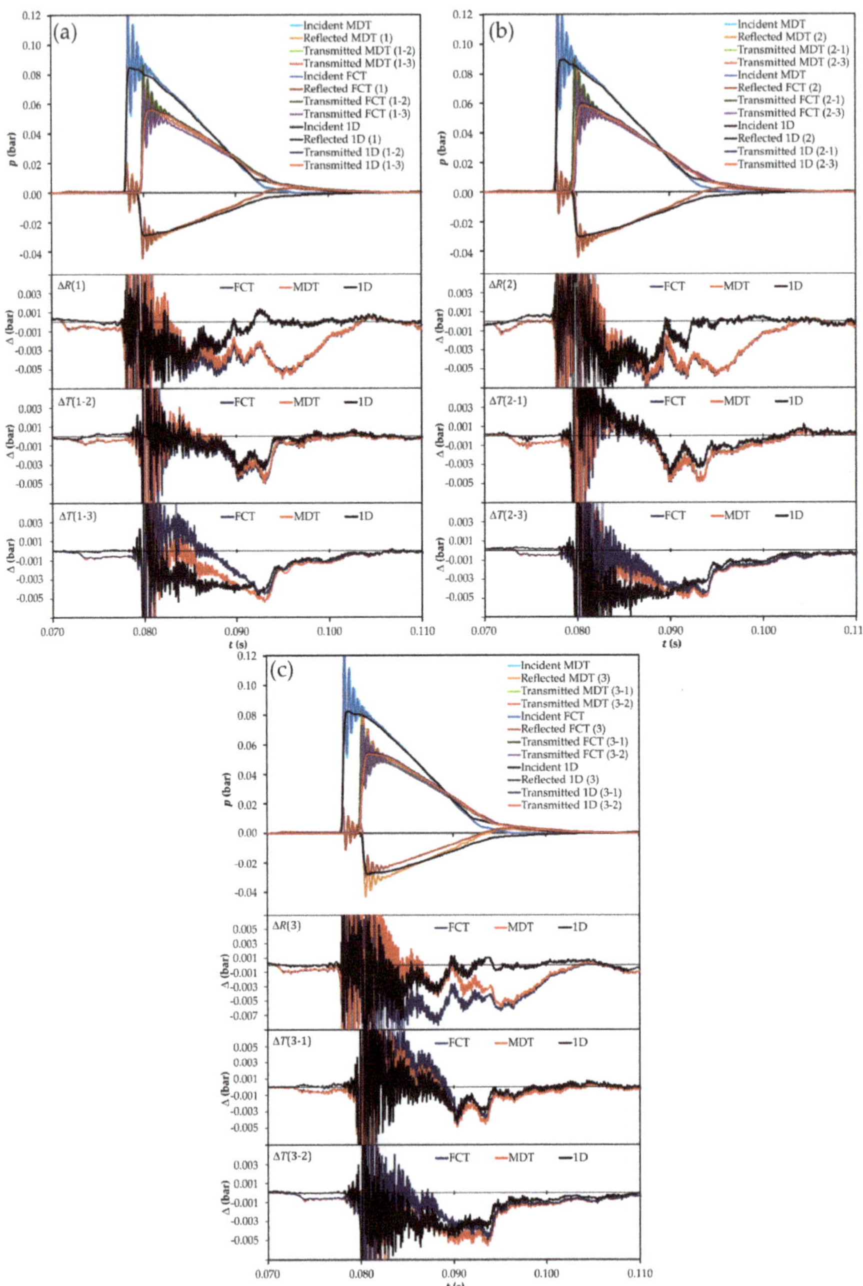

Figure 9. Comparison between the different models considered for the Y-junction (time domain): (**a**) excitation at port 1; (**b**) excitation at port 2; (**c**) excitation at port 3. Ports are denoted as in Figure 1. $\Delta R(i)$: difference in the pulse reflected at port i; $\Delta T(i - j)$: difference in the pulse transmitted between ports i and j. FCT: flux-corrected transport; MDT: momentum diffusion term; 1D: conventional pressure-loss model.

3.2.2. Frequency Domain Assessment

As already detected when describing the experimental results, it is in the frequency domain where the benefits of the staggered mesh finite volume method are more apparent. Consider first the results corresponding to the T-junction, shown in Figure 10 in the case of the reflection coefficients. Here, using either MDT or FCT as flux limiter, the staggered mesh finite volume method produces results for the reflection coefficients which overestimate dynamic effects when the excitation is at port 1, but produces a suitable approximation up to 1000 Hz when the excitation is at port 3 and the FCT flux limiter is used. In comparison with this, it is apparent that the conventional pressure loss model (again labeled as 1D in the figure) is unable to fully capture the dynamic features of the results, while still providing a sort of suitable average value, even if all the dynamic issues are lost, as an unavoidable consequence of the quasi-steady assumption underlying the calculation.

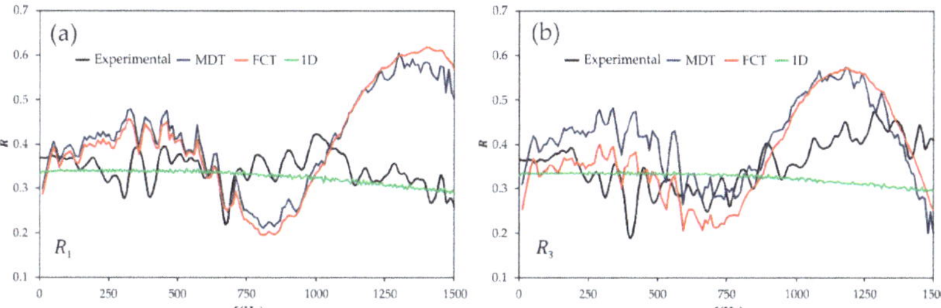

Figure 10. Comparison between the experimental results and the different models for the reflection coefficients of the T-junction (frequency domain): (**a**) excitation at port 1; (**b**) excitation at port 3. Ports are denoted as in Figure 1.

The corresponding transmission coefficients are shown in Figure 11, where it can be observed that the staggered mesh finite volume method produces results that follow the overall trend of the experimental results, with two exceptions: when the excitation is at port 1 the method underestimates the transmission to port 3, and when the excitation is at port 3 the method is unable to capture the behavior observed between 1000 and 1250 Hz. In the case of the conventional model, it is apparent that in this case it is fully unable to reproduce neither the level nor the dynamic features of the measured data, the only acceptable results being produced when the excitation is at port 1 and that only for very low frequencies.

This essential difference between the two modeling approaches considered is even more apparent in the case of the Y-junction, whose results are shown in Figures 12 and 13 for the reflection and transmission coefficients, respectively. In this case, the results provided by the conventional model are rather similar regardless of the port at which the junction is excited. In all the cases, an acceptable value of the transmission coefficient in the very low frequencies is produced in those propagation paths with smaller change in direction, and also a suitable average value for the reflection coefficient as seen from any of the exciting ports. However, differences in transmission between the two propagation paths are not reproduced in any case and, moreover, the results start to decrease monotonically at about 200 Hz and reach totally unrealistic values for frequencies above 750 Hz in all the cases.

On the contrary, the staggered mesh finite volume method reproduces quite fairly the overall dependency with frequency, but tends to overestimate the influence of the change in direction of the propagation path on the transmission coefficients (and thus to underestimate the value of the corresponding coefficient). With this geometry, this effect is especially evident in the results obtained with the FCT flux limiter for $|T_{13}|$, $|T_{23}|$, $|T_{31}|$ and $|T_{32}|$, i.e., all the cases in which the side branch (port 3) is involved. On the contrary, the results of the FCT method are affected by a certain overestimation when transmission through the main branch is considered ($|T_{12}|$ and $|T_{21}|$). Accordingly, with the description given in Appendix B, this difference in behavior between the FCT and the MDT methods can only be due to the effect of the application to the junction itself of the different ways used to handle the information of the neighboring volumes when limiting the flow.

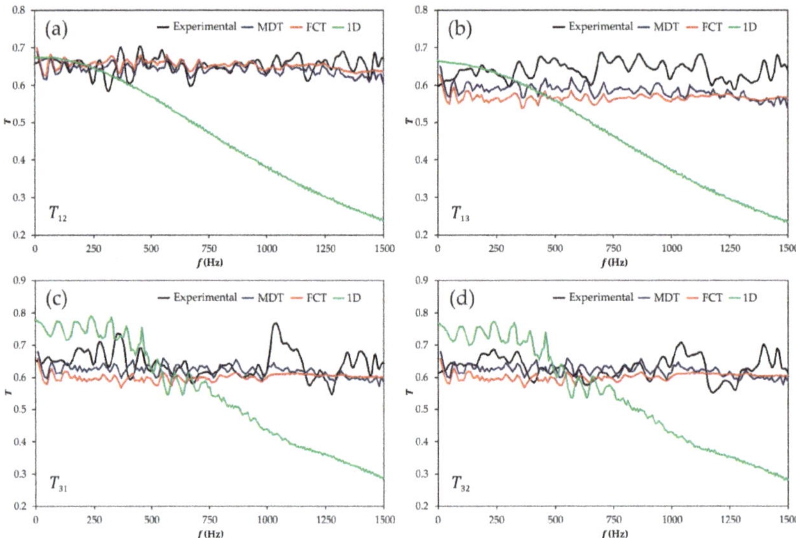

Figure 11. Comparison between the experimental results and the different models for the transmission coefficients of the T-junction (frequency domain): (**a**) excitation at port 1, transmission through port 2; (**b**) excitation at port 1, transmission through port 3; (**c**) excitation at port 3, transmission through port 1; (**d**) excitation at port 3, transmission through port 2. Ports are denoted as in Figure 1.

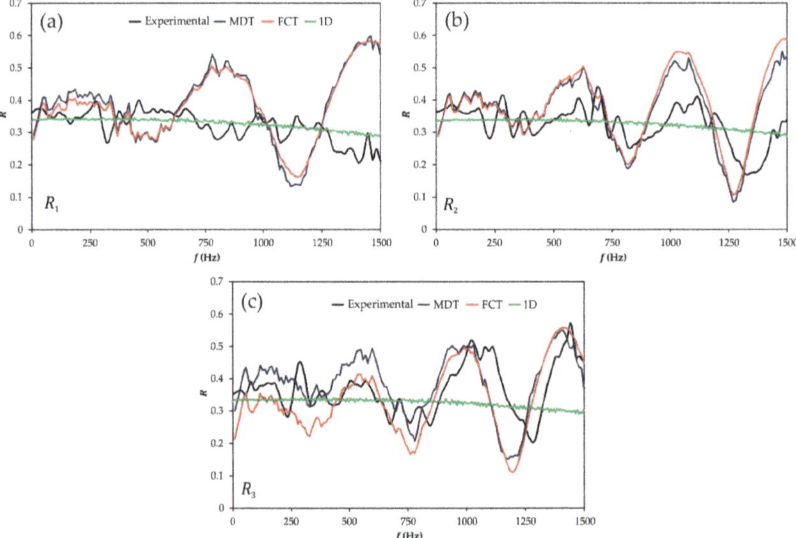

Figure 12. Comparison between the experimental results and the different models for the reflection coefficients of the Y-junction (frequency domain): (**a**) excitation at port 1; (**b**) excitation at port 2; (**c**) excitation at port 3. Ports are denoted as in Figure 1.

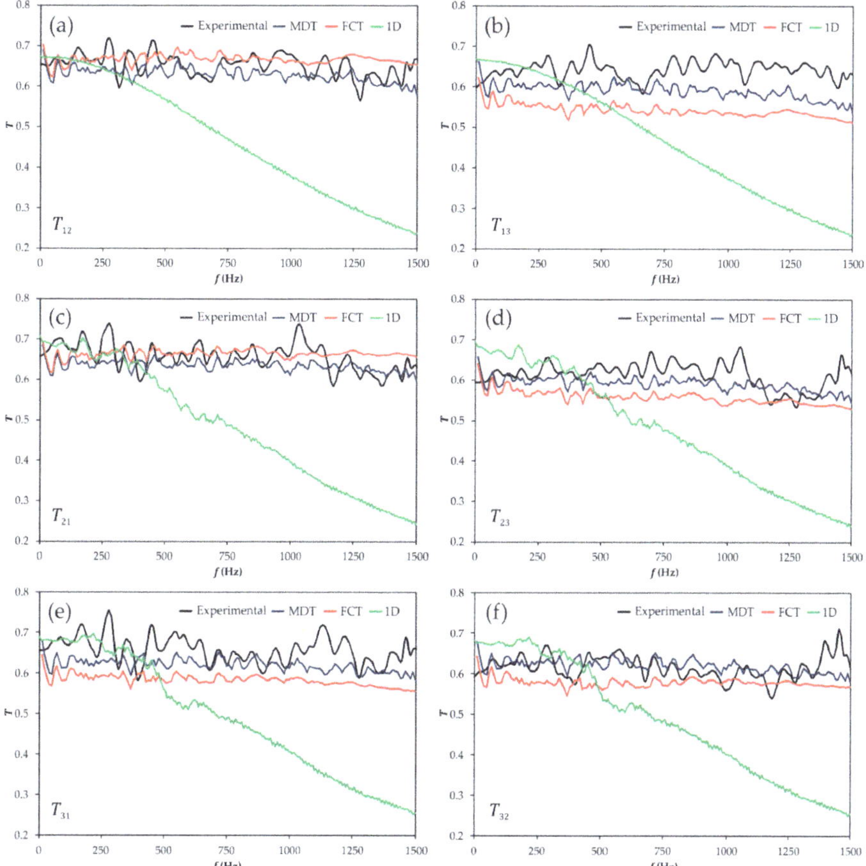

Figure 13. Comparison between the experimental results and the different models for the transmission coefficients of the Y-junction (frequency domain): (**a**) excitation at port 1, transmission through port 2; (**b**) excitation at port 1, transmission through port 3; (**c**) excitation at port 2, transmission through port 1; (**d**) excitation at port 2, transmission through port 3 (**e**) excitation at port 3, transmission through port 1; (**f**) excitation at port 3, transmission through port 2. Ports are denoted as in Figure 1.

In the case of the reflection coefficients, the overestimation of the junction dynamics already observed in the T-junction is also present here when the junction is excited at port 1, but the measured dynamics are quite successfully reproduced when the junction is excited at ports 2 and 3. The characteristic frequencies governing the reflection coefficient are not exactly captured, but the overall amplitude and the influence of the exciting port are reproduced by the numerical results.

3.3. Assessment of a Modelling Approach with a Quasi-3D Description of the Junction

In order to explore the additional potential offered by the staggered mesh method regarding the approximate solution of the three-dimensional flow field inside the junction, such an approach (commonly referred to in the literature as quasi-3D) was finally considered. In Figure 14 the mesh used is shown together, for reference, with that used in the previous subsections. The four volumes at the endpoints of the part shown are then connected to a single volume, thus providing the connection with the one-dimensional computation at the ducts. The mesh chosen is relatively modest, in order to keep the computation time at reasonable values, but sufficient to show any potential advantages of this description.

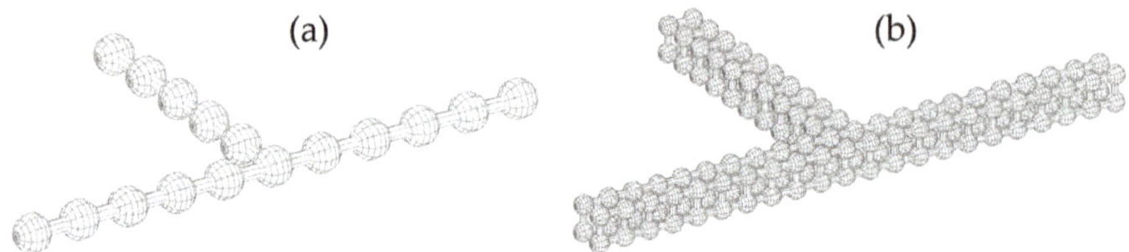

Figure 14. Meshes used in the staggered-grid method: (**a**) a 0D description of the junction; (**b**) a quasi-3D description.

Additionally, in view of the previous results, only the MDT method will be used as a flow limiter, since overall it has appeared to be more robust and consistent, and only the case of the T-junction will be analyzed in the following, as no new qualitative issues have been identified in the Y-junction that were not present also in the T-junction.

3.3.1. Time Domain Assessment

In Figure 15 the results obtained with the quasi-3D description of the junction (labeled MDT Q3D) are compared with those previously shown in Figure 7 corresponding to the MDT with 0D description of the junction. Again the plots on top represent the raw results, whereas the bottom plots show the differences with respect to the experimental values. It can be seen that, in all the cases, a certain improvement is achieved when using the quasi-3D junction, improvement which is more apparent when the junction is excited at port 3, this is, at the side branch, which is intuitively reasonable.

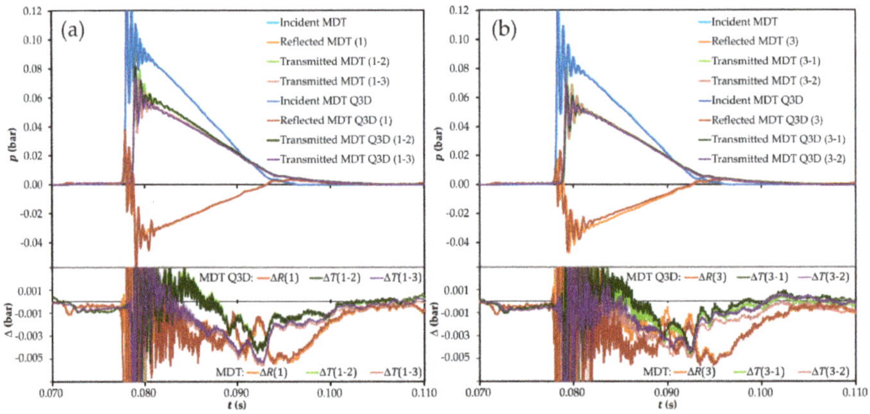

Figure 15. Influence of the description of the junction in the time domain, MDT method: raw data (top) and differences with measurement (bottom). (**a**) Excitation at port 1; (**b**) excitation at port 3. Ports are denoted as in Figure 1.

Again, the mean quadratic errors were computed, as shown in Table 3.

These results indicate that, while the reflection coefficients exhibit a similar mean error, there is a substantial improvement in the transmission coefficients, thus confirming the previous analysis.

However, the improvement achieved is not sufficient to produce results comparable to those shown in Figure 8 for the conventional model in the case of the reflection coefficient, and the differences in the transmission coefficients are clearly significant only when the junction is excited at the side branch. Therefore, while it appears that further refinement of the mesh at the junction should improve further the quality of the results, this might produce in turn an unacceptable increase in the computation time.

Table 3. Values of the mean quadratic error: T-junction.

Path	MDT	MDT Q3D
$R(1)$	1.514×10^{-4}	1.513×10^{-4}
$R(3)$	1.102×10^{-4}	1.809×10^{-4}
$T(1\text{–}2)$	1.005×10^{-4}	1.018×10^{-4}
$T(1\text{–}3)$	1.609×10^{-4}	1.469×10^{-4}
$T(3\text{–}1)$	1.121×10^{-4}	1.044×10^{-4}
$T(3\text{–}2)$	1.554×10^{-4}	1.249×10^{-4}

Abbreviations: MDT Q3D: momentum diffusion term with quasi-3D description of the junction.

3.3.2. Frequency Domain Assessment

Quite unexpectedly, the improvements just commented do not have a translation in the frequency domain. In the case of the reflection coefficients, shown in Figure 16, the influence of the new description is apparent, but the new results do not provide any improvement in the reproduction of the experimental trend, showing even a certain degree of degradation in the quality of the results, with the abnormally high values achieved around 1300 Hz when the junction is excited at port 1.

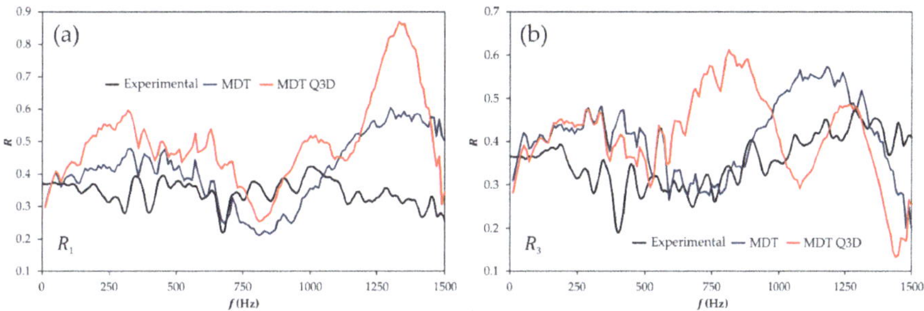

Figure 16. Influence of the description of the junction on the reflection coefficients (frequency domain), MDT method. (**a**) Excitation at port 1; (**b**) excitation at port 3. Ports are denoted as in Figure 1.

The same comments apply to the transmission coefficients shown in Figure 17. The strange behavior around 1300 Hz is again present in those coefficients associated with the excitation at port 1, while when the junction is excited at port 3 no significant improvement can be detected.

The only possible explanation for the behavior observed, this is, something that produces minor but evident improvements in the time domain, but that induces a rather severe degradation in the quality of the results in the frequency domain, would be that some spurious high-frequency oscillations are being generated at the interface between the quasi-3D junction and the 1D elements of the adjacent ducts, due to the virtual merging of four volumes into a single one.

In order to clarify this point, the whole system was meshed as shown in Figure 14b for the junction, and the results are shown in Figure 18, only for the case in which the junction is excited at port 1. It is apparent that a dramatic improvement in the quality of the transmission coefficients is achieved, now showing a more realistic influence of the change in direction. In the case of the reflection coefficients the improvement is not so apparent, but the amplitude of the oscillations is smaller, what indicates that further refining of the mesh could lead to substantially improved results. However, that would be impractical, since the computation time increases substantially when the whole system is meshed in this way.

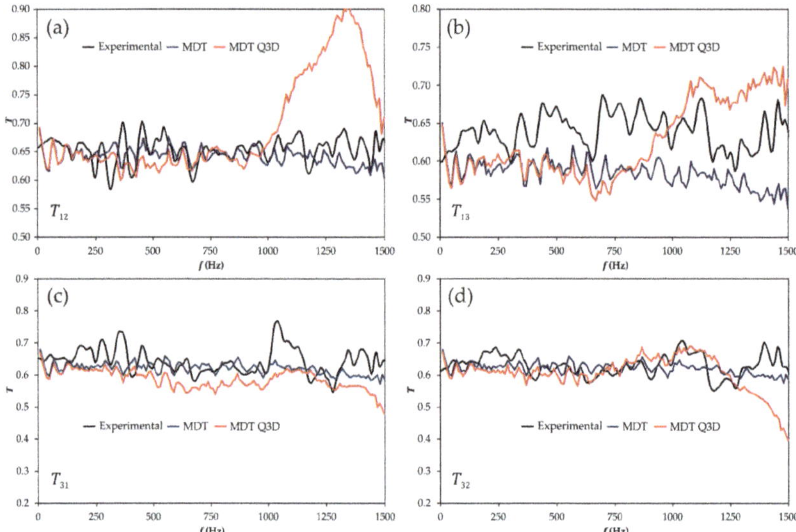

Figure 17. Influence of the description of the junction on the reflection coefficients (frequency domain), MDT method. (**a**) excitation at port 1, transmission through port 2; (**b**) excitation at port 1, transmission through port 3; (**c**) excitation at port 3, transmission through port 1; (**d**) excitation at port 3, transmission through port 2. Ports are denoted as in Figure 1.

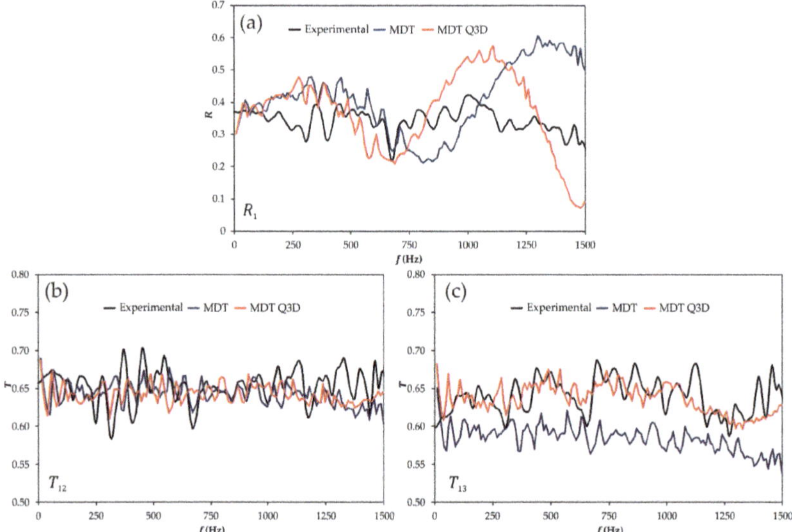

Figure 18. Influence of using a quasi-3D approach for the whole system on the reflection and transmission coefficients (frequency domain). MDT method, excitation at port 1. (**a**) reflection at port 1; (**b**) transmission through port 2; (**c**) transmission through port 3. Ports are denoted as in Figure 1.

4. Conclusions

The objective of the work presented was to establish the potential of staggered mesh finite volume models as a way to improve the description of the effect of simple duct junctions on an otherwise one-dimensional flow system, as the intake or exhaust of an internal combustion engine.

With that purpose, specific experiments were performed making use of a modified impulse method, in which two different junctions were characterized as a multi-port, and that provided precise and reliable results on the propagation of pressure pulses across junctions.

The results obtained were then compared to numerical results obtained from different methods, both in the time and the frequency domains. First, methods assuming a zero-dimensional description of the junction were assessed, including the staggered mesh finite volume method with different flux limiters and, as a reference for comparison, a more conventional pressure loss-based model. Then, the potential of using the staggered mesh finite volume method in order to produce a quasi-3D description of the junction, to be coupled with the one-dimensional description of the adjacent ducts, was explored.

As an overall conclusion of the results found, one may state that none of the modeling approaches considered is able to reproduce totally the observed behavior. However, the performance of the different models is such that a suitable choice seems to be possible depending on which is the actual focus of the problem under study: situations in which a suitable time domain description may be sufficient may be addressed either with the conventional quasi-steady pressure loss model (most notably when the focus is on the reflection properties of the junction) or with the staggered mesh model with quasi-3D junction description (in this last case, when the main interest is on transmission, and given that the lengths involved in the problem will not be as long as to give rise to spurious oscillations due to the dispersive character of the method).

When the focus is on the frequency domain and on the dynamic behavior of the junction, it is the staggered mesh method the one that provides the most suitable results, at least from a qualitative point of view, as a consequence of the fact that momentum conservation across the junction is accounted for. However, due to spurious oscillation arising from the method used to couple a quasi-3D junction to the 1D ducts, suitable results with an acceptable computation time have been obtained only either with the zero-dimensional description of the junction or with a full quasi-3D description of the whole system, this last option being unacceptable in practice. It is thus clear that further work is needed in this case in order to find the optimal settings for the calculation, most notably in the connection between the quasi-3D and the 1D regions.

Finally, it should be recalled that no empirical information has been included in the staggered mesh method used; the incorporation of such information in terms of effective sections and characteristic lengths and the evaluation of their potential could be additional topics for further research.

Acknowledgments: Manuel Hernández is partially supported through contract FPI-S2-2015-1064 of Programa de Apoyo para la Investigación y Desarrollo (PAID) of Universitat Politècnica de València. The authors wish to thank Adolfo Guzmán for manufacturing the junctions and for his technical support during the tests.

Author Contributions: A.J. Torregrosa conceived the structure of the work, contributed to the analysis and wrote a large part of the article; A. Broatch supervised the experiments, wrote Appendix A and contributed to the discussion; L.M. García-Cuevas implemented the conventional model, ran the related simulations, wrote Appendix C and contributed to the discussion; M. Hernández implemented the staggered mesh method, ran the related simulations, wrote Appendix B and contributed largely to the discussion.

Conflicts of Interest: The authors declare no conflict of interest. The founding sponsors had no role in the design of the study; in the collection, analyses.

Appendix A. Experimental Procedure

Making reference to the notation in Figure 2, the determination of the transmission and reflection coefficients defined in Equation (1) requires the three following measurements:

- Excitation in duct 1, with anechoic terminations in ducts 2 and 3, so that $A_1 \neq 0$ and $A_2 = A_3 = 0$, and thus,

$$R_1 = B_1/A_1; \quad T_{12} = B_2/A_1; \quad T_{13} = B_3/A_1 \,. \tag{A1}$$

- Excitation in duct 2, with anechoic terminations in ducts 1 and 3, so that $A_2 \neq 0$ and $A_1 = A_3 = 0$; then,

$$R_2 = B_2/A_2; \quad T_{21} = B_1/A_2; \quad T_{23} = B_3/A_2 \,. \tag{A2}$$

- Excitation in duct 3, with anechoic terminations in ducts 1 and 2, so that $A_3 \neq 0$ and $A_1 = A_2 = 0$, so that,

$$R_3 = B_3/A_1; \quad T_{31} = B_1/A_3; \quad T_{32} = B_2/A_3 \,. \tag{A3}$$

In order to perform the above-indicated tests, the modified version of the impulse method described in [28] was used, since pressure components, which all the previous developments are based upon, are directly obtained in the time domain with a simple procedure, and the consideration of three-port elements is straightforward. In Figure A1 both the experimental setup used and the relevant pressure waves recorded are illustrated.

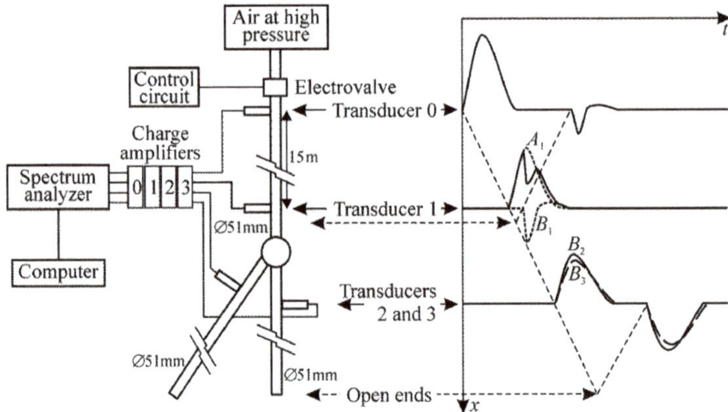

Figure A1. Experimental setup used.

The incident pulse is generated by means of a high speed electrovalve that controls the discharge from a high-pressure tank. A proper choice of the opening time ensures that the spectrum associated with the incident pulse is essentially flat. The length of the ducts placed between the valve and transducer 0, transducer 0 and the junction, and the junction and the open ends is chosen so that no windowing is necessary in order to isolate the incident, the reflected, and the transmitted pulses, as indicated in the figure. Transducer 0 was located 15 m away from both the valve and the junction, and transducers 2 and 3 were placed 0.15 m downstream of the junction, and 15 m away from their corresponding open end.

At the position indicated for transducer 1 in Figure 2, it is clear that this transducer records the addition of the incident and the reflected pulses, as illustrated in the figure. In order to surpass this difficulty, the solution adopted is to estimate the pulse incident on the junction at section 1 (whose Fourier transform will give the complex amplitude of the A_1 component) from an additional test performed without any element, using the pressure recorded by transducer 0 only to check the comparability of the excitations used in both types of tests (with and without junction).

Appendix B. Staggered-Grid Finite-Volume Approach

As described by Torregrosa et al. [34], the quasi-3D model uses a staggered grid with two different basic elements: volumes and connectors. Volumes have associated scalar information such as pressure, density or temperature, as well as the cell volume. Connectors contain vector information, like flow velocity, momentum or the orientation of the connector in space, together with its area. With this configuration, a volume might have attached as many connectors as needed, but a connector will always connect only two volumes. In Figure A2 two volumes and a connector are shown schematically (volumes and connectors actually have an undefined shape).

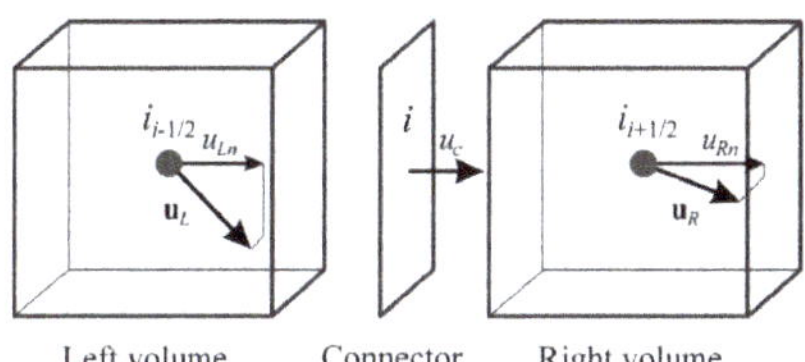

Figure A2. Basic mesh elements, definition of velocity projections and notation of volumes and connectors.

The method is based on the solution of the 3D Euler conservation equations without source term:

$$\partial \rho / \partial t + \nabla \cdot (\rho \mathbf{u}) = 0 \tag{A4}$$

$$\partial (\rho \mathbf{u}) / \partial t + \nabla \cdot (\rho \mathbf{u}\mathbf{u}) = -\nabla p \tag{A5}$$

$$\partial (\rho e_0) / \partial t + \nabla \cdot [(\rho e_0 + p)\mathbf{u}] = 0 \tag{A6}$$

together with the perfect gas equation of state. Here, ρ is the density, p is the pressure, $\mathbf{u}$ is the velocity vector, $\nabla \cdot$ indicates the divergence, and e_0 is the specific stagnation internal energy, whose expression for a perfect gas is:

$$e_0 = c_v T + u^2 / 2 \tag{A7}$$

where c_v is the specific heat capacity at constant volume and T is the fluid static temperature.

However, in the context of a finite volume method on a staggered grid, the key issue is how and where these equations are solved. The mass conservation equation, being scalar, is solved in the volumes. Upon discretization it becomes:

$$\rho^{n+1} = \rho^n + \frac{\Delta t}{V} \sum_{c=1}^{N_c} \rho_c^n u_c^n A_c \tag{A8}$$

where u is the flow velocity. Superscript n indicates the time step, Δt is the time interval, V the volume of the cell, N_c the number of connectors, A is the cross section, and subscript c indicates that the variable is taken at a connector (otherwise it is assumed that the variable is taken at the volumes).

Following a similar procedure, the discretized energy equation is written as:

$$(\rho e_0)^{n+1} = (\rho e_0)^n + \frac{\Delta t}{V} \sum_{c=1}^{N_c} \rho_c^n e_0^n u_c^n A_c + \frac{\Delta t}{V} \sum_{c=1}^{N_c} p_c^n u_c^n A_c \tag{A9}$$

The momentum equation in a 3D case consists of three coupled equations. In this context, the momentum equation is solved in the connectors and only in the direction orthogonal to its surface. This is achieved by projecting the flow velocity of the two adjacent volumes into that direction, as shown in Figure A2, where the velocity u_c in the connector, and the projections of the volume flow velocity from its left and right, u_{Ln} and u_{Rn}, are shown. As a result, the momentum in the connector follows a one-dimensional momentum equation, whose discretization gives:

$$(\rho_c u_c A_c)^{n+1} = (\rho_c u_c A_c)^n + (\Delta t/\Delta L)\left[\left(\rho u_n^2 + p\right)_L + \left(\rho u_n^2 + p\right)_R\right] A_c \tag{A10}$$

Here, u_n is the velocity projection onto the direction orthogonal to the connector surface and subscripts R and L denote the variables taken from the volumes at the right and left of the connector, respectively. With this simplification, a single equation is solved for each connector instead of three coupled equation for each volume. While usually the number of connectors is higher than the number of volumes in a 3D mesh, the fact that the momentum equations for each direction are not coupled reduces drastically the computation time when compared to a regular 3D method.

Once the momentum is calculated, its value is used to compute the mass and energy conservation equations in the next time step. It is worth noticing that some terms in the momentum and energy equations, like density or pressure, must be evaluated in the connectors, but are only calculated in the volumes. To solve this, an upwind approach is adopted, the required values being taken from the right or left volumes, depending on the flow direction.

Finally, the momentum in the volumes is computed by distributing the momentum calculated in the connectors among the volumes they connect, taking into account their size. In a uniform 1D mesh, half the momentum is thus assigned to each volume. As the orientation of the connectors in space is known, the resulting momentum vector of a volume can be calculated from the vector sum of the momentum in the connectors:

$$(\rho_c \mathbf{u} V)_v^{n+1} = \frac{1}{2} \sum_{c=1}^{N_c} (\rho \mathbf{u}_c A_c \Delta L)_c^{n+1} \tag{A11}$$

The resulting method is a second-order accuracy method based on an explicit scheme with a staggered grid, as shown in Figure A3.

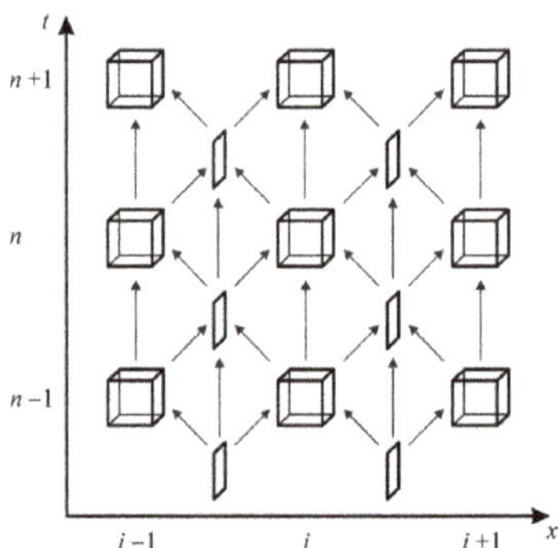

Figure A3. Scheme of the staggered mesh and the associated time marching.

The fact that the resulting scheme offers second-order accuracy, together with the simplifications adopted in the momentum equation, results in nonphysical oscillations, especially in the vicinity of significant pressure gradients. In order to mitigate such overshoots, two different flux limiters have been used here: the momentum diffusion term (MDT) proposed in [35] and the flux-corrected transport (FCT) methodology proposed in [34].

In the case of the MDT, the main goal is to add a diffusion term to the momentum equation so that the mass flux computed at that connector is conveniently limited. With this purpose, the momentum flux density tensor used in the momentum Equation (A5) can be modified, in a way similar to that used for incorporating viscosity effects, as follows:

$$\partial(\rho\mathbf{u})/\partial t + \nabla\cdot(\rho\mathbf{u}\mathbf{u} + \mathbf{D}) = -\nabla p \tag{A12}$$

where the tensor $\mathbf{D}$ is assumed to depend linearly on the local momentum gradients, i.e.,

$$\mathbf{D} = \epsilon\nabla(\rho\mathbf{u}) \tag{A13}$$

The scalar quantity ϵ has the dimensions of a kinematic viscosity and can thus be interpreted as a momentum diffusion coefficient. With this prescription, the contribution of the diffusion term $\nabla\cdot\mathbf{D}$ will only be relevant if significant gradients exist, and any resulting spurious oscillations will be damped.

Adding the projection of Equation (A12) onto the direction of a connector to the discretized momentum equation one gets:

$$(\rho_c u_c A_c)^{n+1} = (\rho_c u_c A_c)^n + \left(\frac{\Delta t}{\Delta L}\right)\left\{\left[\left(\rho u_n^2 + p\right)_L + \left(\rho u_n^2 + p\right)_R\right]A_c + \left(\widetilde{D}_{Ln} - \widetilde{D}_{Rn}\right)\right\} \tag{A14}$$

where $\widetilde{D}_L$ and $\widetilde{D}_R$ are the projections onto the connector direction of tensor $\widetilde{\mathbf{D}} = \epsilon\nabla(\rho\mathbf{u}A)$ computed in the two adjacent volumes. Following [35], the momentum diffusion coefficient is computed by relating the mesh size and the time step to the local flow velocity at the volume, as:

$$\epsilon = \frac{|\mathbf{u}|}{2}(\Delta L - |\mathbf{u}|\Delta t) \tag{A15}$$

and the gradient of mass flow $\nabla(\rho\mathbf{u}A)$ is also computed from the projections of the mass flows of the adjacent connectors onto each direction.

Considering now the FCT technique, it was initially conceived to be used in a finite differences scheme, but it was adapted to a finite volume staggered grid in [34]. In the FCT technique three stages can be identified: a transport stage based on the scheme considered, a diffusion stage where the numerical dispersion is reduced, and an anti-diffusion stage in which the accuracy of the scheme where the solution is smooth is restored. The diffusion operator is defined as:

$$D_j(W) = \theta\left(W_{j+1/2}\right) - \theta\left(W_{j-1/2}\right) \tag{A16}$$

with,

$$\theta\left(W_{j+1/2}\right) = \vartheta/4(W_{j+1} - W_j) \tag{A17}$$

where W_j is the conservative variable computed at cell j in the transport stage, $j \pm 1/2$ indicates that the conservative variable is evaluated in the midpoint between j and $j \pm 1$, and the factor ϑ is a positive real number with $\vartheta \geq 1/2$. Calculating the diffusion via damping, as suggested in [34], the new variable $\overline{W}_j$ is obtained as:

$$\overline{W}_j^{n+1} = W_j^{n+1} + D_j(W^n) \tag{A18}$$

When considering a staggered mesh finite volume method, the FCT is only applied to the momentum equation, since this is the main source of oscillations. Therefore, all the calculations in the volumes are made using $W_j^n = (\rho u_c A_c)_j^n$ as the conservative variable, whereas the calculations in the midpoints make use of the variables evaluated at the volumes adjacent to the connector.

Finally, the non-linear anti-diffusion operator A_j is defined as:

$$A_j(W) = \Psi\left(W_{j-1/2}\right) - \Psi\left(W_{j+1/2}\right) \tag{A19}$$

Making use of the anti-diffusive limited flow defined in [37] gives:

$$\Psi\left(W_{j+1/2}\right) = s\max\left[0, \min\left((5/8)s\,\Delta W_{j-1/2}, (1/8)\left|\Delta W_{j+1/2}\right|, (5/8)s\,\Delta W_{j+3/2}\right)\right] \tag{A20}$$

where $s = \text{sign}\left(\Delta W_{j-1/2}\right)$, $\Delta W_{j-1/2} = W_j - W_{j-1}$, $\Delta W_{j+1/2} = W_{j+1} - W_j$ and $\Delta W_{j+3/2} = W_{j+2} - W_{j+1}$. Then, according to [34], the phoenical form should be used, so that:

$$\overline{\overline{W}}_j^{\,n+1} = \overline{W}_j^{\,n+1} + A_j\left(W^{n+1}\right) \tag{A21}$$

The anti-diffusion stage equations can be adapted to the staggered grid mesh finite volume method in a way similar to that used with the diffusion stage.

Appendix C. 1D Method with Pressure Loss-Based Junction Model

In this case, a collocated one-dimensional finite volume method is used for all the calculations inside the pipes. The Euler equations of fluid dynamics simplified for one-dimensional flow in a straight uniform duct can be expressed as:

$$\frac{d\overline{W}_i}{dt} = \frac{d}{dt}\begin{pmatrix}\rho \\ \rho u \\ \rho e_0\end{pmatrix}_i = \frac{A\left(F_{i-1,i} - F_{i,i+1}\right)}{V_i} \tag{A22}$$

Here, $\overline{W}_i$ is the cell-averaged state vector of cell i, $F_{i-1,i}$ and $F_{i,i+1}$ are the inter-cell fluxes between cells $i-1$ and i and between i and $i+1$, respectively, and the other symbols refer to the same magnitudes as in Appendix B, with u being now the axial velocity of the flow. The inter-cell fluxes are computed by an approximate solution of the Riemann problem as described by Toro et al. [38]. The state vector is extrapolated to the cell boundaries to compute the fluxes by means of a Monotonic Upstream-Centered Scheme for Conservation Laws (MUSCL) approach as described in [39], while the solution is propagated in time using Heun's method, leading to a second order in time and space, total variation diminishing scheme.

While the main one-dimensional flow inside the ducts is simulated, the effects of the geometry of the junction are modelled. The connections of the ducts to the junction are solved using an auxiliary small zero-dimensional element. Each of the one-dimensional branches is connected to that zero-dimensional element making use of the Riemann variables to compute the fluxes at their corresponding boundary. At each connection, it is assumed that a certain amount of stagnation pressure is lost, depending on the angle of the junction and of the ratio between the outflow mass flow $\dot{m}_{out}$ passing through the branch of interest and the inflow mass flow $\dot{m}_{in}$. The pressure loss coefficients are computed following the expressions given in [15,16], and are defined as the ratio of the difference in stagnation pressure between the outflow branch and the inflow branch to the dynamic pressure $(\rho u^2/2)$ of the inflow branch. Finally, the total pressure loss coefficient K for a three-branch junction with the same section in all the branches can be estimated as:

$$K = \left(\frac{\dot{m}_{out}}{\dot{m}_{in}}\right)^2 - \frac{3}{2}\frac{\dot{m}_{out}}{\dot{m}_{in}} + \frac{1}{2} \tag{A23}$$

when the branch is collinear with the inflow branch, and,

$$K = \left(\frac{\dot{m}_{out}}{\dot{m}_{in}}\right)^2 + 1 - 2\frac{\dot{m}_{out}}{\dot{m}_{in}}\cos\left(\frac{3}{4}\theta\right) \tag{A24}$$

for the lateral branch, when the flow is split between a collinear and a lateral branch. In this case, θ is the angle between the lateral branch and the axial outflow branch, so that $0°$ means that both outflow branches are parallel. The same expression applies when the inflow branch is not parallel to any of the outflow branches: in that case, the angle θ is measured between the inflow branch and the other outflow branch.

In the auxiliary zero-dimensional element, the gas is again considered as a perfect gas, and the mass and energy equations are solved:

$$\frac{\mathrm{d}m}{\mathrm{d}t} = \sum \dot{m} \tag{A25}$$

$$\frac{\mathrm{d}(mc_v T)}{\mathrm{d}t} = \sum \dot{m}h_0 \tag{A26}$$

where m is the mass trapped in the zero-dimensional element, $\dot{m}$ is the mass flow, positive when it enters the element, and h_0 is the specific stagnation enthalpy associated with the mass moving inside or outside of the element. These two equations set an additional limitation to the maximum possible time step.

1. Winterbone, D.E.; Pearson, R.J. *Design Techniques for Engine Manifolds*, 3rd ed.; Professional Engineering Pub. Ltd.: London, UK, 1999.
2. Payri, F.; Reyes, E.; Galindo, J. Analysis and modelling of the fluid-dynamic effects in branched exhaust junctions of I.C.E. *J. Eng. Gas Turbines Power* **2001**, *123*, 197–203. [CrossRef]
3. Tang, S.K. Sound transmission characteristics of Tee-junctions and the associated length corrections. *J. Acoust. Soc. Am.* **2004**, *115*, 218–227. [CrossRef] [PubMed]
4. Harrison, M.F.; De Soto, I.; Rubio-Unzueta, P.L. A linear acoustic model for multi-cylinder IC engine intake manifolds including the effects of the intake throttle. *J. Sound Vib.* **2004**, *278*, 975–1011. [CrossRef]
5. Karlsson, M.; Abom, M. Quasi-steady model of the acoustic scattering properties of a T-junction. *J. Sound Vib.* **2011**, *330*, 5131–5137. [CrossRef]
6. Karlsson, M.; Abom, M. Aeroacoustics of T-junctions—An experimental investigation. *J. Sound Vib.* **2010**, *329*, 1793–1808. [CrossRef]
7. Desantes, J.M.; Torregrosa, A.J.; Broatch, A. Experiments on flow noise generation in simple exhaust geometries. *Acta Acust. United Acust.* **2001**, *87*, 46–55.
8. Benson, R.S. *The Thermodynamics and Gas Dynamics of Internal-Combustion Engines;* Clarendon Press: Oxford, UK, 1982; Volume 1.
9. Corberán, J.M. A new constant pressure model for N-branch junctions. *Proc. Inst. Mech. Eng. D* **1992**, *206*, 117–123. [CrossRef]
10. Schmandt, B.; Herwig, H. The head change coefficient for branched flows: Why "losses" due to junctions can be negative. *Int. J. Heat Fluid Flow* **2015**, *54*, 268–275. [CrossRef]
11. Shaw, C.T.; Lee, D.J.; Richardson, S.H.; Pierson, S. Modelling the effect of plenum-runner interface geometry on the flow through an inlet system. *SAE Tech. Pap. Ser.* **2000**. [CrossRef]
12. Pérez-García, J.; Sanmiguel-Rojas, E.; Hernández-Grau, J.; Viedma, A. Numerical and experimental investigations on internal compressible flow at T-type junctions. *Exp. Therm. Fluid Sci.* **2006**, *31*, 61–74. [CrossRef]
13. Naeimi, H.; Domiry Ganji, D.; Gorji, M.; Javadirad, G.; Keshavarz, M. A parametric design of compact exhaust manifold junction in heavy duty diesel engine using computational fluid dynamics codes. *Therm. Sci.* **2011**, *15*, 1023–1033. [CrossRef]
14. Sakowitz, A.; Mihaescu, M.; Fuchs, L. Turbulent flow mechanisms in mixing T-junctions by Large Eddy Simulations. *Int. J. Heat Fluid Flow* **2014**, *45*, 135–146. [CrossRef]
15. Bassett, M.D.; Winterbone, D.E.; Pearson, R.J. Calculation of steady flow pressure loss coefficients for pipe junctions. *Proc. Inst. Mech. Eng. C* **2001**, *215*, 861–881. [CrossRef]
16. Hager, W.H. An approximate treatment of flow in branches and bends. *Proc. Inst. Mech. Eng. C* **1984**, *198*, 63–69. [CrossRef]
17. Paul, J.; Selamet, A.; Miazgowicz, K.D.; Tallio, K.V. Combining flow losses at circular T-junctions representative of intake plenum and primary runner interface. *SAE Tech. Pap. Ser.* **2007**. [CrossRef]

18. Pérez-García, J.; Sanmiguel-Rojas, E.; Viedma, A. New coefficient to characterize energy losses in compressible flow at T-junctions. *Appl. Math Model.* **2010**, *34*, 4289–4305. [CrossRef]

19. Wang, W.; Lu, Z.; Deng, K.; Qu, S. An experimental study of compressible combining flow at 45° T-junctions. *Proc. Inst. Mech. Eng. C* **2015**, *229*, 1600–1610. [CrossRef]

20. Peters, B.; Gosman, A.D. Numerical simulation of unsteady flow in engine intake manifolds. *SAE Tech. Pap. Ser.* **1993**. [CrossRef]

21. Bingham, J.F.; Blair, G.P. An improved branched pipe model for multi-cylinder automotive engine calculations. *Proc. Inst. Mech. Eng. Part D* **1985**, *199*, 65–77. [CrossRef]

22. William-Louis, M.J.P.; Ould-El-Hadrami, A.; Tournier, C. On the calculation of the unsteady compressible flow through an N-branch junction. *Proc. Inst. Mech. Eng. C* **1998**, *212*, 49–56. [CrossRef]

23. Bassett, M.D.; Pearson, R.J.; Fleming, N.P.; Winterbone, D.E. A multi-pipe junction model for one-dimensional gas-dynamic simulations. *SAE Tech. Pap. Ser.* **2003**. [CrossRef]

24. Pearson, R.J.; Bassett, M.D.; Batten, P.; Winterbone, D.E.; Weaver, N.W.E. Multi-dimensional wave propagation in pipe junctions. *SAE Tech. Pap. Ser.* **1999**. [CrossRef]

25. Bassett, M.D.; Winterbone, D.E.; Pearson, R.J. Modelling engines with pulse converted exhaust manifolds using one-dimensional techniques. *SAE Tech. Pap. Ser.* **2000**. [CrossRef]

26. Montenegro, G.; Onorati, A.; Piscaglia, F.; D'Errico, G. Integrated 1D-multiD fluid dynamic models for the simulation of I.C.E. intake and exhaust systems. *SAE Tech. Pap. Ser.* **2007**. [CrossRef]

27. Onorati, A.; Montenegro, G.; D'Errico, G.; Piscaglia, F. Integrated 1D-3D fluid dynamic simulation of a turbocharged Diesel engine with complete intake and exhaust systems. *SAE Tech. Pap. Ser.* **2010**. [CrossRef]

28. Montenegro, G.; Onorati, A.; Della Torre, A. The prediction of silencer acoustical performances by 1D, 1D-3D and quasi-3D non-linear approaches. *Comput. Fluids* **2013**, *71*, 208–223. [CrossRef]

29. Morel, T.; Silvestri, J.; Goerg, K.; Jebasinski, R. Modeling of engine exhaust acoustics. *SAE Tech. Pap. Ser.* **1999**. [CrossRef]

30. Sapsford, S.M.; Richards, V.C.M.; Amlee, D.R.; Morel, T.; Chappell, M.T. Exhaust system evaluation and design by non-linear modeling. *SAE Tech. Pap. Ser.* **1992**. [CrossRef]

31. Montenegro, G.; Della Torre, A.; Onorati, A.; Fairbrother, R.; Dolinar, A. Development and application of 3D generic cells to the acoustic modelling of exhaust systems. *SAE Tech. Pap. Ser.* **2011**. [CrossRef]

32. Payri, F.; Desantes, J.M.; Broatch, A. Modified impulse method for the measurement of the frequency response of acoustic filters to weakly nonlinear transient excitations. *J. Acoust. Soc. Am.* **2000**, *107*, 731–738. [CrossRef] [PubMed]

33. Torregrosa, A.J.; Broatch, A.; Fernández, T.; Denia, F.D. Description and measurement of the acoustic characteristics of two-tailpipe mufflers. *J. Acoust. Soc. Am.* **2006**, *119*, 723–728. [CrossRef]

34. Torregrosa, A.J.; Broatch, A.; Arnau, F.J.; Hernández, M. A non-linear quasi-3D model with Flux-Corrected-Transport for engine gas-exchange modelling. *J. Comput. Appl. Math.* **2016**, *291*, 103–111. [CrossRef]

35. Montenegro, G.; Della Torre, A.; Onorati, A.; Fairbrother, R. Nonlinear quasi-3D approach for the modeling of mufflers with perforated elements and sound-absorbing material. *Adv. Acoust. Vib.* **2013**, *2013*, 546120. [CrossRef]

36. OpenWAM. CMT—Motores Térmicos, Universitat Politècnica de València. Available online: http://www.openwam.org/ (accessed on 20 March 2017).

37. Ikeda, T.; Nakagawa, T. On the SHASTA FCT algorithm for the equation $\partial \rho / \partial t + (\partial / \partial x)(v(\rho)\rho) = 0$. *Math. Comput.* **1979**, *33*, 1157–1169. [CrossRef]

38. Toro, E.F.; Spruce, M.; Speares, W. Restoration of the contact surface in the HLL-Riemann solver. *Shock Waves* **1994**, *4*, 25–34. [CrossRef]

39. Van Leer, B. Towards the ultimate conservative difference scheme. V. A second-order sequel to Godunov's method. *J. Comput. Phys.* **1979**, *32*, 101–136. [CrossRef]

 *applied sciences*

Article

Electric Turbocharging for Energy Regeneration and Increased Efficiency at Real Driving Conditions

Pavlos Dimitriou [1,*], Richard Burke [1], Qingning Zhang [1], Colin Copeland [1] and Harald Stoffels [2]

[1] Department of Mechanical Engineering, University of Bath, Bath BA2 7AY, UK; R.D.Burke@bath.ac.uk (R.B.); Q.Zhang@bath.ac.uk (Q.Z.); C.D.Copeland@bath.ac.uk (C.C.)

[2] Powertrain Research & Advanced, Ford-Werke GmbH, Köln-Merkenich, Cologne D-50725, Germany; hstoffel@ford.com

* Correspondence: P.Dimitriou@bath.ac.uk; Tel.: +44-1225-38-3313

Academic Editor: Jose Ramon Serrano
Received: 23 January 2017; Accepted: 14 March 2017; Published: 1 April 2017

Abstract: Modern downsized internal combustion engines benefit from high-efficiency turbocharging systems for increasing their volumetric efficiency. However, despite the efficiency increase, turbochargers often lack fast transient response due to the nature of the energy exchange with the engine, which deteriorates the vehicle's drivability. An electrically-assisted turbocharger can be used for improving the transient response without any parasitic losses to the engine while providing energy recovery for increasing overall system efficiency. The present study provides a detailed numerical investigation on the potential of e-turbocharging to control load and if possible replace the wastegate valve. A parametric study of the optimum compressor/turbine sizing and wastegate area was performed for maximum torque, fast response time and energy regeneration across the real driving conditions speed/load area of the engine. The results showed that the implementation of a motor-generator could contribute to reducing the response time of the engine by up to 90% while improving its thermal efficiency and generating up to 6.6 kWh of energy. Suppressing the wastegate can only be achieved when a larger turbine is implemented, which as a result deteriorates the engine's response and leads to energy provision demands at low engine speeds.

Keywords: turbocharger; e-turbo; boosting; electrically-assisted; turbo-compound; energy regeneration; internal combustion engines; 1D simulation

1. Introduction

The demand for low fuel consumption and CO_2 generation vehicles over the last few years has popularly increased the necessity of downsizing and increasing the overall thermal efficiency of Internal Combustion (IC) engines. Downsizing is the process of reducing the volumetric capacity of an engine for reduced throttling and friction losses while its boosting capabilities need to be increased for higher specific heat. This can be achieved by the implementation of a boosting device (turbocharger or supercharger) for increased air pressure at the intake of the engine and therefore higher volumetric efficiency.

A turbocharger is a device that recovers the waste energy from the engine's exhaust gasses and uses it to compress the air at the engine's intake. The level of compression is directly linked to the amount of air passing through the turbine, and it can be controlled by either bypassing part of the flow through a Wastegate (WG) or by changing the nozzle position of the turbine (VGT, Variable-geometry turbocharger). The proportion of the waste-gated flow can be up to 50% for high speed and load conditions, which imply a vast amount of unexploited energy. The drawback of this device is that due to the nature of the energy exchange between the engine and the turbocharger (filling of the intake and exhaust manifolds and low exhaust energy at low speeds/loads), the transient performance during engine load increase is relatively poor, which deteriorates the drivability of the vehicle [1]. On the other hand, a supercharger is a device mechanically driven by the engine to increase its volumetric efficiency. It has a very fast response time in transient conditions, but the power required for its operation is a parasitic loss for the engine; therefore, it is not widely used in recent technologies.

The parallel use of a turbocharger and a supercharger could potentially improve the transient performance of the system during load increase while reducing the engine losses compared to a solely supercharger boosted system. However, this will increase the complexity, as a two-stage boosting system is required [2,3]. An electrically-assisted turbocharger of a larger size and with higher efficiency could be used in case a simpler one-stage boosting system is needed. The motor can provide the electricity required at the periods of load increase for a faster response, while the turbocharger works as a conventional system during steady-state and tip-out conditions.

Katrasnik et al. [4] investigated the influence of an electric motor attached to the turbo shaft on the transient response of a diesel engine. It was found that the time required to perform transient power increase was reduced from 3.9 s for the original conventional turbocharger down to 1.7 s. Ibaraki et al. [5] tested a hybrid turbo developed by Mitsubishi Heavy Industries under transient operating conditions. An improved engine peak torque and enhanced transient response compared to a conventional system was demonstrated. Torque was enhanced by 18% at low engine speeds, while the response time was reduced by 70% when 2 kW of turbo motor assistance was provided. Millo et al. [6] investigated the potential of an electrically-assisted turbocharger for a heavy-duty diesel engine to evaluate the turbo-lag reductions and the fuel consumption savings that could be obtained in an urban bus at different operating conditions. The system allowed fuel consumption reductions of 6% to 1%, depending on the driving cycle, with lower values corresponding to congested traffic conditions.

Burke [7] applied various electric boosting systems to a gasoline and a diesel engine and evaluated their steady state and transient performance from the perspective of the air path. The author compared the performance characteristics of an electrically-assisted turbocharger with those of a two-stage system with an electrically-driven compressor placed before or after the main waste-gated turbocharger. He found that under steady-state operating conditions, there was significant system efficiency to installing the electric compressor downstream of the turbocharger's compressor. The author also concluded that an electrically-assisted turbocharger is not ideal as a replacement for the second compressor in a two-stage system, as it will push the compressor into surge. However, it was mentioned that this could be overcome through re-matching of the compressor. Bumby et al. [8,9] investigated the technical problems in selecting an appropriate machine to use with an electrically-assisted turbocharger. The authors also demonstrated that the required time for accelerating an electrically-assisted turbocharger from 40 to 110 krpm could be around 50% less than a conventional turbocharger.

However, albeit an electrically-assisted turbocharged engine requires less energy for its operation than a solemnly supercharged engine, the power required is still a parasitic loss for the engine, which results in an increased fuel consumption. Divekar et al. [10] proposed an electrical supercharging and a turbo-generation system integrated in a diesel engine for overcoming the issue of parasitic losses. The proposed system showed distinct transient response improvement benefits over a conventional turbocharged system and 7% improvement in fuel consumption over the Federal Urban Driving Schedule (FUDS) cycle. Furthermore, during transients and high load operation, the proposed system did not build up exhaust backup pressure in order to accelerate the supercharger, and as such, no additional pumping losses were incurred. However, the authors commented that the electrical energy required by the supercharger can be only partially obtained by the turbo-generation system, which still leads to additional energy losses.

Panting et al. [11] was one of the first research groups to implement the idea of a motor-generator electrical turbocharger for a 5.2 L truck diesel engine in a theoretical study. Adding a directly-coupled motor-generator offers tremendous advantages to the operation of the turbocharger. It abolishes the requirement of the turbine and compressor power to be matched under steady-state conditions while it assists the acceleration and deceleration of the shaft during transients with no engine energy needs.

Over the last few years, keen interest has been shown in the numerical and experimental investigation of the motor-generator technology application in high-duty engines by several research groups. Terdich and Martinez-Botas [12] experimentally characterized a variable geometry turbocharger with a motor-generator technology. The authors found that the motor-generator is capable of delivering a maximum shaft power of 3.5 kW in motoring mode and 5.4 kW in generating mode. The peak electrical efficiency was more than 90% in both modes and occurred at 120,000 revs/min. Airse et al. [13] developed a comprehensive powertrain model to evaluate the benefits of an electric turbo-compound, working in both generator and motor mode, in reducing CO_2 emissions from small diesel passenger cars. The simulations showed a reduction in CO_2 and fuel consumption of over 4% for the New European Driving Cycle (NEDC). Algrain [14] developed an advanced control system for a motor-generator fitted in a heavy-duty diesel engine for improving the overall fuel efficiency. The simulation results showed that at the rated power, the fuel consumption of a Class-8 on-highway truck engine could be reduced by almost 10%, while the overall reduction in fuel consumption was estimated to be around 5%. Pasini et al. [15] focused on the evaluation of the benefits resulting from the application of an Electric Turbo Compound (ETC) to a small-sized twin-cylinder Spark-Ignition (SI) engine and to a four-cylinder Compression-Ignition (CI) engine with the same power rating. They found that by absorbing electrical energy from the battery, the ETC can lead to significant Brake-Specific Fuel Consumption (BSFC) reductions of up to 4% at the highest engine speeds and loads for the SI engine and up to 6% for the CI engine at 4000 rpm half load. Furthermore, calculations have shown that in the case of the CI engine, the maximum electric power that can be recovered by the ETC is around 4 kW at 4000 rpm full load, while in the case of the SI engine, the maximum power is 1.5 kW.

Tavcar et al. [16] presented a comprehensive study on engine performance improvement attributable to the application of different electrically-assisted turbocharger topologies, including a single-stage turbocharger, an electrically-assisted turbocharger, a turbocharger with an additional electrically-driven compressor and an electrically-split turbocharger (supercharger and turbo-generator). The results revealed that all of the electrically-assisted turbocharger topologies improve the transient response of the engine and, thus, the drivability of the vehicle. However, no electrically-assisted turbocharger topology could clearly be favoured in general.

In this paper, a detailed investigation of the potential of e-turbocharging to control load while providing energy recovery for increasing the overall system efficiency and if possible replace the wastegate boost control is provided. The current approach of e-turbocharging requires larger turbine systems that do not build up backup pressure and provide electrical assistance at low engine speeds. However, with this configuration, the energy regeneration occurs only at high speed and load areas of the engine, as shown in Figure 1a. The proposed study focusses on shifting the energy regeneration towards the low-speed area, which represents more realistic driving conditions, as shown in Figure 1b. Energy assistance is provided to the engine when the maximum power characteristics of the engine need to be met. The study is performed on a 2.0 L turbocharged SI engine under steady-state and transient driving conditions. The research work focuses on the availability of energy and the effects of component sizing, the transient behaviour for eliminating turbo lag and the overall system energy balance during various driving conditions.

After a comprehensive introduction in this section, Section 2 outlines the engine model used for the study and any relatively small modifications occurring for the purpose of electrifying the turbocharger. A detailed model validation against experimental data is also presented in Model Validation. Section 3 describes the methodology followed for all of the simulation studies conducted under steady-state and transient conditions. The simulation results for all of the studies performed are analysed and discussed in Section 4. Finally, Section 5 summarizes the main findings of this work and proposes future work in the area of the electrification of turbocharging systems.

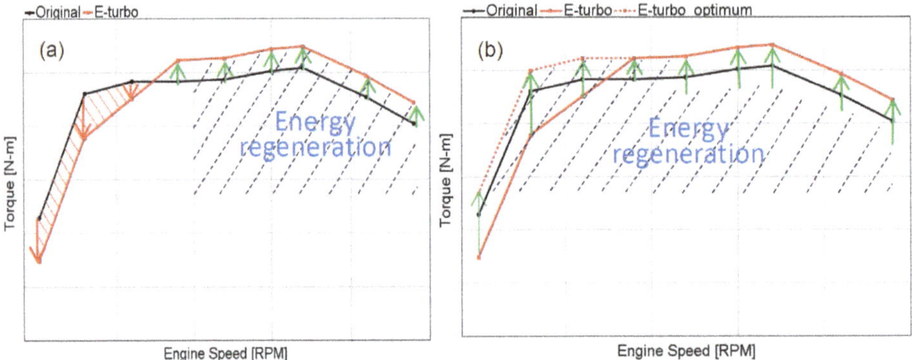

Figure 1. Energy regeneration areas over the load/speed map of an engine: (**a**) current approach; (**b**) desired operation.

2. Computational Model

The present numerical study was performed on a 1D gas dynamic environment using the GT-Power software tool (v7.5.0, Gamma Technologies, LLC., Westmont, IL, USA, 2014). The model used for this study, shown in Figure 2, represents a 2.0 L spark-ignition turbocharged engine. The model, previously used in [17], was provided by the engine supplier, and it has been validated against experimental data in both steady-state and transients by POWERTECH Engineering (see Model Validation).

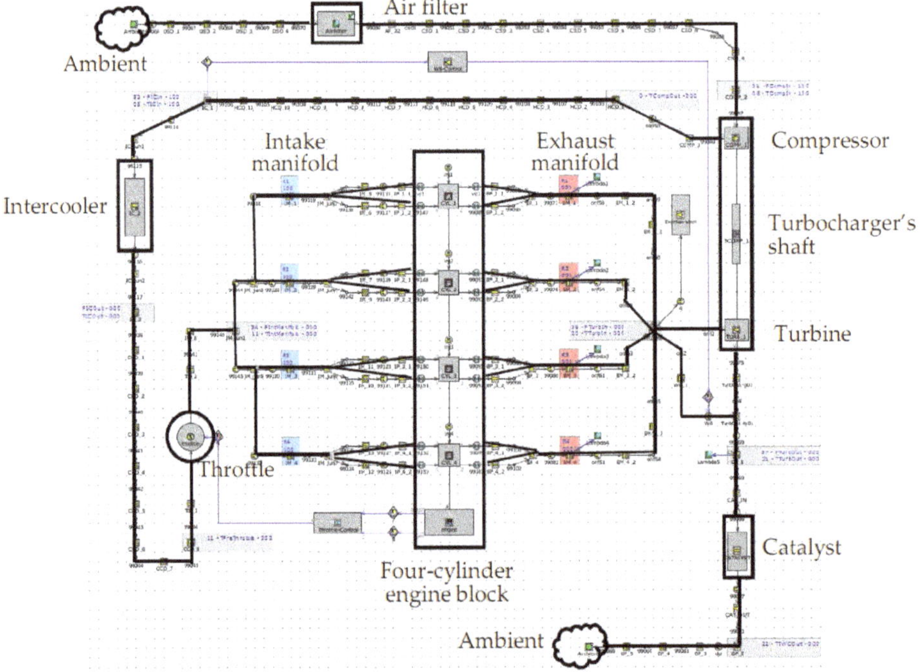

Figure 2. 1D model of the baseline 2.0 L turbocharged engine used in the present study.

The baseline 2.0 L engine model was used only for the Phase 1 study of the steady-state simulations, which included a wastegate enthalpy loss study for quantifying the energy availability across the speed/load map of the engine.

For the Phase 2 and Phase 3 of the steady-state simulations, as well as the transient simulations included in this paper, the model was modified by implementing a motor/generator (M/G) component model. The motor/generator was connected directly to the turbocharger's shaft, as shown in Figure 3.

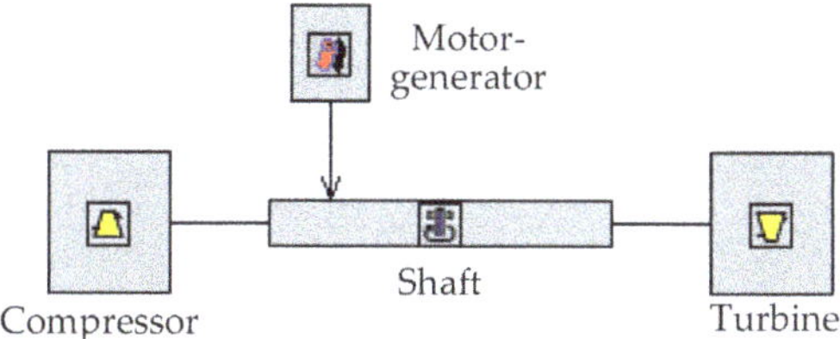

Figure 3. Motor/generator component model used for converting the model to an e-turbo engine.

The original model was fitted with a wastegate/boost pressure and a throttle/brake mean effective pressure proportional-integral-derivative (BMEP PID) controllers for achieving the targeted boost pressure and BMEP. The target values for different engine speeds and loads were provided to the model using look-up tables. For the electrically-assisted model, two new PID controllers were introduced or replaced the original controllers as necessary, depending on the type of study. These are:

- A motor/generator (M/G)/boost pressure controller for achieving the desired intake pressure
- A WG/pre-turbine pressure controller for achieving the desired pre-turbine pressure

The compressor and turbine mass flow multipliers were modified and spanned in the range of 0.7 to 1.3 for the purpose of undertaking parametric studies with different component sizing, as shown in Equation (1). By restricting or enhancing the mass flow rate within a component, the size of the component can be simulated.

$$\text{Average mass flow rate } = \text{Multiplier } \times \frac{\int \dot{m}\, dt}{\int dt} \tag{1}$$

where $\dot{m}$ is the mass flow rate through the part (in the case of the turbine, the wastegate mass flow is excluded).

Although this approach would not promise a high level of accuracy, it is deemed to be reliable to predict the trend of the overall system's behaviour when the turbine or compressor sizing is increased or decreased compared to the original components based on a highly-calibrated engine. Furthermore, the wastegate area was modified from completely closed to diameters larger than the baseline engine. However, due to the change of the size of critical components, the operation of the engine and its main parameters needs to be closely monitored to ensure realistic performance. For this reason, the parameters limits shown in Table 1 were set and established, which were not violated during the parametric studies.

Table 1. Maximum limits set for the parametric studies.

Parameter	Unit	Value
Engine torque	(Nm)	304
Turbine inlet temperature	(°C)	1000
Turbocharger speed	(krpm)	200
Pre-turbine pressure	(bar)	3
Post-compressor temperature	(°C)	200

Model Validation

A high fidelity engine model validated against experimental data for a conventional layout, in both steady-state and transients, has been applied in this study. The calibration of the combustion model (SI Wiebe) was performed by isolating Cylinder 1 of the engine and performing a Three Pressure Analysis (TPA). Figure 4 represents the comparison between measurements and simulation results for all engine cylinders at low loads.

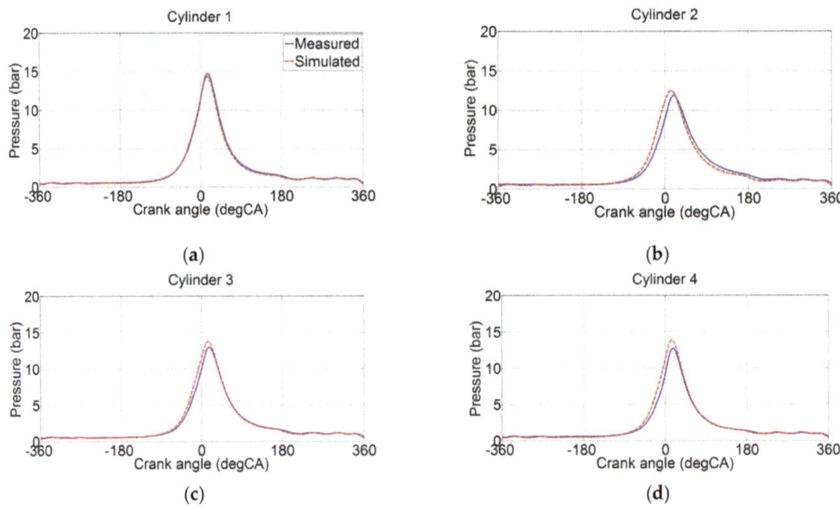

Figure 4. Simulated cylinder pressure from Three Pressure Analysis (TPA) and measured pressure at low loads. (**a**) Cylinder 1; (**b**) Cylinder 2; (**c**) Cylinder 3; (**d**) Cylinder 4.

The small cylinder-to-cylinder variations observed in the measured data were difficult to capture in the 1D simulation. These variations may be a result of several factors, such as 3D air and gas flow behaviours at the intake and exhaust or different thermal conditions. Although there were slight variations, a good agreement between measured and simulated mass air flow and averaged indicated mean effective pressure (IMEP) (error of less than 2%) was observed for full load conditions, as shown in Table 2.

Table 2. Measured data and simulation results at full loads. IMEP, indicated mean effective pressure.

Attribute Value	Cylinder 1	Cylinder 2	Cylinder 3	Cylinder 4
Measured total mass air flow (kg/h)		258.69		
Simulated total mass air flow (kg/h)		263.65		
Error in total mass air flow (%)		1.92		
Measured net IMEP (bar)	20.71	21.19	20.75	18.29
Simulated net IMEP (bar)	20.44	20.76	20.30	18.02
Absolute error in net IMEP prediction (bar)	0.39	0.43	0.45	0.27
Error in net IMEP prediction (%)	1.87	2.05	2.15	1.47
Averaged absolute error in net IMEP (bar)		0.36		
Averaged error in net IMEP (%)		1.89		

For part-load cases, the discrepancy between measured and simulated results was slightly larger, as can be found in Table 3. However, despite the 9.37% error in the averaged IMEP for the low-load case, the averaged absolute error in net IMEP was below 0.3 bar. Therefore, the modelling of the scavenging system and combustion system at low engine load is satisfactory.

Table 3. Summary of combustion model validation at 2000 rpm.

Attribute Value	Low-Load	Medium-Load	Full-Load
Error in total mass air flow prediction (%)	0.69	3.26	1.92
Averaged error in net IMEP prediction (bar) (Cylinder 1)	0.19	0.39	0.39
Averaged error in net IMEP prediction (%) (Cylinder 1)	7.39	2.56	1.87

The overall shape of the port pressures at three operating points was captured. Figure 5 outlines the measured and simulated instantaneous pressure at the exhaust port of Cylinder 1 for medium engine loads. As can be seen from the figure, the magnitude of the pressure wave reflections was well predicted.

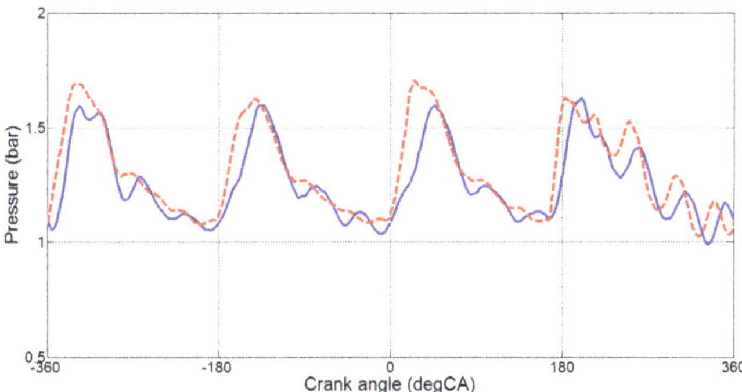

Figure 5. Instantaneous pressure at exhaust port of Cylinder 1, 2000 rpm medium-load.

The validation of the transient performance of the model was also performed against experimental data collected from transient tests performed with various wastegate positions. The calibration was performed for three different boost pressures and the comparison between measured and simulated mass air flow is presented in Figure 6. The experimental and simulation results show a good agreement with small variations (5%) that could be a result of the scavenging model in the 1D engine simulation.

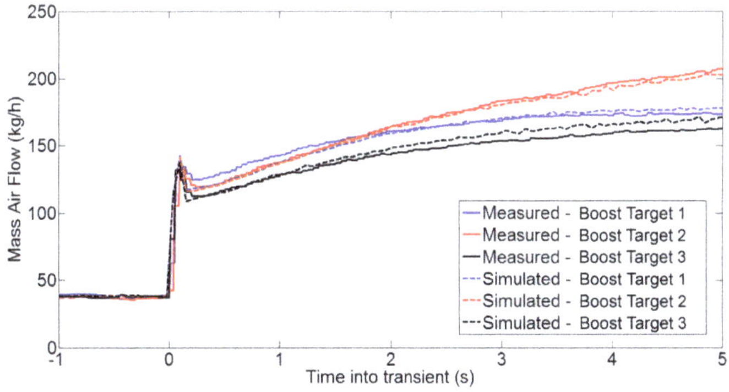

Figure 6. Comparison between measurement and simulation results of transient performance for different boost targets with imposed turbocharger rotational speed.

3. Methodology

The simulations performed in this paper are divided into two main subsections of steady-state and transient analysis. Each of the subsections is further divided into smaller segments to represent different types of studies. A detailed flowchart of all of the simulations performed in this paper is shown in Figure 7.

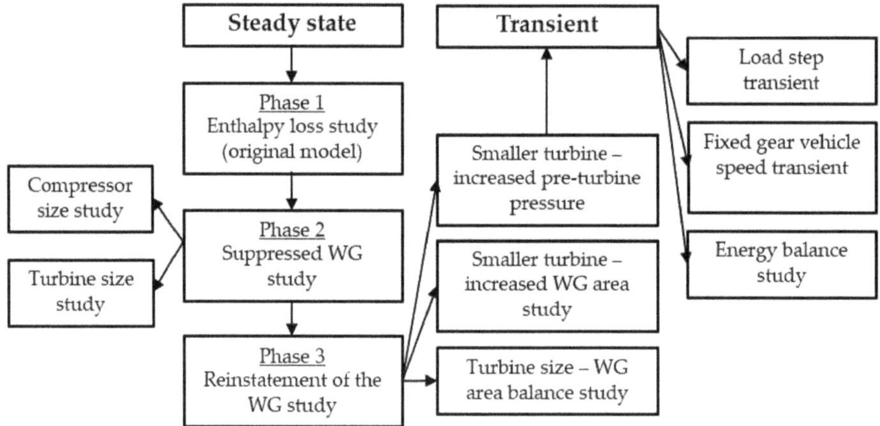

Figure 7. Flowchart of the simulation process. WG, wastegate.

Finally, it needs to be highlighted that all of the studies presented in this paper show the potential available energy in the system. Any electrical losses such as alternator, converter and battery losses have not been considered at this point.

3.1. Steady-State Simulations

The steady state analysis of the model was performed in the area of three axes, compressor and turbine size and WG area, as shown in Figure 8. The purpose of the study was to understand the effects of the compressor and turbine sizing on the amount of energy that needs to be provided/harvested by the motor/generator and to identify any regions where the e-turbo can replace the wastegate for load control.

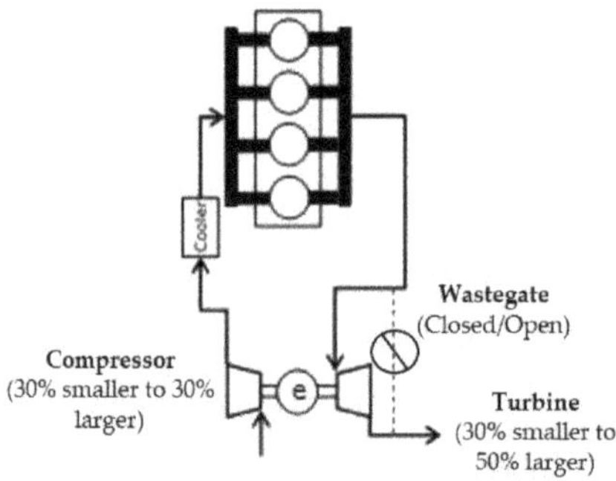

Figure 8. Area of investigation for the steady-state simulations.

The steady-state simulations are divided into three phases based on the type of investigation.

- Phase 1: Assessment of the amount of enthalpy loss for the baseline engine across the speed/load map and how this is affected by the compressor and turbine's size.
- Phase 2: Investigation of the potential of suppressing WG and using e-turbo to control boosting.
- Phase 3: Reinstatement of the WG to control exhaust manifold pressure and explore the trade-off between WG and turbine size.

3.1.1. Phase 1: WG Enthalpy Loss Study

The first phase of the study was performed to calculate the amount of energy that is lost through the wastegate of a modern 2.0 L turbocharged gasoline engine. The amount of power loss was calculated using the equation:

$$Q = m \times c_{\text{p}} \times \Delta T, \tag{2}$$

where Q is the amount of heat to the system (power), m is the mass flow rate through the wastegate, c_{p} is the specific heat of the gas and ΔT is the temperature difference.

A Design of Experiment (DoE) analysis was performed to evaluate the effects of the compressor and turbine's size on the amount of enthalpy loss, combustion limiting parameters and maximum engine power. The nine cases studied are described in Table 4.

Table 4. Phase 1: Design of Experiment (DoE) analysis for different compressor and turbine sizes.

Case	Turbine Size Multiplier	Compressor Size Multiplier
1	0.7	0.7
2	0.7	0.85
3	0.7	1
4	0.85	0.7
5	0.85	0.85
6	0.85	1
7	1	0.7
8	1	0.85
9	1	1

3.1.2. Phase 2: Suppressing WG and Using e-Turbo to Control Boosting

Phase 2 of the steady-state simulations section involves the application of a motor/generator, which is directly linked to the shaft of the turbocharger. The purpose of the motor/generator is to provide or harvest energy, as needed, to/by the compressor for achieving the boosting demands of the engine. For this phase, the WG has been completely suppressed and the model run once with the original size of the compressor and turbine. However due, to the shut off of the WG valve, extremely high pre-turbine pressures violating the limits at high loads were noticed.

A sweep study was performed for five different turbine sizes (10% to 50% larger) to evaluate the optimum turbine size for the model without a WG valve. Then, the optimum turbine was tested for different compressor sizes (20% smaller to 20% larger) to evaluate its effect on the energy harvesting/provision requirements.

3.1.3. Phase 3: Reinstating WG to Control Exhaust Manifold Pressure

Phase 3 of the steady-state simulations involves the reinstatement of the WG valve for reducing the required size of the turbine and enhancing the energy performance across the low speeds area of the engine's speed/load map. This section is divided into four studies, as shown below:

- Turbine size: WG area balance study
- Smaller turbine: increased WG area study
- Smaller turbine: increased pre-turbine pressure study (smaller WG area)
- Smaller turbine: variable pre-turbine pressure study (smaller and larger WG area)

3.2. Transient Simulations

The findings in the steady-state simulations section (see the Results Section) demonstrated that a model fitted with a motor-generator and a turbine 10% smaller than the original could provide energy harvesting and thermal efficiency gain across the speed/load map of the engine. The transient behaviour of the model with this configuration, an original size compressor and increased pre-turbine pressures was tested by performing three different types of simulations, as shown below:

- Load step transient
- Fixed gear vehicle speed transient
- Energy balance study for various driving cycles and real driving conditions

The purpose of the transient study was to reveal the response time improvement under different levels of energy provision to the compressor, as well as to perform an energy balance review. This is required in order to identify whether the e-turbocharger can operate without energy consumption from the engine and if any excess on the total energy generated.

3.2.1. Load Step Transient Simulations

The load tip-in study was performed for two engine speeds of interest, 1600 and 1100 rpm, close to the surge limit of the compressor. The load tip-in step requested was from 2 bar BMEP to full load BMEP (19 bar). However, due to the engine speeds at which the tests were performed, the requested BMEP could not be achieved.

The study was initially performed for the model with all of the PID controllers. However, for ensuring that the results are not affected by the tuning of the controllers, the study was repeated in an open-loop environment (no PID controllers). The PID controllers in the model were removed and replaced by look-up tables for controlling the boost pressure and throttle position of the engine, as shown in Table 5. This allows a hardware capability investigation to be performed by eliminating any effects of the PID controllers. The energy provided to the compressor was divided into four sections of 0.25 seconds, and different levels of power (1, 3 and 5 kW) were provided in each section.

Table 5. Control of basic process parameters for the model with proportional-integral-derivative (PID) controllers and the open-loop model. BMEP, brake mean effective pressure; M/G, motor/generator.

Process Parameter	Model with PIDs (Control Variables)	Open-Loop Model
M/G Power	-	Provided power profiles
Boost pressure	M/G power	Speed/load look-up table
Throttle position	Engine speed/BMEP	Speed/load look-up table
Pre-turbine pressure	WG position	WG position (PID)

3.2.2. Fixed Gear Vehicle Speed Transient Simulations

The purpose of this study was to represent the vehicle's driving conditions and test the model for a fixed gear (third) vehicle speed transient. For this reason, a driving resistance load of 30 Nm was applied to represent the rolling resistance along with the flywheel's inertia given by the following formula:

$$I_{eq} = I_{eq,eng} + I_{eq,prop} + I_{eq,axle} + I_{eq,veh}, \tag{3}$$

where I_{eq} is the total inertia (kg·m^2), $I_{eq,eng}$ is the engine's inertia, $I_{eq,prop}$ is the prop shaft inertia, $I_{eq,axle}$ is the axle inertia and $I_{eq,veh}$ is the vehicle mass equivalent inertia, as shown in Figure 9.

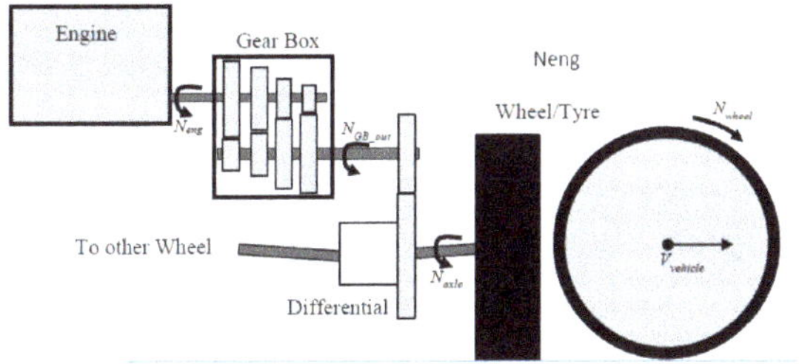

Figure 9. Schematic of total inertia applied to the engine.

The study was performed for four different levels of targeted BMEP, as shown in Figure 10, representing the aggressiveness with which a vehicle's pedal can be pressed. The tests were repeated twice for different pre-turbine pressures, which were controlled by implementing a pre-turbine pressure/WG PID controller.

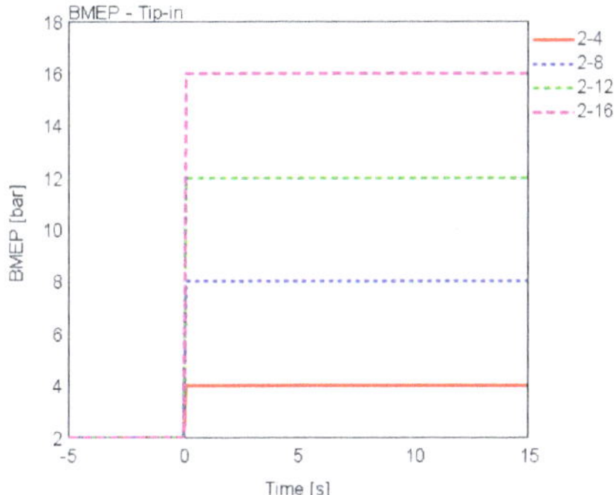

Figure 10. Fixed gear vehicle speed BMEP tip-ins.

3.2.3. Energy Balance Simulations

The studies performed in the previous sections highlight the significant improvement on the transient response time of the engine during load and speed tip-in conditions (see Results Section). The work performed in this section investigates whether the power harvested by the generator is enough to cover the motor demands. The energy balance study for a specific vehicle was conducted for the following driving cycles with the pre-turbine pressure target set at 15% higher than the baseline engine for the low to medium loads and 5% for the full load conditions:

1. New European Driving Cycle (NEDC): designed to represent the typical usage of a car in Europe, but often criticized for delivering unrealistic economy figures.
2. Worldwide harmonized Light vehicles Test Cycle (WLTC): a harmonized driving cycle representing realistic driving conditions data in different regions around the world, combined with suitable weighting factors.
3. US06: representing aggressive, high-speed and/or high acceleration driving behaviour, rapid speed fluctuations and driving behaviour following startup.
4. Combined driving conditions: including equal amounts of various realistic driving conditions, such as low speed, start-stop, highway, motorway, uphill, downhill and dangerous overtaking conditions.

The engine speed and BMEP demands for the four driving cycles are shown in Figure 11.

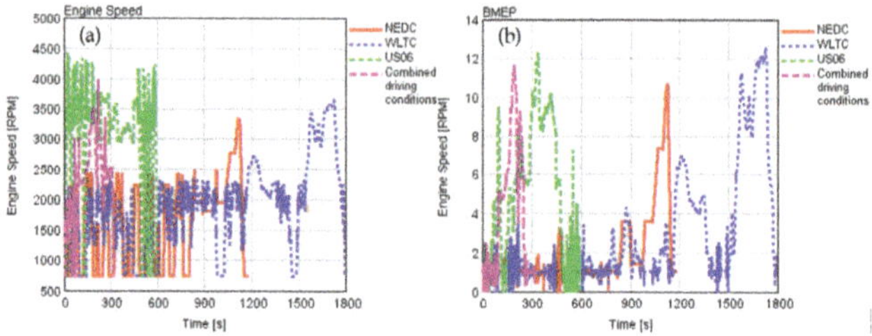

Figure 11. Driving cycles conditions: (**a**) engine speed; (**b**) engine BMEP. NEDC, New European Driving Cycle; WLTC, Worldwide harmonized Light vehicles Test Cycle.

4. Results and Discussion

The results of the simulation studies performed in this paper are categorized in a similar manner to that of the Methodology Section.

4.1. Steady-State Analysis

4.1.1. Phase 1: WG Enthalpy Loss Study (Baseline Engine)

The results in Figure 12 show that the energy loss through the wastegate for the model with the original size of compressor and turbine (Case 9) can reach the amount of 5 kW at high engine speeds and loads. The amount of the potential energy dismissed at medium load and engine speed conditions is of lower magnitude, up to 2 kW. At low loads and speeds, there is no wasted energy, as the WG is completely closed.

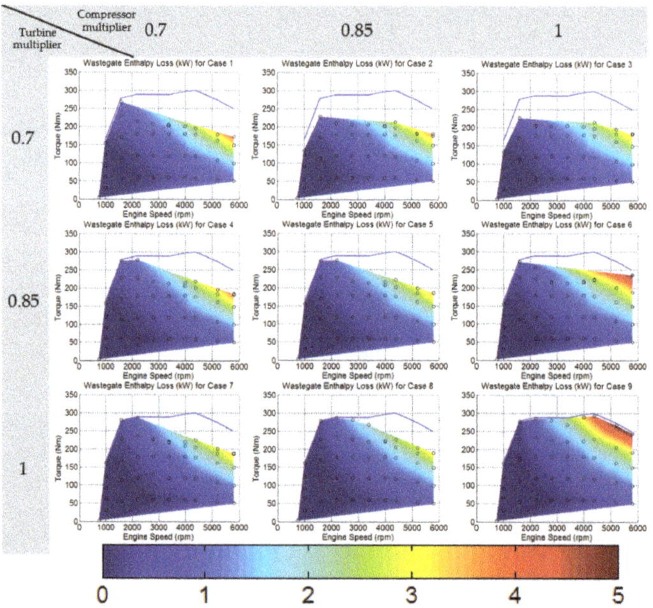

Figure 12. Phase 1: DoE analysis of the waste-gated enthalpy loss for the different compressor and turbine sizes; the blue line represents the maximum torque line of the baseline engine's model as provided by the manufacturer ($\lambda < 1$); the $\bigcirc$ symbols indicate data points; the $\otimes$ symbol indicates a limit violation at specific points.

Figure 12 highlights the effect of a smaller turbine and compressor on the amount of waste-gated flow and the maximum power of the engine. As can be seen, a smaller compressor (Cases 7 and 8) cannot provide the boosting requirements needed to achieve maximum engine's power at medium and high engine speeds. On the other hand, a smaller turbine (Cases 3 and 6) can slightly increase the amount of waste-gated flow at medium engine speed and load conditions. However, a smaller turbine leads to increased pre-turbine pressure, in some cases violating the maximum limit set, and reduces the maximum torque of the engine.

4.1.2. Phase 2: Suppressing WG and Using e-Turbo to Control Boosting

Suppressing the wastegate of an engine leads to extremely high pre-turbine pressures at medium to high loads. Figure 13 presents the pre-turbine pressures and peak engine torques achieved for various turbine sizes and the wastegate valve closed. As shown from the figure, a turbine 30% larger than the original size can provide or even increase the maximum power targets of the baseline engine without violating the pre-turbine pressure limit. A torque deficit occurs at 1000 rpm, but this is mainly due to a poor model convergence at this specific engine speed near the surge line and not the incapacity of the engine to achieve the targeted torque.

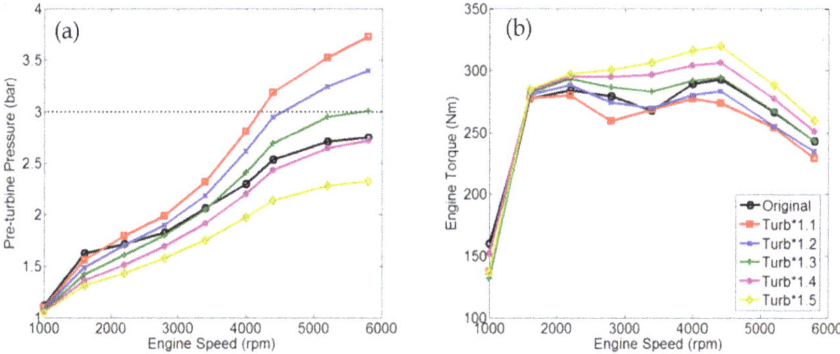

Figure 13. Engine's performance results for various turbines sizes and the wastegate completely shut: (**a**) pre-turbine pressure; (**b**) engine torque.

The results in Figure 14 show that with this configuration, extremely high amounts of energy, more than 15 kW, can be harvested at high engine speeds and loads. The system can harvest energy up to 10 kW at medium to high engine speeds. However, as can be seen, the larger turbine results in considerable amounts of energy that need to be provided by the motor for meeting, if possible, the targeted maximum torque at low engine speed conditions. This is happening due to the lower pre-turbine pressures and increased turbocharger's inertia at a non-boosting area of the engine's map.

The next stage in this phase was to investigate the potential of reducing the amount of energy needs to be provided by changing the compressor's size. For this reason, five different compressors (multipliers of 0.8 to 1.2) were tested, and the results are presented in Figure 15.

It is clearly shown in Figure 15 that a small compressor could reduce the power level needs to be provided by the motor from 0.6 down to 0.2 kW at 1600 rpm with no effect on the engine's maximum torque. However, a smaller compressor would also reduce the amount of energy that can be harvested (from 18.5 down to 17 kW) at high engine speeds; but foremost, it would struggle to meet the maximum engine's torque demands.

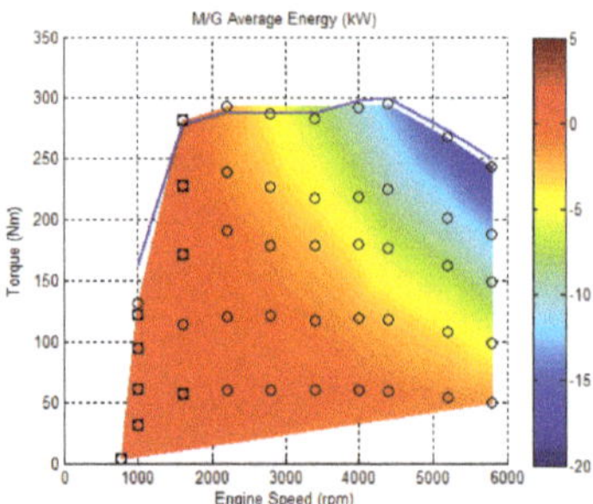

Figure 14. Phase 2: Motor-generator average energy for the case with original compressor, 30% larger turbine and the WG valve completely shut; negative values indicate energy harvesting; positive values indicate the need for energy provision; the blue line represents the maximum torque line of the baseline engine's model as provided by the manufacturer ($\lambda < 1$); the $\otimes$ symbol indicates limit violation; $\square$ indicates energy provision area.

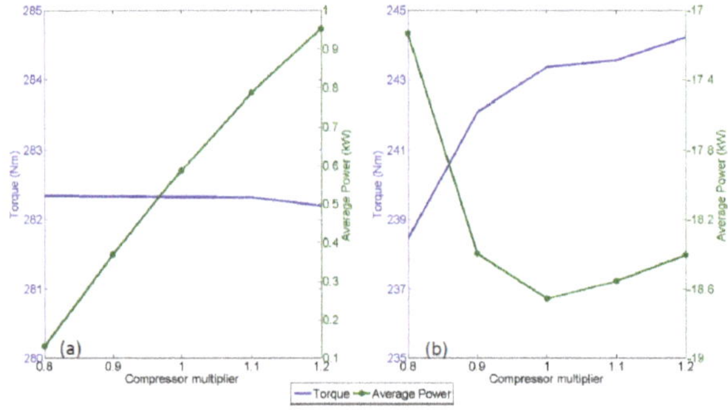

Figure 15. Effect of compressor's size on average power and torque: (**a**) 1600 rpm; (**b**) 5800 rpm.

4.1.3. Phase 3: Reinstating WG to Control Exhaust Manifold Pressure

Turbine Size: WG Area Balance Study

The first part of this phase includes a DoE study for investigating the benefits on the engine's maximum power and energy recovery for a smaller turbine (multipliers less than 1.3) and the WG valve open at different positions (smaller than the original model), as shown in Table 6.

Table 6. Phase 3: DoE analysis for different turbine sizes and WG areas.

Case	Turbine Size Compared to the Original Model	WG Area Compared to the Original Model
Turb×1.1–WG/2	10% larger	50% smaller
Turb×1.1–WG/3	10% larger	67% smaller
Turb×1.2–WG/2	20% larger	50% smaller
Turb×1.2–WG/3	20% larger	67% smaller

The results in Figure 16 show that with a 20% larger turbine (rather than 30%) and the WG valve open at half the size as the baseline engine, maximum engine torque can be achieved. The smaller turbine and the open WG reduced the maximum amount of energy harvested at high speeds from 18 down to 12 kW. However, the benefit on the low-speed side of the map was relatively low, as the amount of energy that needs to be provided by the motor at 1600 rpm went from 0.6 down to 0.45 kW.

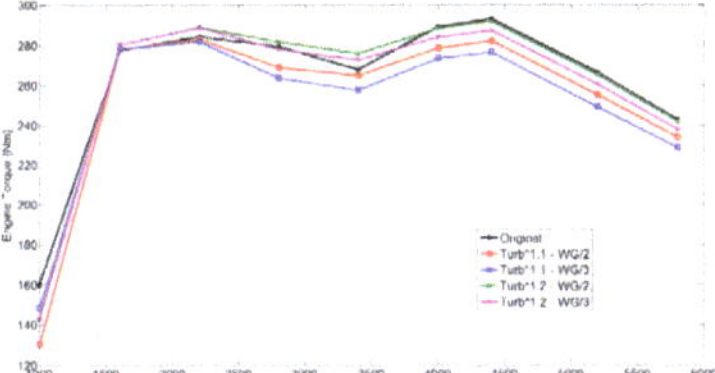

Figure 16. Balance study for different turbine sizes and wastegate areas for meeting the baseline engine's maximum torque.

Smaller Turbine: Increased WG Area Study

Despite the fact that a large turbine allows energy harvesting within the pre-turbine pressure limit at the high speed and load conditions, it also leads to low speeds and loads' poor performance. This could be theoretically resolved by implementing a small turbine and controlling the pre-turbine pressure limits by increasing the wastegate area. The following DoE study results show the effects of three compressors smaller than the original (multipliers of 0.7 to 0.9) and the WG area open at values 10% to 50% larger (multipliers of 1.1 to 1.5) than in the baseline engine.

The results in Figure 17 illustrate that a small turbine can lead to lower energy demands and power generation at low speed/load conditions. However, the smaller the turbine, the higher the pre-turbine pressures, which deteriorate the engine's power output at full load conditions. By increasing the WG area, the pre-turbine pressure drops, and therefore, the full load performance of the engine increases. However, even a 10% increase in the WG area leads to high levels of energy requirements for meeting the baseline engine's power characteristics.

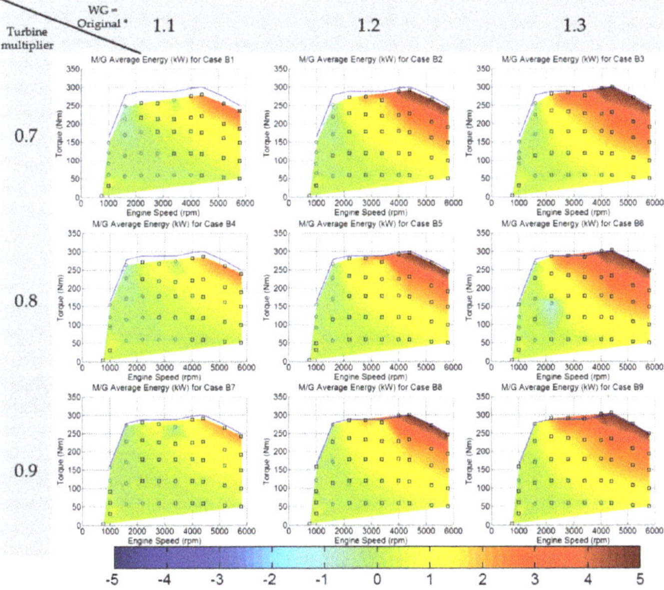

Figure 17. Phase 3: DoE analysis of the motor-generator average energy for different turbine sizes and WG areas; negative values indicate energy harvesting; positive values indicate need for energy provision; blue line represents the maximum torque line of the baseline engine's model as provided by the manufacturer ($\lambda < 1$); $\otimes$ symbol indicates limit violation; $\square$ indicates energy provision area.

Smaller Turbine: Increased Pre-Turbine Pressure Study

The next study in phase 3 focuses on the effects of a smaller turbine with a reduced WG area compared to the baseline engine for benefiting from a good energy balance across the low and high load areas of the engine's speed/load map. For this study, the WG area is controlled indirectly by setting up a PID controller between the WG valve and the pre-turbine pressure. The targeted value is the pre-turbine pressure, which is set to increased values of 5%, 10% and 15% (multipliers of 1.05 to 1.15) compared to the baseline engine, but always within the set pre-turbine pressure limit.

Figure 18 shows that for a 10% smaller turbine (multiplier of 0.9) than the baseline engine, the e-turbo can harvest energy (up to 4 kW) at most points of the map, except the 1000 rpm speed, where the results are not highly trusted due to non-convergence of the model. Although, the increased pre-turbine pressure leads to increased energy harvesting levels, it also, as expected, reduces the maximum power output of the engine. A 5% pre-turbine pressure increase leads to a 5% penalty on the engine's maximum torque, while for the case with the pre-turbine pressure set at 15% higher, the torque penalty is around 10%.

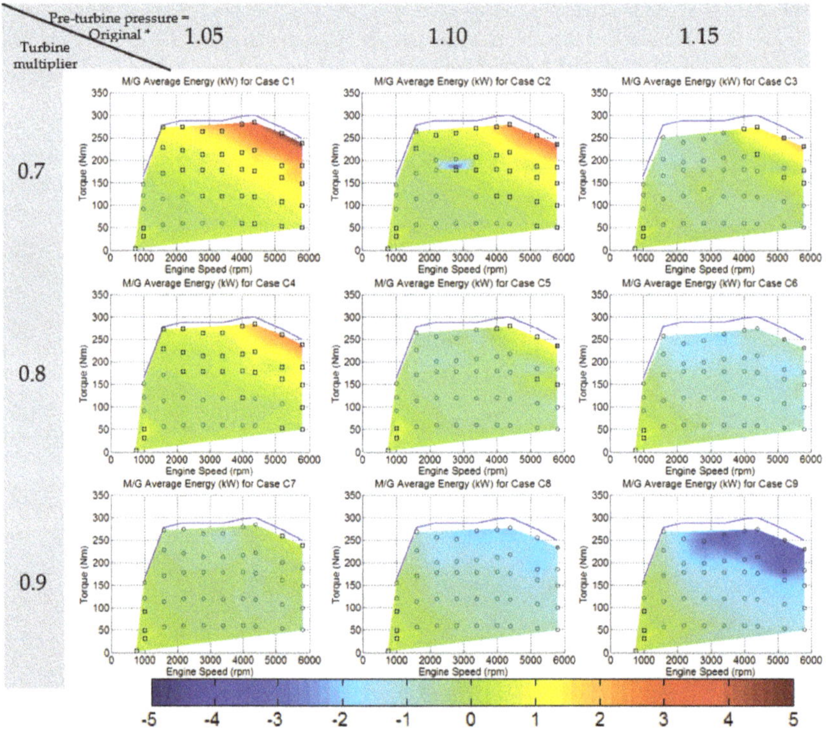

Figure 18. Phase 3: DoE analysis of the motor-generator average energy for different turbine sizes and pre-turbine pressures; negative values indicate energy harvesting; positive values indicate the need for energy provision; the blue line represents the maximum torque line of the baseline engine's model as provided by the manufacturer ($\lambda < 1$); the $\otimes$ symbol indicates limit violation; $\square$ indicates energy provision area.

Smaller Turbine: Variable Pre-Turbine Pressure Study

The previous study showed that with the right component sizing, energy harvesting could be achieved across most of the speed/load map area of the engine. However, this leads to torque sacrifice at full load conditions. This penalty can be reduced or even eliminated by adjusting the targeted pre-turbine pressure of the engine when running at full load conditions. The present study shows the effects of various pre-turbine targeted values, larger and smaller than the baseline engine, to the maximum power output of the engine and the average power harvested or that needs to be provided by the motor-generator.

As is shown in Figure 19, at full load conditions, the maximum power output of the baseline engine can be met by running the engine at the original pre-turbine pressures or slightly lower. However, this leads to energy provision demands by the e-turbo of up to 2 kW. Higher power outputs, which often work as selling points for automotive manufacturers, can be achieved by running at lower pressures and providing significant amounts of power to the compressor.

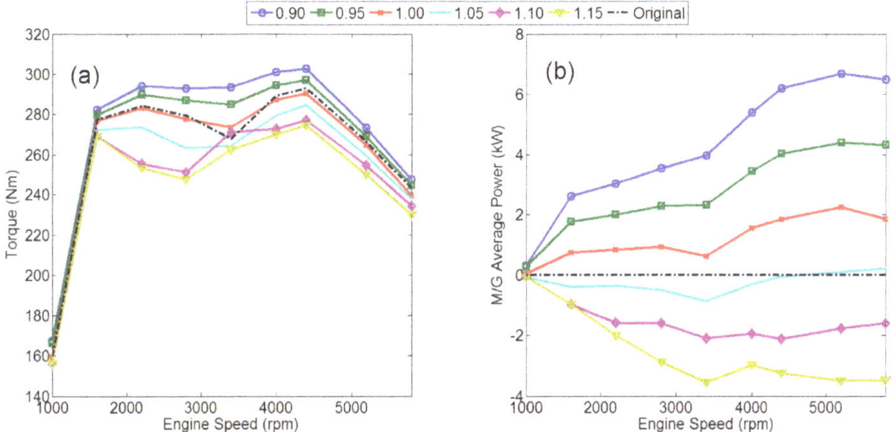

Figure 19. Effect of pre-turbine pressure % change on: (**a**) engine's torque at full load; (**b**) M/G performance at full load; negative values indicate energy harvesting.

4.1.4. Overall Engine Efficiency

The overall efficiency of the baseline engine, which is calculated using the brake power achieved over the total fuel power, can reach levels higher than 35% at medium to high load conditions. On the other hand, the overall efficiency of the electrically-assisted turbocharged engine is given by the brake power achieved plus the energy harvested or provided by the e-turbo divided by the total fuel power. Figure 20 shows the comparison charts between the overall efficiency of the baseline engine and four cases studied in Phase 2 and Phase 3 of the steady-state simulations.

It is evident in Figure 20 that the implementation of a motor-generator linked to the turbocharger leads to an overall engine efficiency increase at medium and high loads and engine speeds for all of the cases. The cases with a reduced turbine size (multiplier of 0.9) benefit from an increased overall efficiency across the whole map. Finally, the high pre-turbine pressures show a significant effect on the overall efficiency gain of the model.

By assessing the work presented in the three phases of the steady-state simulations, the ideal engine layout can be summarized as shown in Figure 21.

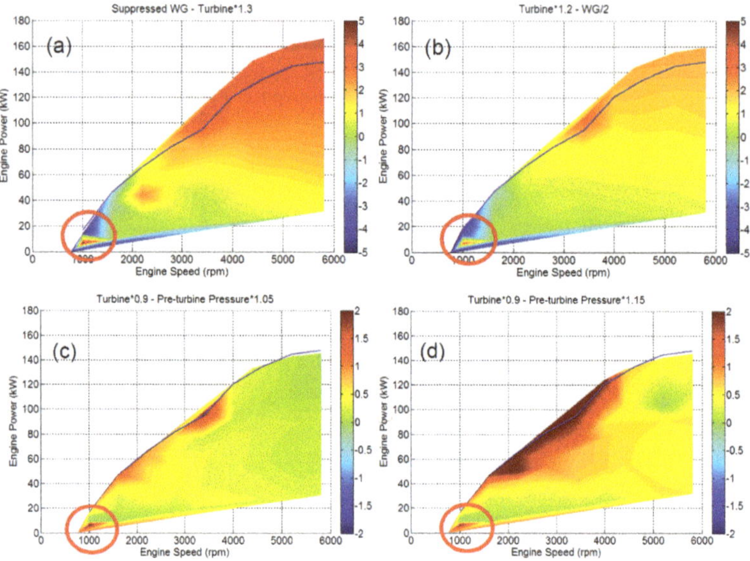

Figure 20. Absolute difference in overall efficiency between four different e-turbo configurations and the baseline engine; negative values indicate efficiency loss; positive values indicate efficiency gain; the area in red circle indicates non-convergence of the model: (**a**) model with 30% larger turbine and WG completely shut; (**b**) model with 20% larger turbine and WG half opened compared to the baseline engine; (**c**) model with 10% smaller turbine and 5% increased pre-turbine pressure compared to the baseline engine; (**d**) model with 10% smaller turbine and 15% increased pre-turbine pressure compared to the baseline engine.

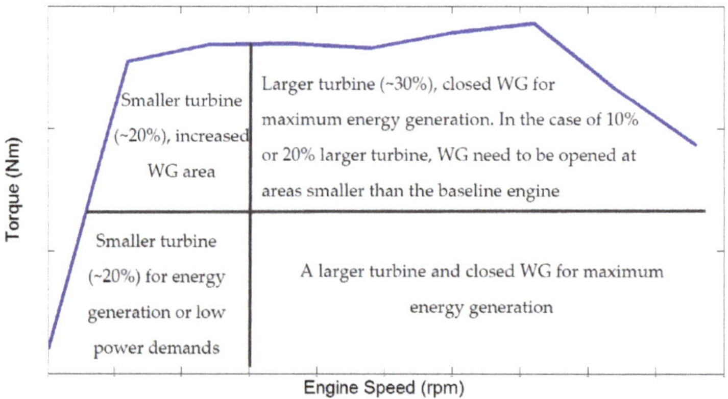

Figure 21. Ideal component sizing for maximum power and energy harvesting.

4.2. Transient Analysis

The results in the steady-state simulations section demonstrated that a model fitted with a motor-generator and a turbine 10% smaller than the original could provide energy harvesting and thermal efficiency gain across the speed/load map of the engine. The transient behaviour of the model with this configuration, an original size compressor and increased pre-turbine pressures were tested by performing three different types of simulations.

4.2.1. Load Step Transient

Model with PID Controllers

The load step transient study was repeated five times for 1600 and 1100 engine speeds by limiting the amount of power (from 1 to 5 kW) provided as electrical assistance to the compressor, as shown in Figure 22.

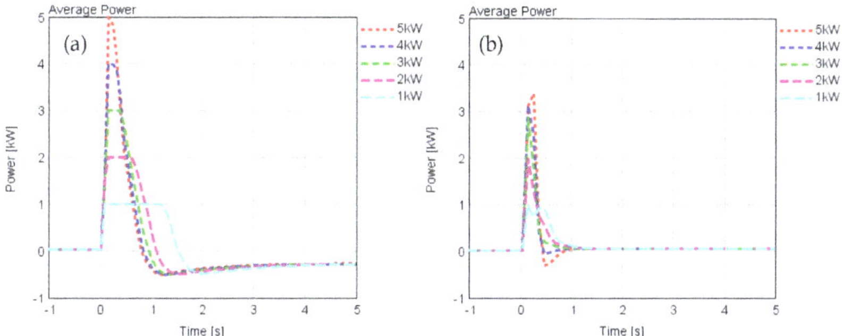

Figure 22. Limits and level of power provided to the compressor during tip-in transient: (**a**) 1600 rpm; (**b**) 1100 rpm.

It is clear that the maximum power limit set for the 1100 rpm case exceeds the model requirement, based on the current PID tuning, for achieving the fastest possible response.

The results in Figure 23 reveal that an electrical assistance of 1 kW can reduce the transient response time of the engine by 70% while higher levels of power can lead to a reduction of more than 85%. As shown in Figure 23a, power levels of more than 3 kW have almost a negligible effect on the response time of the engine. It is obvious by looking at Figure 23b that the improvement in response time is due to the rapid increase of the turbocharger's shaft speed. The fact that the initial shaft speed is higher, as a result of the smaller turbine, also contributes to a fast transient change. Moreover, Figure 23c shows that the electrical assistance provided to the turbocharger pushes the operation of the compressor towards the right side of the mass flow/pressure ratio map, which leads to a more efficient operation of the compressor.

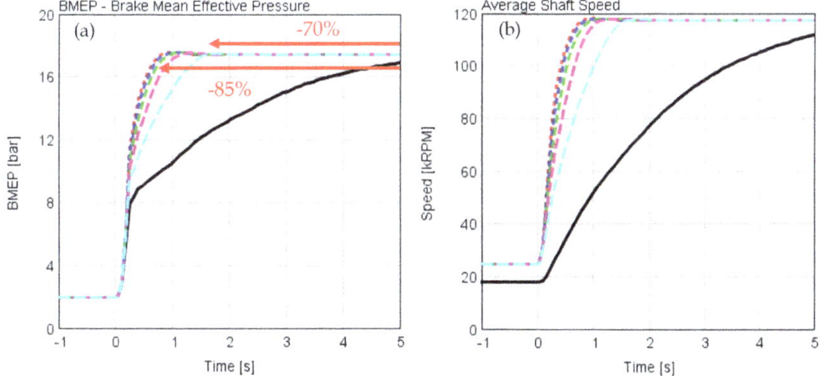

Figure 23. *Cont.*

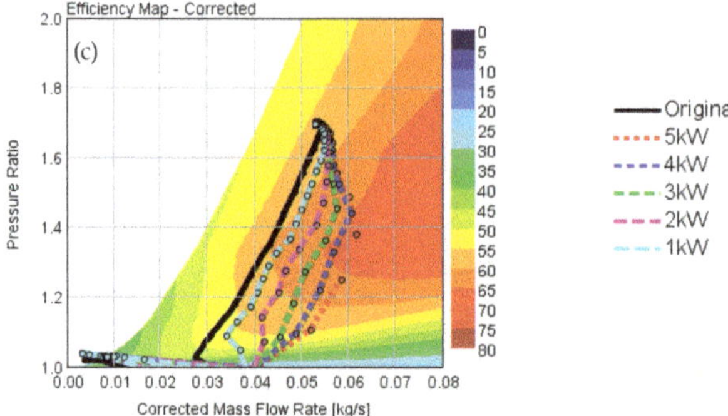

Figure 23. Load tip-in results at 1600 rpm engine speed: (**a**) BMEP response time; (**b**) turbocharger speed response time; (**c**) compressor performance.

On the other hand, the transient improvement at very low engine speeds is of higher magnitude, as shown in Figure 24a,b. This happens due to the very low pre-turbine pressures occurring at 1100 rpm. Moreover, as illustrated in Figure 24c, the electrical assistance provided to the turbocharger pushes the operation of the compressor towards the surge line of the map.

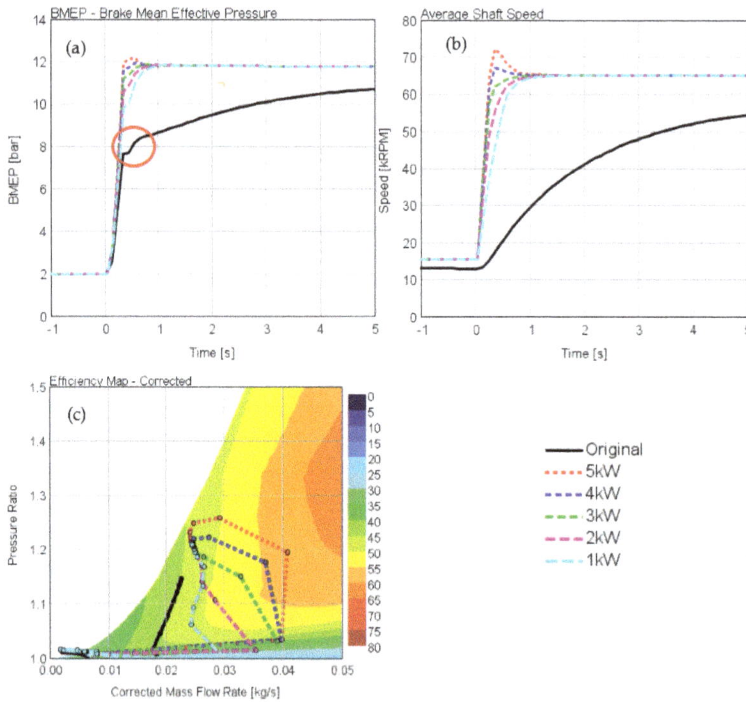

Figure 24. Load tip-in results at 1100 rpm engine speed; the two peaks highlighted in the red circle are a result of the mass flow rate fluctuation when the throttle valve opens: (**a**) BMEP response time; (**b**) turbocharger speed response time; (**c**) compressor performance.

Open-Loop Study (No PID Controllers)

The open-loop study was performed for investigating the hardware capabilities without any effects from the tuning of the PID controllers. Figure 25 reveals the importance of the initial level of power, as well as the average power provided to the compressor. Cases with the maximum amount of power provided during the first section benefit from the fastest response and highest BMEP levels. On the other hand, Figure 25b shows that the power distribution is equally important as the average power provided to the compressor for a fast transient response time.

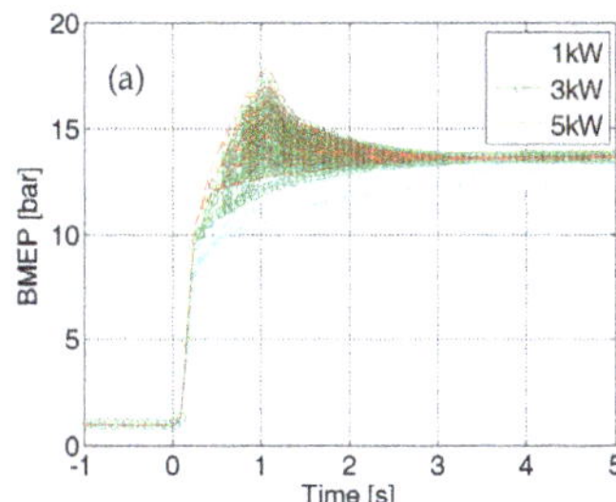
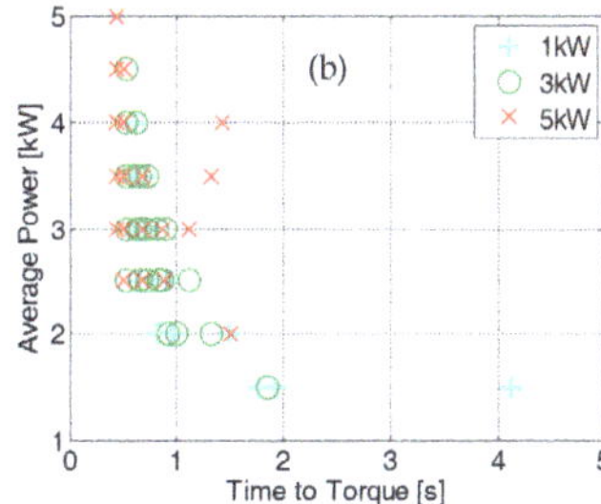

Figure 25. Open-loop load tip-in results at 1600 rpm engine speed: (**a**) BMEP vs. time; (**b**) average power vs. time to torque; colour and symbol categorization indicates the amount of power provided during the first section.

4.2.2. Fixed Gear Vehicle Speed Transient

The results in Figure 26 illustrate the required time for increasing the vehicle's speed from 1100 to 3000 rpm. As is expected, an aggressive tip-in leads to a fast vehicle speed transient. The pre-turbine pressure has almost a negligible effect on the required time. The comparison between the e-turbo and the baseline engine for the 2–16 BMEP shows the improvement in transient response for the electrically-assisted model. On the other hand, the comparison for the 2–4 BMEP tip-in shows no difference at all due to operating at the non-boosted area of the engine. Figure 26b demonstrates the amount of energy provided or harvested by the motor during the transient period of interest. It is clear that during the first phase of the tip-in, the motor provides a high amount of power (limited to 5 kW) to the compressor for meeting the BMEP demands. Once the targeted boost pressure is achieved, the e-turbo starts harvesting energy for most of the cases. The overall average energy demand is negative (energy generation) for most of the cases, except the most aggressive one, as shown in Figure 26b, where the BMEP is increasing for the whole duration of the transient as presented in Figure 26c. The Brake-Specific Fuel Consumption (BSFC) of the engine is slightly deteriorated when higher pre-turbine pressures than the baseline model are applied, as shown in Figure 26d.

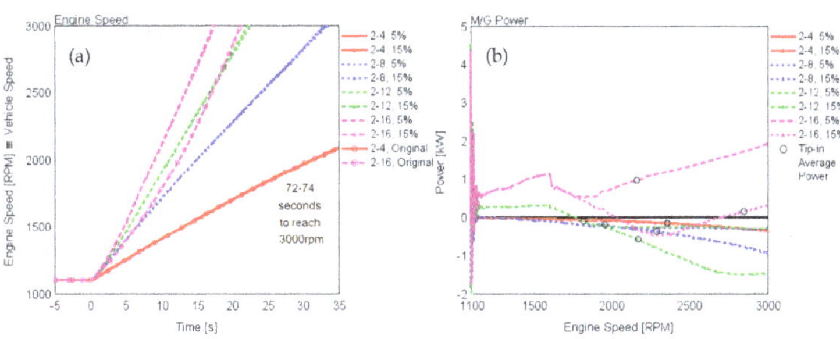

Figure 26. *Cont.*

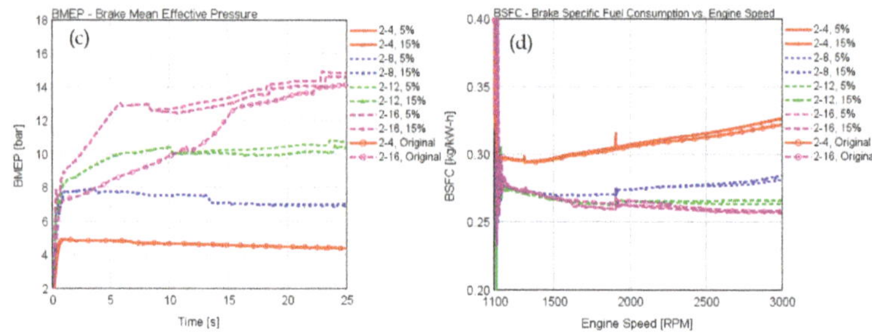

Figure 26. Fixed gear vehicle speed transient results; % values represent pre-turbine pressure increase compared to the baseline engine: (**a**) engine speed vs. time; (**b**) motor-generator power vs. engine speed; (**c**) BMEP vs. time; (**d**) Brake-Specific Fuel Consumption (BSFC) vs. engine speed.

4.2.3. Energy Balance Study

The energy balance study was performed for the NEDC, WLTC, US06 cycles, as well as real driving conditions cycles shown in the Methodology Section.

Figure 27 reveals that all of the driving cycles tested provide a negative average power, which indicates energy harvesting. Energy is provided to the compressor during transient load and speed conditions. The motor-generator is almost inactive when running at low to medium load steady-state conditions. Finally, at extra-urban driving conditions, the generators can harvest a high amount of energy. The average amount of energy harvested is highly dependent on the driving behaviour. The combined driving conditions and the US06 cycle, which represents aggressive driving conditions, demonstrated the maximum amounts of energy gain (6.6 and 5.9 kWh, respectively). The NEDC and WLTC cycles that represent mild driving behaviour provided energy gains between 0.4 and 0.9 kW.

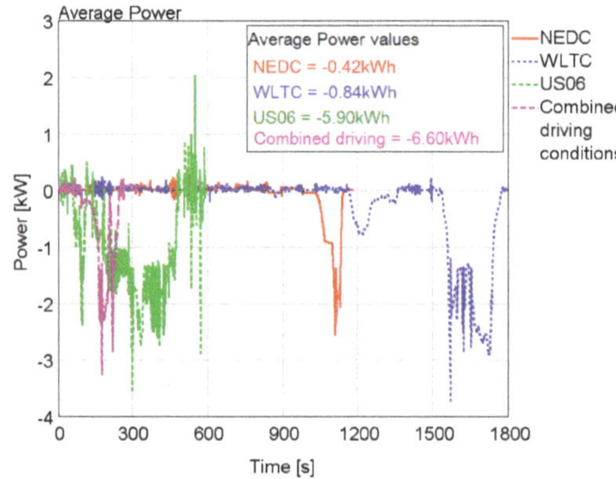

Figure 27. Average power provided/harvested by the motor-generator; negative values indicate energy harvesting.

The fuel consumption of the model in transient mode cannot be highly trusted, as it may be affected by the tuning of the different PID controllers used for the baseline and the electrically-assisted models, as shown in Figure 28.

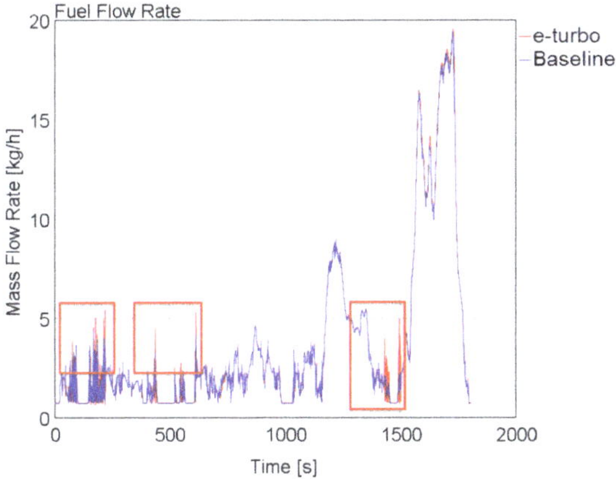

Figure 28. Fuel flow mass flow rate comparison between e-turbo and baseline model for the WLTC cycle; highlighted area indicates the disturbance of PID tuning on the accuracy of the results.

The obtained results for the transient cycles show that with the implementation of the e-turbocharger, the fuel economy of the engine is deteriorated by up to 1.8% (Table 7) due to the increased pre-turbine pressures. However, this fuel consumption increase is compensated by the generated power of the e-turbocharger. The combined driving cycle demonstrates the maximum net gain of 5.5 kWh, followed by the US06 cycle with 3.8 kWh. The WLTC is the only cycle that provides a marginally negative energy gain when considering the increased fuel consumption.

Table 7. Fuel consumption difference between baseline and e-turbo models for various cycles; energy gain = average power values − heat loss (Q) due to increased fuel consumption, Q (kJ) = Calorific value of fuel (CV) (kJ/kg) × fuel (kg), CV of gasoline = 47,300 kJ/kg; * considering the increased fuel consumption.

Cycle	Difference (g/h)	Difference (%)	Energy Gain * (kWh)
NEDC	−3.7	−0.12	+0.37
WLTC	−66.4	−1.34	−0.03
US06	−161.3	−1.80	+3.78
Combined driving	−81.8	−1.26	+5.52

5. Conclusions

The present paper provides a detailed theoretical analysis on the potential of e-turbocharging to control load while providing energy recovery for increasing the overall system efficiency and if possible replacing wastegate boost control. The initial results show that at medium to high loads, significant amounts of enthalpy, up to 5 kW, are lost through the WG valve of a 2.0 L turbocharged spark-ignition engine at medium to high loads. A significant amount of this enthalpy loss can be recovered by the implementation of a motor-generator directly linked to the shaft of the turbocharger.

The study reveals that replacing the wastegate is only achievable when a 30% larger turbine is used for non-violating the pre-turbine pressure limit of the engine. However, a large turbine leads to very poor performance and energy provision needs at the low loads and speeds. A 10% smaller turbine when combined with increased pre-turbine pressures was found to provide efficiency gains and energy harvesting conditions across the whole area of the engine's load/speed map. However, this energy harvesting and efficiency gain come with a 5% penalty on the maximum power output of the engine. Increased power outputs are still achievable by reducing the pre-turbine pressure (opening the WG) and providing extra energy to the compressor.

The transient response time of the electrically-assisted engine showed an improvement between 70% and 90% depended on the engine speed and the power provided to the compressor (~1 to 5 kW). Testing the system under various driving cycles showed that the average amount of energy generated exceeds by up to 1 kW on average (6.6 kWh) the energy required by the motor during the transient events. The e-turbocharged engine demonstrated a fuel deterioration of up to 1.8% during the transient due to the increased pre-turbine pressures. However, this was compensated by the additional generated power. The maximum net energy gain of 5.5 kWh when considering the fuel consumption increase was for the combined driving cycle followed by the US06 cycle with a net gain of 3.8 kWh.

Finally, it needs to be mentioned again that none of the results presented in this study take into account any electrical losses, such as alternator, converter and battery losses, which will obviously reduce the net amount of harvested energy. Future work will include investigations on gasoline and multi-turbo systems with a comprehensive turbine/compressor matching and the electrification of VGT turbocharging systems. Finally, the simulation studies can be supported by experimental investigations with actual hardware testing and e-turbocharger emulation studies (using an externally-boosted engine transient facility located at the University of Bath) for a comprehensive in-cylinder combustion analysis.

Acknowledgments: The authors would like to acknowledge Ford Motor Company and Jaguar Land Rover for their funding support for this research work.

Author Contributions: Pavlos Dimitriou, Richard Burke and Harald Stoffels conceived of and designed the experiments. Pavlos Dimitriou performed the experiments, analysed the data and wrote the paper. Qingning Zhang and Colin Copeland contributed with their experience on electrically-assisted turbochargers and simulation tools.

Conflicts of Interest: The authors declare no conflict of interest. The founding sponsors have agreed to publish the findings of this work.

1. Winterbone, D.E.; Benson, R.S.; Mortimer, A.G.; Kenyon, P.; Stotter, A. Transient response of turbocharged diesel engines. *SAE Technical Paper* **1977**. [CrossRef]

2. Turner, J.W.G.; Popplewell, A.; Marshall, D.J.; Johnson, T.R.; Barker, L.; King, J.; Martin, J.; Lewis, A.G.J.; Akehurst, S.; Brace, C.J.; et al. SuperGen on ultraboost: Variable-speed centrifugal supercharging as an enabling technology for extreme engine downsizing. *SAE Int. J. Engines* **2015**, *8*, 1602–1615. [CrossRef]

3. Rose, A.J.M.; Akehurst, S.; Brace, C.J. Investigation into the trade-off between the part-load fuel efficiency and the transient response for a highly boosted downsized gasoline engine with a supercharger driven through a continuously variable transmission. *Proc. Inst. Mech. Eng. Part D* **2013**, *227*, 1674–1686. [CrossRef]

4. Katrašnik, T.; Rodman, S.; Trenc, F.; Hribernik, A.; Medica, V. Improvement of the dynamic characteristic of an automotive engine by a turbocharger assisted by an electric motor. *J. Eng. Gas Turbines Power* **2003**, *125*, 590–595.

5. Ibaraki, S.; Yamashita, Y.; Sumida, K.; Ogita, H.; Jinnai, Y. Development of the "hybrid turbo", an electrically-assisted turbocharger. *Mitsubishi Heavy Ind. Tech. Rev.* **2006**, *43*, 1–5.

6. Millo, F.; Mallamo, F.; Pautasso, E.; Ganio Mego, G. The potential of electric exhaust gas turbocharging for HD diesel engines. *SAE Tech. Pap.* **2006**. [CrossRef]

7. Burke, R.D. A numerical study of the benefits of electrically-assisted boosting systems. *J. Eng. Gas Turbines Power* **2016**, *138*, 092808. [CrossRef]

8. Bumby, J.R.; Spooner, E.S.; Carter, J.; Tennant, H.; Ganio Mego, G.; Dellora, G.; Gstrein, W.; Sutter, H.; Wagner, J. Electrical machines for use in electrically-assisted turbochargers. In Proceedings of the Second International Conference on Power Electronics, Machines and Drives, Edinburgh, UK, 31 March–2 April 2004; IET: Stevenage, Hertfordshire, UK, 2004; Volume 1, pp. 344–349.

9. Bumby, J.; Crossland, S.; Carter, J. Electrically-assisted turbochargers: Their potential for energy recovery. In Proceedings of the IET The Institution of Engineering and Technology Hybrid Vehicle Conference, Coventry, UK, 12–13 December 2006; IET: Stevenage, Hertfordshire, UK, 2006; pp. 43–52.

10. Divekar, P.S.; Ayalew, B.; Prucka, R. Coordinated electric supercharging and turbo-generation for a diesel engine. *SAE Tech. Pap.* **2010**. [CrossRef]

11. Panting, J.; Pullen, K.R.; Martinez-Botas, R.F. Turbocharger motor-generator for improvement of transient performance in an internal combustion engine. *Proc. Inst. Mech. Eng. Part D* **2001**, *215*, 369–383. [CrossRef]
12. Terdich, N.; Martinez-Botas, R. Experimental efficiency characterization of an electrically-assisted turbocharger. *SAE Tech. Pap.* **2013**. [CrossRef]
13. Arsie, I.; Cricchio, A.; Pianese, C.; De Cesare, M.; Nesci, W. A comprehensive powertrain model to evaluate the benefits of electric turbo compound (ETC) in reducing CO_2 emissions from small diesel passenger cars. *SAE Tech. Pap.* **2014**. [CrossRef]
14. Algrain, M. Controlling an electric turbo compound system for exhaust gas energy recovery in a diesel engine. In Proceedings of the 2005 IEEE International Conference on Electro information Technology, Lincoln, NE, USA, 22–25 May 2005; IEEE: New York, NY, USA, 2005; p. 6.
15. Pasini, G.; Frigo, S.; Marelli, S. Numerical comparison of an electric turbo compound applied to a SI and a CI engine. In Proceedings of the ASME 2015 Internal Combustion Engine Division Fall Technical Conference (ICEF2015), Houston, TX, USA, 8–11 November 2015.
16. Tavčar, G.; Bizjan, F.; Katrašnik, T. Methods for improving transient response of diesel engines—Influences of different electrically-assisted turbocharging topologies. *Proc. Inst. Mech. Eng. Part D* **2011**, *225*, 1167–1185. [CrossRef]
17. Tang, H. Application of Variable Geometry Turbine on Gasoline Engine and the Optimisation of Transient Behaviours. Ph.D. Dissertation, University of Bath, Bath, UK, 2015.

 *applied sciences*

Article

A Study on the Optimal Actuation Structure Design of a Direct Needle-Driven Piezo Injector for a CRDi Engine

Sangik Han [1], Juhwan Kim [1] and Jinwook Lee [2,*]

[1] Department of Mechanical Engineering, Graduate School, Soongsil University, Seoul 06978, Korea; shdow88@naver.com (S.H.); jwan0502@gmail.com (J.K.)
[2] Department of Mechanical Engineering, Soongsil University, Seoul 06978, Korea
* Correspondence: immanuel@ssu.ac.kr; Tel.: +82-2-820-0929

Academic Editor: Jose Ramon Serrano
Received: 13 January 2017; Accepted: 14 March 2017; Published: 24 March 2017

Abstract: Recently, the high-pressure fuel injection performance of common-rail direct injection (CRDi) engines has become more important, due to the need to improve the multi-injection strategy. A multiple injection strategy provides better emission and fuel economy characteristics than a normal single injection scheme. The CRDi engine performance changes with the type of high-pressure electro-mechanical injector that is used and its injection response in a multi-injection scheme. In this study, a direct needle-driven piezo injector (DPI) was investigated, to optimize its actuation components, including the plate length, number of springs, and the elasticity of the spring between the injector needle and the piezo stack. Three prototype DPIs were proposed by this research. They were classified as Type 1, 2, and 3, depending on whether the injector needle was hydraulic or mechanical. Then, the optimal prototype was determined by conducting four evaluation experiments analyzing the maximum injection pressure, injection rate, spray visualization, and real engine combustion application. As a result, it was found that the Type 3 DPI prototype, with several pan-springs and plates, had the highest injection pressure, a steady injection rate, and the fastest spray speed. It also demonstrated the most effective emission reduction for a two-stage rapid spray injection in a single-cylinder CRDi engine. The Type 3 DPI displays an increased elasticity from its hydraulic needle that provides a synergy effect for improving DPI actuation.

Keywords: direct needle-driven piezo injector; high-pressure injection pressure; injection rate; spray visualization; CRDi engine

1. Introduction

The combustion process in an automotive engine is never perfect and small amounts of more harmful emissions are also produced in internal combustion engines. In order to reduce engine emissions, modern automotive engines carefully control the amount of fuel that they burn. The thermal efficiency of a combustion engine is defined as the relationship between the total energy contained in the fuel, and the amount of energy used to perform useful work or generate kinetic energy. Carbon dioxide (CO_2) emissions, as a product of combustion by bonding the carbon in the fuel with the oxygen in the air, are by far the largest amount of emissions produced by internal combustion engine vehicles. Until recently, they were thought to pose no immediate threat to the environment and the health of human beings. Therefore, CO_2 has not been regulated in the same way as HC, CO, and NO_x. However, due to recent concerns about the increasing production of greenhouse gases and the increasing use of fossil fuels, regulators have begun attempts to limit the release of CO_2 and other greenhouse gases into the atmosphere. The ICCT (International Council on Clean Transportation) [1] says that most developed countries have set CO_2 emissions targets for new vehicles. Compared to the EU's target of 95 g CO_2/km by 2020/2021, the US's 97 g CO_2/km by 2025 for passenger cars, Canada's 97 g CO_2/km by 2025, China's 117 g CO_2/km by 2020, and South Korea's 97 g CO_2/km by 2020, illustrate similar targets.

Recently, combustion engine vehicles with a high fuel economy have also been included as eco-friendly vehicles that have less harmful impacts on the environment. In general, modern practical engines are always compromised by trade-offs between different properties, such as efficiency, weight, power, exhaust emissions, or NVH (noise, vibration, harshness). Sometimes, the economy not only plays a role in the engine manufacturing cost, but also in the manufacturing and distributing of the fuel. Increasing the engine's efficiency produces a better fuel economy. Reduced exhaust emissions have been strengthened in passenger cars and light duty trucks after EURO 1 was initiated in 1992. Europe adopted EURO 6 in 2013, and it was applied in South Korea in 2014. Nitrogen oxides (NO_x) of 80% and particulate matters (PM) of 60% are now more tightened than EURO 5. Some countermeasure technologies have been developed for satisfying this regulation. Among them, clean diesel vehicles and combustion engine-based hybrid electric vehicles have been studied, requiring a great deal of high-tech concentration. Specifically, a common-rail direct injection system is at the core of clean diesel technology. This makes it possible to pressurize above 200 MPa and then spray fuel into the combustion chamber through an electro-hydraulic injector. This presents accurate and precise fuel spraying using electronic control and is a great development for clean diesel engines.

Typical solenoid-driven and piezo-driven fuel injectors can generate a favorable response and provide an effective injection in a common-rail direct injection system. These effects reduce exhaust emissions and improve the combustion performance in a clean diesel engine. Therefore, a rapid injector needle response and multiple injections are needed to obtain a high degree of freedom spray strategy, in order to improve diffusion combustion with a higher air to fuel (A/F) mixing ratio, reduce exhaust emissions, and increase engine performance. Research [2,3] has shown that five to seven multiple injections and rapid On/Off switching of the injector hole by the needle can freely control the shape of the injection rate. Daiji Ueda, etc., announced G4P (4th generation diesel piezo) injector concepts, structures, and hydraulic performances, using a rectangular injection rate and shorter multiple injection intervals. Also, they showed engine performance improvement results with G4P [2]. Arpaia, A., etc., reported that a numerical model for simulating a piezo indirect acting fuel injection system under a steady state and transient operating conditions was conducted through a commercial code [3]. Other papers [4], referring to the above result, showed that an injection mechanism with a high-pressure performance reduces not only harmful PM exhaust emissions, but also the auto-ignition combustion performance. However, the design of the piezo injector needs further improvement to increase the response of the injector needle.

In this study, a direct needle-driven piezo injector (DPI) based on the Delphi product [5] was considered to improve injection characteristics with a maximum injection pressure, injection response, and manufacturing simplicity. For investigation, three steps were followed in this study: spray visualization based on Mie-scattering, an injection rate measurement, and a compression ignition (CI) combustion experiment based on a single-cylinder CRDi engine. DPI prototype design concepts focused on an internal DPI constitution, depending on whether or not a hydrodynamic needle was used. This study investigates the injection characteristics and combustion performance of three prototype DPIs, to obtain practical results. The effects of the internal design structure of the direct needle-driven piezo-injector and its optimal design parameters on the spray formation and CI combustion are discussed experimentally.

2. Experimental Apparatus and Procedure

2.1. Design Concept of Prototype DPIs

By using a solenoid coil or a piezo stack as an actuator, diesel injectors have recently been designed in response to stringent emission regulations. A solenoid-driven injector, operated by peak and hold current control, is still commonly used for its cost effectiveness, reliability, and smaller unit size. However, this type of injector is likely to exhibit an injection delay. Therefore, a number of studies [6,7] have focused on the characteristics and injection flexibility of a piezo-driven injector.

A DPI directly utilizes a longer piezo stack to lift the needle for injections. The longer piezo stack generates enough displacement for the lifting using an adapting amplifier located between the needle and the piezo stack. The elongation generated by the piezo stack is used to directly control the needle. During DPI operation, the piezo stack shrinks when its electric energy is discharged. This causes the needle to move upward and reveal previously blocked injector holes. Then, high-pressure fuel is injected through the injector holes. This allows the DPI to have a low fuel consumption by controlling various injection rates. Therefore, as shown in Figure 1, the prototype DPI design concepts in this study focused on the internal DPI construction with three types. Type 1 uses a needle body, a cylindrical shell, and a spring. This type is simple to make, produces a good response, and reduces the unit cost. Type 2 consists of pan-springs, plates, a needle body, a needle cylindrical shell, and a small spring. There is no cap to decrease the hydraulic influence and increase the needle response. Type 3 is constructed without limiting the number of pan-springs and plates. It is designed using a needle that has a needle body, a cylindrical shell, a spring, and cap. This design allows hydraulics to be used. This Type 3 design is used to analyze and study the injector influence by varying the number of plate springs and pressure plates.

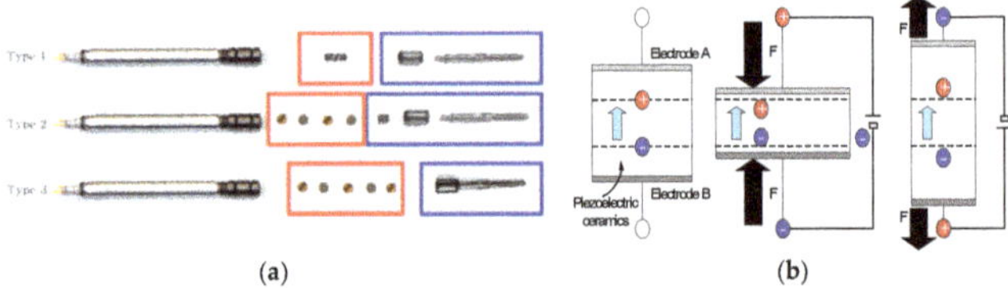

(a)   (b)

Figure 1. DPI proposed by this study; (**a**) Three prototype DPIs, (**b**) Principle of the inverse piezo-electric effect.

2.2. Experimental Apparatus and Methods

2.2.1. Maximum Injection Pressure Measurement

A 1.5 kW class 3-phase motor (Hyosung Co., Seoul, Korea) was used as a high-pressure injection pump for a common-rail system. It can pressurize up to 200 MPa, in order to inject at the maximum possible injection pressure, as listed in Table 1. A temperature sensor is placed inside the fuel tank to automatically turn on and off the radiator and a cooling fan is used to keep the temperature constant. As shown in Figure 2, the injection period and the injection pressure were controlled through the common rail controller (Zenobalti Co., Model:ZB-9013, Daejeon, Korea) and the universal piezo injector driver (Zenobalti Co., Model:ZB-6200, Korea).

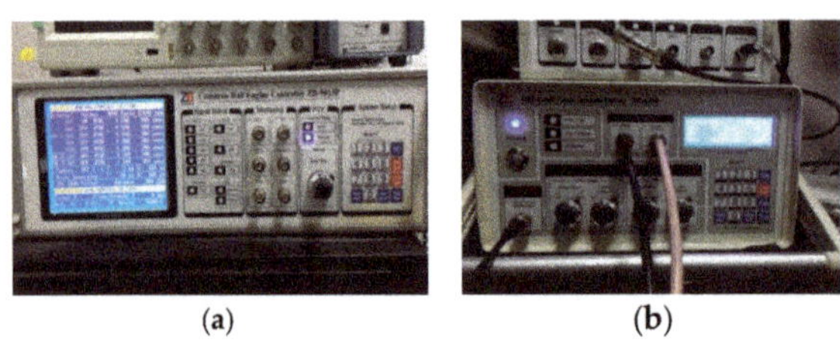

(a)   (b)

Figure 2. Fuel injection control system; (**a**) Common-rail controller, (**b**) Injector driver.

Table 1. Experimental conditions for maximum injection pressure measurement.

Item	Specification
Fuel	ULSD(Ultra-Low-Sulfur Diesel)
Injection duration (μs)	1000
Injection pressure (MPa)	30~180

2.2.2. Injection Rate Measurement

There are two methods which can be used to measure the injection rate. The Zeuch method obtains the injection rate after injecting fuel into the pressure vessel. With the Bosch-Tube method, fuel is injected into the pipe to detect the pressure change in the injection rate meter. Then, the injection rate is measured. In this study, the Bosch-Tube method was used [5]. The Bosch-Tube measurement principle involves feeding the fuel into the pipe with a cross-sectional area A and selecting the fuel control volume in the pipe flowing at the speed of sound as c, when the fuel in the pipe moves at a speed of u. The characteristics of the fuel moving inside can be expressed as shown in Figure 3. By knowing the sonic velocity and fuel density in the pipe and ascertaining the pressure change in the chamber, the injection rate can be calculated.

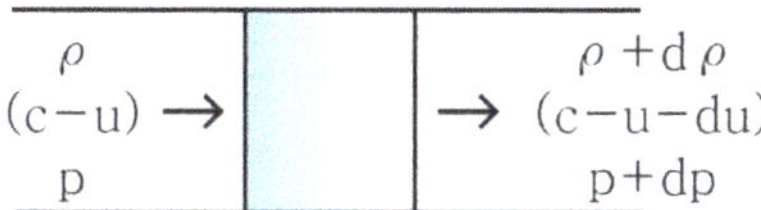

Figure 3. Control volume in the pipe to measure the injection rate.

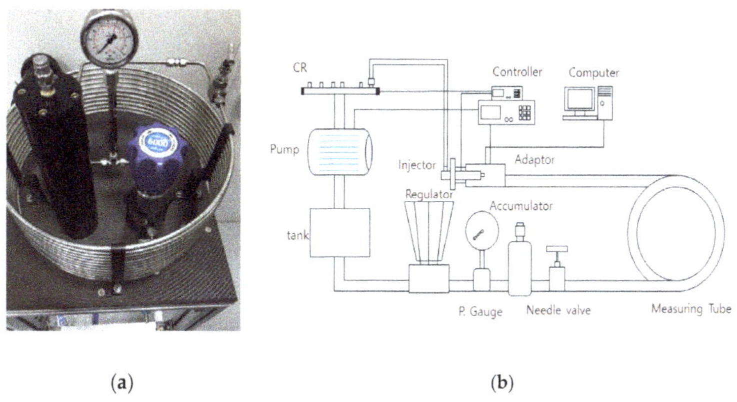

(a)   (b)

Figure 4. Injection rate measurement; (**a**) Injection rate measuring device, (**b**) Schematic diagram.

Table 2. Experimental conditions for measuring injection rate.

Properties	Specification
Fuel	ULSD
Tube length (m)	12.607
Tube area (m^2)	1.6409×10^{-5}
Injection pressure (MPa)	30~180
Duration (μs)	1000

As shown in Figure 4, the injection rate measurement system consists of a measuring tube with a length of about 12.6 m, needle valve, accumulator, pressure gauge, and regulator. The injector is installed on the adapter in the injection rate measuring device and is connected to the common-rail to supply high-pressure fuel. When fuel injection starts, the injection rate is measured by passing the fluid through the pressure sensor (Kistler, Model:6052c, Winterthur, Switzerland) that is mounted on the adapter. The experimental conditions for the injection rate are expressed in Table 2.

2.2.3. Spray Visualization Measurement

Spray visualization experiments were carried out in order to directly confirm the injection speed and behavior for each prototype. In this study, a static chamber with a volume of 885 cc was used, and images were taken with a high speed camera through a quartz glass window with a diameter of 108 mm and a thickness of 50 mm. This experiment was conducted at a normal temperature and pressure, and used a Bosch common-rail system. Since the prototype injector used in the experiment was a piezo-driven type, it was controlled using a universal piezo actuation driver (Zenobalti Co., Model:ZB-6200, Korea). Detailed specifications of the high-speed camera (Vision Research, Model:Phantom V7.3, Wayne, NJ, USA) used in this study are shown in Table 3. Direct lighting spray images were obtained using an 80 W LED light source. The experimental conditions of the spray visualization experiment are shown in Table 4. Figure 5 shows a schematic and actual picture of the injection test apparatus used in this experiment.

Table 3. Specifications of the high-speed camera.

Properties	Specification
Model	Phantom V7.3
Max. resolution	800×600
Sensor type	16 bit sensor
Max. PPS	500,000
Trigger source	TTL signal
Memory	8 G DDR RAM

Table 4. Experimental conditions for spray visualization.

Properties	Specification
Injection duration (μs)	1000
Injection Pressure (MPa)	120
Resolution (pixels)	320×320
Sample rate (frame per second)	25,000
Exposure (μs)	40
Ambient condition	1 atm. 300 K

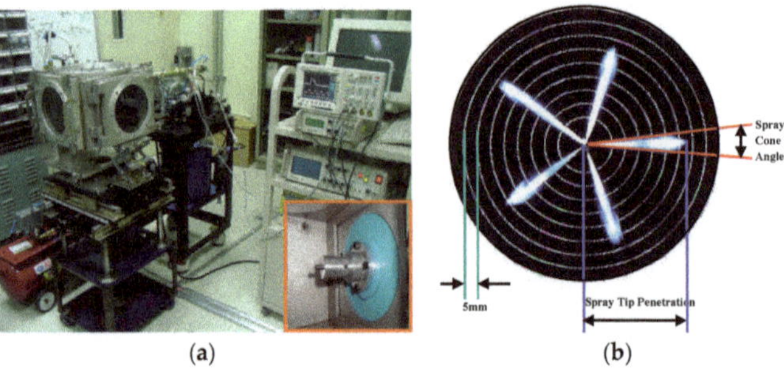

Figure 5. Spray visualization measurement; (**a**) Visualization apparatus, (**b**) Definition of spray parameter.

2.2.4. Real Engine Combustion System

The engine used in this study is a direct-injection type single-cylinder CI engine (Daedong Co., Model:ND10DE, Kyeongsan, Korea). Figure 6 shows the single-cylinder diesel engine system and its specifications are given in Table 5. Table 6 shows the combustion test conditions for the single-cylinder diesel engine. The emission performance of the most optimized DPI type and the conventional type was analyzed by measuring the exhaust gas through an exhaust gas measurement system (Horiba, Model:MEXA-554JKNO$_x$, Kyoto, Japan). The main specifications of the exhaust gas measurement system are shown in Table 7.

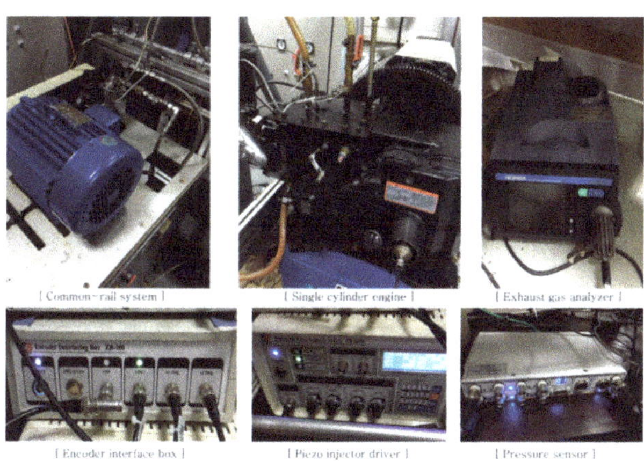

Figure 6. Experiment system of single-cylinder diesel engine.

Table 5. Main specifications of the single-cylinder diesel engine.

Properties	Specification
Engine type	Horizontal DI diesel engine
Bore × Stroke (mm)	95 × 95
Displacement (cc)	673
Compression ratio	18.0
Valve type	SOHC (Single OverHead Camshaft) 2 valve
Combustion chamber type	Reentrant bowl-in-piston

Table 6. Engine combustion test conditions.

Item	Specification
Engine speed (rpm)	1200
Injection quantity (mg/stroke)	3, 7
Injection timing (BTDC)	5, 15, 20

Table 7. Specifications of the exhaust gas measuring device.

Item	CO	HC	CO$_2$	NO$_x$	O^2
Measuring rage	0.00~10.00%	0~10,000 ppm	0.00~20.00%	0~5000 ppm	0.00~25.00%
Repeatability	± 0.06	± 12	± 0.5	± 7	-
Measurement methods	NDIR	NDIR	NDIR	CLD	CLD
Responsibility	up to 10 s, 90% response				

3. Results and Discussion

3.1. Design Complexity and Specialized Knowledge in Needed for Each DPI Prototype

3.1.1. Type 1 DPI

In Type 1, the spring (70C hardened SW-C steel wire) was fabricated according to the wire thickness, outer diameter, number of spring revolutions, and length. It was inserted into the injector inner-body, and then the needle body and needle cylinder were also inserted. Experimental results showed that there is no problem in spraying up to the available injection pressure of 120 MPa. However, at 130 MPa, normal injection becomes intermittent. For this reason, the injector normally injected only 20 of the total injection signals.

As shown in Figure 7, each graph is divided into steel wire thicknesses of 0.7 mm and 0.8 mm. It is shown that the production spring constant (K) increases toward the right of the X-axis, and the actual number of injections continually increases. The reason for this is that the number of actual injections is expected to increase as the spring stiffness and K increases. The spring with high K becomes harder than the lower K springs, such that the pressure inside the injector provides the force to lift the injector needle. It was found that the best Type 1 performance was achieved by the spring with a steel wire thickness of 0.8 mm, outer diameter of 4.8 mm, five spring revolutions, and a length of 6 mm.

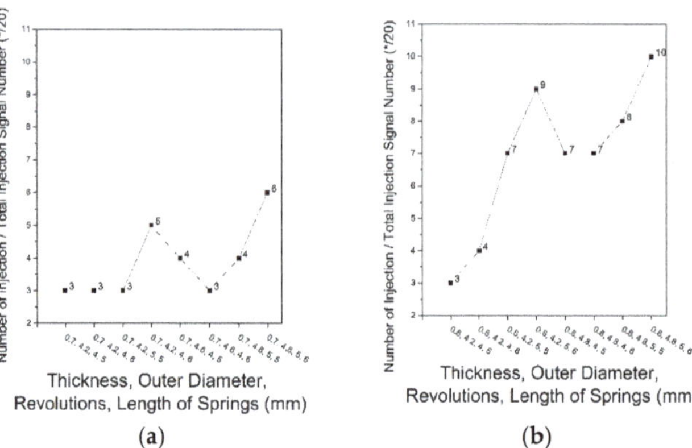

Figure 7. Actual number of injections according to the thickness of the steel wire; (a) 0.7 mm, (b) 0.8 mm.

3.1.2. Type 2 DPI

As shown in Figure 8, the total number of leaf springs and pressure plates is specified as a design parameter, and the needle part is assembled by inserting the needle body, cylinder, and spring. The number of leaf springs was reduced by one and three cases were analyzed. As shown in Figure 9, the Type 2 needle configuration allows fluid to move more freely than the conventional type. When hydraulic pressure is used in the conventional type, the needle moving distance is amplified and the needle can move with a small pressure force. However, the injection responsiveness is considered to be slow. Therefore, the design purpose of Type 2 is to increase the injection responsiveness by allowing pressured fluid to flow freely without hydraulic pressure, by taking advantage of the mechanical characteristics.

Case 2-1 in Figure 8 has an available maximum injection pressure of 130 MPa, as shown in Figure 10. The injection pressures of Case 2-2 and Case 2-3 were about 100 bar lower than Case 2-1. It was found that the available maximum injection pressure can be obtained, depending on the number of leaf springs. If one leaf spring in Type 2 is absent, the force for lifting the needle under CR high pressure is insufficient. However, when two leaf springs are provided, it becomes possible to lift the needle with a greater force, due to the increase in the elastic force. It is considered that the elastic force in the design of Type 2 is determined by the number of leaf springs.

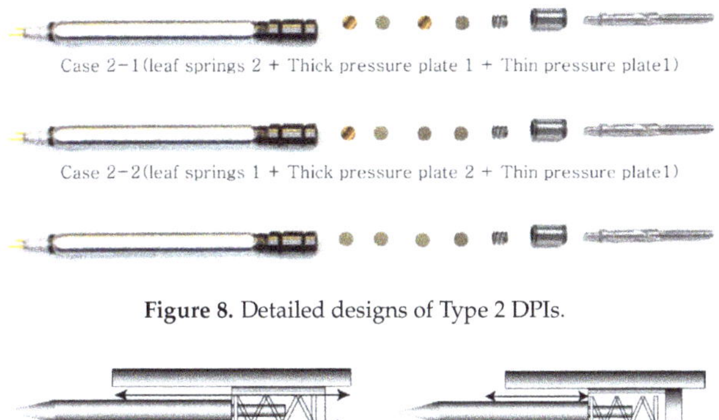

Figure 8. Detailed designs of Type 2 DPIs.

Figure 9. Comparison of needle configuration; (**a**) Type 2, (**b**) Conventional Type DPI.

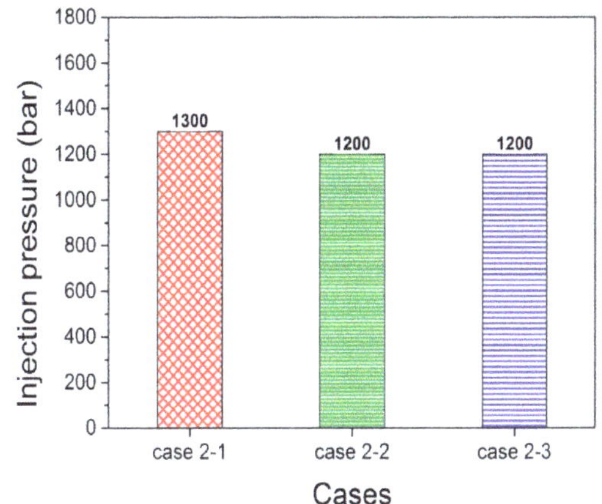

Figure 10. Comparison of the maximum injection pressure of Type 2 DPIs.

3.1.3. Type 3 DPI

As shown in Figure 11, the Type 3 internal structure was analyzed by eight cases with different combinations of pressure plates and leaf springs, including the same nozzle configuration as a conventional type. In Type 3 DPIs, it was found that the available maximum injection pressure was 180 MPa in all cases, except Cases 3-3 and 3-6 in Figure 12. In Case 3-3 with 800 bar and Case 3-6, the pressured fuel leaked and the fuel spray could not be measured. It was thought that Case 3-3 did not inject more than 800 bar, due to the absence of a plate spring for amplifying the piezo stack force, as seen in the Type 2 DPI. Unlike Case 2-3 of Type 2, the reason for not injecting under 120 MPa is that there is not enough elasticity to use the needle's hydraulic pressure. This indicates that when there is not a large enough elastic force, it is advantageous to spray only when using the needle mechanically, without any hydraulic pressure. In Case 3-6, the difference between Case 3-6 and Case 3-4 is the thickness of the pressure plate. The thickness of the thick pressure plate is 0.65 mm, and the thickness of the thin pressure plate is 0.3 mm. The total length of the variables, excluding the length of the needle and piezo stack, is 1.95 mm for Case 3-4 and 1.6 mm for Case 3-6. Therefore, it was found that the minimum length inside the injector must exceed 1.6 mm.

For Type 3 DPI Cases that have an injection performance of up to 180 MPa, spray images were acquired by a high-speed camera and the spray speed was compared, as shown in Figure 13. As known, the spray speed was derived by the definition of the spray tip penetration and spray cone angle, with the edge defined as a line of 80% transmittance of back-light in the raw spray images. It was found that Case 3-8 had the best spray performance with an average speed of 107.3 m/s, an observed-maximum speed of 144 m/s, and an SOI (start of injection) value of 80 μs. The observed-maximum speed in this study means that the spray has not reached the real maximum speed because the bore size of the test engine used in this study is 95 mm, as shown in Table 5, which is about the same as the 100 mm diameter disk shown in Figure 5b. In other words, the high speed camera shoots the fuel stream until it reaches the end of the disk and identifies the highest speed. When the fuel is injected into the cylinder, it is blocked by the inner wall of the cylinder and the maximum speed is not fully achieved in the real spray chamber.

In some Type 3 DPIs, as the elasticity increases, the needle movement is amplified and its moving speed increases. Therefore, it can be expected to have a good effect on two- or multi-injection schemes that are favorable for enhancing the air-fuel mixing rate of the diffusion combustion in the CI engine.

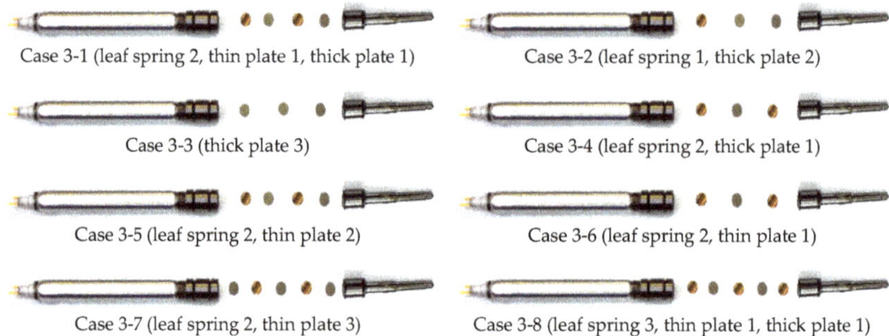

Figure 11. Detailed designs of Type 3 DPIs.

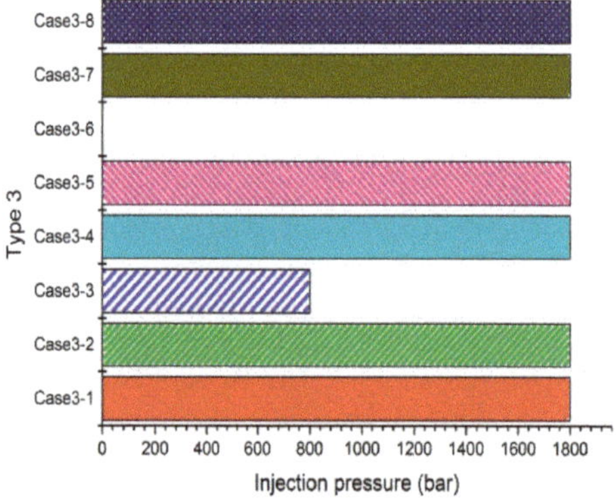

Figure 12. Comparison of the maximum injection pressures of Type 3 DPIs.

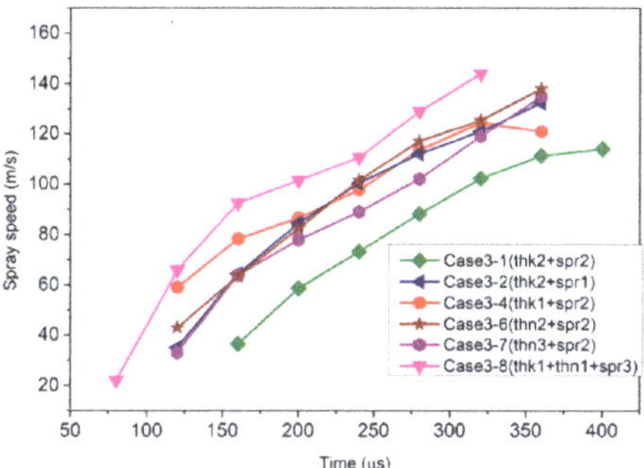

Figure 13. Comparison of the spray speed results in Type 3 DPIs.

3.2. Comparison and Evaluation of Prototype DPIs

The following prototype DPIs were selected to compare the injection performance, spray, and combustion characteristics. (1) The Type 1 DPI has a 0.8 mm thick steel wire spring with an outer diameter of 4.8 mm, spring rate of five, and a length of 6 mm; (2) For the Type 2 DPI, Case 2-1 (two leaf springs + one thick pressure plate + one thin pressure plate) was selected; (3) For the Type 3 DPI, Case 3-8 (three plate springs + one thick pressure plate + one thin pressure plate) was selected.

3.2.1. Comparison of Maximum Injection Pressure Results

The maximum injection pressure with the representative Types 1, 2, and 3, is shown in Figure 14. The Type 3 DPI showed the highest injection pressure at 180 MPa. It was found that the hydraulic application required to build up needle movement is an essential element to obtain an injection performance at high pressures, over 130 MPa. As a result, it can be said that the Type 3 DPI prototype has a similar injection performance to the conventional type DPI, and it has a better flexibility for designing the internal structure of a direct needle-driven piezo-injector.

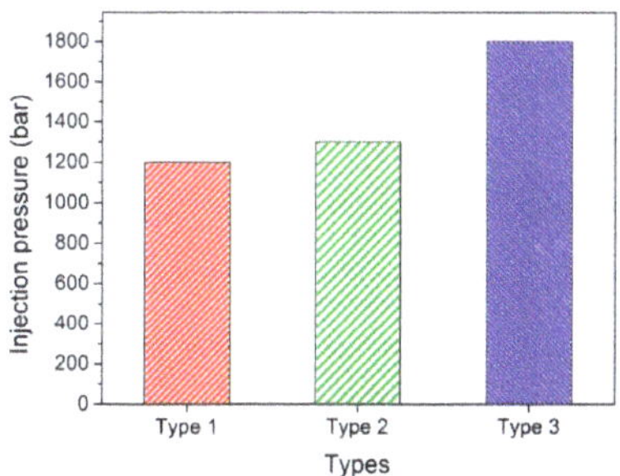

Figure 14. Comparison of the maximum injection pressure for representative Type 1, 2 and 3 DPI.

3.2.2. Comparison of Mie-Scattering Spray Image Results

In this study, LED back illumination for Mie-scattering was applied to make the initial liquid spray to full development spray visible. This enabled an accurate SOI determination and simultaneously measured the initial liquid portion of the spray. Additionally, this Mie scattering technique is suited to the spray characteristics in cold conditions, where there is insignificant vaporization [8,9]. The spray image was taken with an exposure time of 38 μs at a resolution of 320 × 320 pixels in 25,000 pps by a high-speed camera.

Figure 15 and Table 8 show the comparative results of the spray speeds with four different DPIs. The Type 1 and 2 DPI had observed-maximum speeds of 91 m/s and 104 m/s, respectively, with an average speed of 59 m/s and 67 m/s. In comparison to the conventional type, Types 1 and 2 have slower spray speeds and a slow SOI response of 200 µs. It is known that the motion condition of a DPI needle is highly affected by the response parameter [9]. Generally, this shows that the use of hydraulic pressure for needle movement enhances the response performance. In this study, Type 1 and 2 DPIs with the needle only being controlled by a mechanical mechanism, without the use of hydraulic pressure, at about 80 µs, would be slower than the conventional type in terms of the SOI response. In particular, by considering the internal actuation structure of Type 2 and the conventional type DPI, it can be seen that the plate spring and the pressure plate are key factors influencing the determination of the spray characteristics.

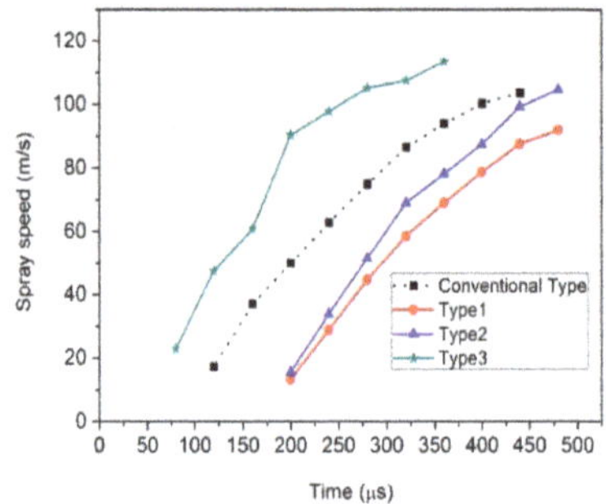

Figure 15. Comparison of spray speed with various DPIs.

Table 8. Comparison of spray characteristics with various DPIs.

Item	Type 1	Type 2	Type 3	Conventional Type
Max. speed (m/s)	91	104	113	103
Ave. speed (m/s)	59	67	80	69
SOI (µs)	200	200	80	120

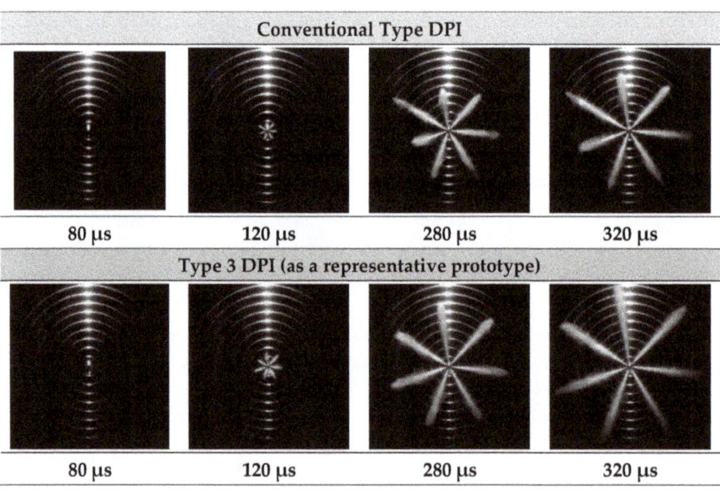

Figure 16. Comparison of Mie-scattering spray image between conventional type and Type 3 DPI.

In the case of the Type 3 DPI, the observed-maximum spray speed is 113 m/s, the average speed is 80 m/s, and the SOI response is 80 µs. This value means that the Type 3 DPI has a fast needle movement, since about 40 µs would be faster than the conventional type DPI. In other words, the Type 3 DPI needle is able to start fuel injection at 80 µs, but the conventional type DPI is still closed, maintaining a needle state until the SOI timing is 120 µs.

As shown in Figure 16, there is an obvious difference in the spray characteristics with time after the start of the injection. As stated previously, the Type 3 DPI had a faster needle opening than the other three DPIs, so it displayed a rapid spray behavior until obtaining a full development spray. It is clear that the actuation structure of a direct needle-driven injector significantly affects the initial spray shape. This is due to hydraulic internal flow dynamics within the DPI nozzle. Therefore, when the needle response is relatively high, the spray tip fuel penetration increases in the liquid phase spray characteristics.

3.2.3. Comparison of Injection Rate Results

Figures 17 and 18 compare the injection rate results between Types 1, 2, 3, and the conventional type DPI. To confirm the reliability of the injection rate meter, the injection rate obtained from the meter used in this study was first validated with reference data from Delphi [10]. It was observed that the injection quantity of the Type 1 DPI is 3.14 mg, which is relatively smaller than Type 2 or Type 3, as shown in Table 9. Since the spring constant K is relatively higher than the leaf spring, the length of the compressed spring is shortened when the needle is lifted, so the distance of travel of the needle in the Type 1 DPI is also shortened, and the injection quantity is expected to be reduced. In the case of Type 2, the injection amount is high after the start of injection, but then falls sharply. Compared to Type 3, the end of the injection (EOI) of the Type 2 DPI occurs early and pressured fuel moves freely in the needle. So, the Type 2 DPI has a higher injection rate than other DPI types, until it reaches the EOI. This is a key cause of decreasing injection quantity due to the quick closing operation.

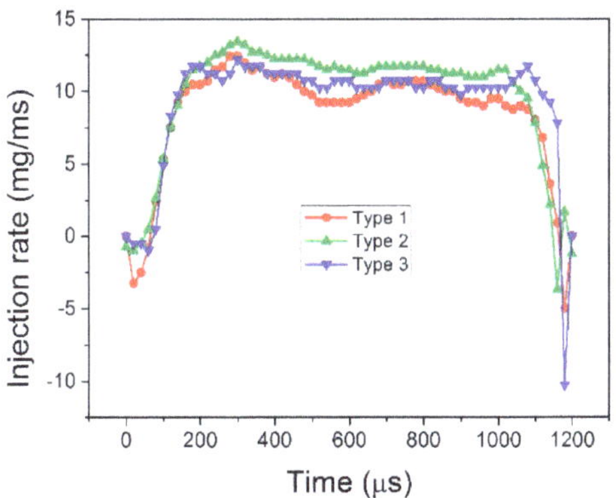

Figure 17. Comparison of the injection rate with various DPIs.

Table 9. Comparison of the injection quantity with various DPIs.

	Type 1	Type 2	Type 3
Quantity (mg)	3.14	3.53	3.56

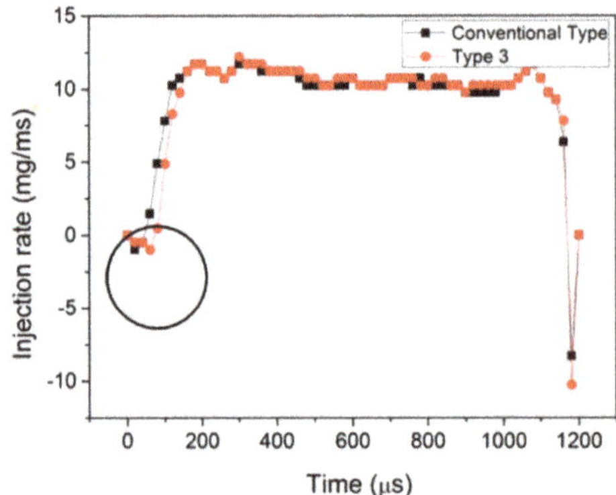

Figure 18. Comparison of the injection rate between conventional type and Type 3 DPI.

However, the Type 3 DPI showed a more stable and sufficient injection rate than Type 1 and 2 DPIs. The shape of the injection rate of the Type 3 DPI is relatively horizontal and the fuel injection lasts for the longest time. This means that the Type 3 DPI has a favorable degree of freedom for controlling the injection amount and the injection shape in real injection mapping processes.

Figure 18 shows a comparison of the injection rates between the conventional type and the Type 3 DPI. As can be seen from this graph, the conventional type shows a shorter start of injection than Type 3. However, as compared with the results of Table 10, the injection rate and injection quantity do not show much deviation. This opinion is based on using a different observational method for the injection rate measurement system and the spray visualization measurement system. Moreover, the first "injection signature" marked with one circle in Figure 18 was detected for the Type 3 DPI. They show an almost equal injection performance, even if the Type 3 DPI has more simple structures in the design processes of the direct needle-driven piezo injector.

Table 10. Comparison of injection quantity between conventional type and Type 3 DPI.

	Conventional Type	Type 3
Quantity (mg)	3.51	3.56

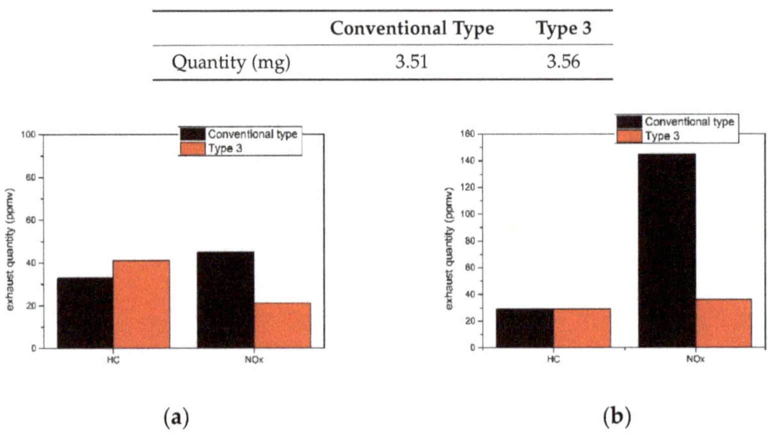

(a)   (b)

Figure 19. Comparison of HC and NO_x emissions between conventional type and Type 3 DPI; (a) 7 mg/stroke (injection quantity) and BTDC 10 CA (injection timing); (b) 3, 7 mg/stroke (pilot/main injection quantity) and BTDC 15, 5 CA (pilot/main injection timing).

The effect of the injection rate on combustion is discussed in Section 3.2.4. It predicts that the Type 2 DPI has a higher NO_x emission due to its higher injection quantity at the beginning of the combustion. It can be inferred that the Type 3 DPI might have the advantage of multiple injections, because the test results indicated that it had a lower injection delay and faster injection start, with a higher initial injection quantity. Therefore, a pilot injection strategy was applied and described, as shown in Figure 19b.

3.2.4. Comparison of Engine HC and NO_x Emission Results

Figure 19 shows the diesel engine experimental results using the same conditions applied in previous chapters. The speed of the engine was maintained at 1200 rpm, and the injection timing was set to BTDC 10 CA in a single injection case and BTDC 15 CA and 5 CA in a two-injection stage (pilot/main) injection case. The fuel injected was 3 mg and 7 mg with the pilot and main injections, respectively. This was done to maintain the total fuel mass with the single injection experiment. By fixing the pilot injection timing at BTDC 15 CA, it was observed that the combustion characteristics and performance were influenced by the injection timing of the main injection. First, it was observed that HC and NO_x emissions for the conventional type and the Type 3 DPI were different, regardless of single and two-injection stages. As shown in Figure 19a, in the case of the Type 3 DPI at the injection quantity of 7 mg/stroke, the HC emission increased by about 8 ppm, but the NO_x value decreased by about 23 ppm, which is more than 50% lower than the conventional type DPI. However when the injection timing of BTDC 10 CA differs for the two-injection stage, as shown in Figure 19b. The NO_x emission value is about 109 ppm, which is about 4.5 times more than the difference. The reason for this is that when the injection angle of the conventional type DPI is small, the injector response is relatively slow and any special distinction between the pilot injection stage and the main injection stage is not clear. Therefore, it is expected to have a similar effect to that of injecting at the injection stroke. Conversely, Type 3 can exhibit low NO_x emission values with a relatively fast injection response and spray speed. This shows that HC and NO_x emissions are significantly reduced in a two-stage injection scheme. The premixed combustion phase and the mixing controlled combustion phase were mostly affected by the injection rate, as stated in Section 3.2.3. It was observed that most of the fuel was injected rapidly following the injection signal at BTDC 15 CA, due to the characteristics of the DPI, a fast SOI, and a high injection rate. Therefore, the premixed combustion phase was promoted. Also, the rest of the remaining fuel was injected after the ignition. Therefore, the combustion was continuous, until the controlled combustion phase. As a result, the air-fuel mixture was saturated, and some fuel in the cylinder was not able to participate in combustion, remaining in an unburned condition. Later, the unburned fuel was used in the late combustion phase. Therefore, a multiple injection strategy could be advantageous in combustion performance assessments. In this sense, a Type 3 DPI specified as Case 3-8 (three plate springs + one thick pressure plate + one thin pressure plate) is a very useful measure for a FIE (fuel injection equipment) system in a newly clean CRDi diesel engine.

4. Conclusions

In this study, a direct needle-driven piezo injector (DPI) based on the Delphi product was considered to improve injection characteristics, including: a maximum injection pressure and injection response, and manufacturing simplicity. The effects of the injection response and spray characteristics on CI combustion emissions were investigated with three concept designs, in order to ascertain the optimal DPI actuation structure for a clean CRDi diesel engine. First, a maximum injection pressure test and a Mie-scattering spray visualization with a high-speed camera were carried out, to compare the realizable criteria for the prototype DPIs. Second, injection rates based on Bosch tube method were applied for each prototype DPI and compared with a conventional type DPI. Finally, diesel engine experiments were carried out to investigate HC and NO_x emissions, and to investigate the engine adaptability and combustion performance of the Type 3 DPI prototype. This approach led to the following conclusions:

(1) In internal design structure for a useful DPI needle response, it is very difficult to apply a high injection pressure over 130 MPa if the hydraulic pressure required to excite the needle's initial state is not applied. From the maximum injection pressure test, it was concluded that a high DPI injection pressure depends on the number of leaf springs. If the leaf spring is not present, the DPI can't spray more than 180 MPa, due to a lack of elasticity. As the number of leaf springs increases, the elasticity increases.

(2) Type 1 and 2 DPI prototypes were conceptually designed with non-hydraulic needles. Therefore, they were able to spray up to a maximum injection pressure of 130 MPa. Type 1, with an internal coil spring, had a high spring constant K, which a shortened needle distance of travel and a reduced injection value. Type 2 had the highest injection rate due to free needle movement. However, the Type 3 DPI, with an optimal number of pan-springs, plates, and needle structure for available hydraulics, had a maximum injection pressure of 180 MPa and advantageous degrees of freedom for controlling the injection value and injection rate with a simpler structure in the design process for a direct needle-driven piezo injector.

(3) In the spray visualization experiment, the Type 3 DPI showed the fastest spraying speed and responsiveness at an average speed, the observed-maximum speed, and the maximum SOI response. As a result, the key factor influencing the spray characteristics is whether the DPI plate spring and the pressure plate of the needle are controlled by a mechanical mechanism only, or by hydraulic pressure.

(4) For the injection rate, the Type 3 DPI had a faster injection rate, which corresponded to the visualization results. Furthermore, by looking at the slope of the injection rate at the beginning and end, the Type 3 DPI had a relatively horizontal slope, and fuel injection lasted the longest.

(5) In diesel engine experiments, relatively high fuel injection rates for the Type 3 DPI prototype generated un-burned emission gases such as HC. However, it had a lower injection delay and faster injection start, with a much higher initial injection quantity. Therefore, it is expected to demonstrate a great advantage for multi-stage injections, where a rapid response and the spraying speed are subdivided.

Acknowledgments: This research was supported by the CEFV (Center for Environmentally Friendly Vehicle) as Global-Top Project of KMOE (Ministry of Environment, Korea).

Author Contributions: Sangik Han, Juhwan Kim and Jinwook Lee conceived and designed the experiments; Sangik Han and Juhwan Kim produced prototype DPI; Sangik Han performed the experiments; Sangik Han and Jinwook Lee analyzed the data and wrote the paper. Jinwook Lee supervised and advised all parts.

Conflicts of Interest: The authors declare no conflict of interest.

1. Jeong, Y.I.; Lee, J.W.; Cho, G.B.; Kim, H.S. *Automotive and Environment*; Publishing Department of Soongsil University: Seoul, Korea, 2010.

2. Ueda, D.; Tanada, H.; Utsunomiya, A.; Kawamura, J.; Weber, J. *4th Generation Diesel Piezo Injector (Realizing Enhanced High Response Injector)*; SAE Technical Paper 2016-01-0846; SAE International: Warrendale, PA, USA, 2016.

3. Arpaia, A.; Catania, A.; Ferrari, A.; Spessa, E. *Development and Application of an Advanced Numerical Model for CR Piezo Indirect Acting Injection Systems*; SAE Technical Paper 2010-01-1503; SAE International: Warrendale, PA, USA, 2010.

4. Husted, H.; Piock, W.; Ramsay, G. Fuel Efficiency Improvements from Lean, Stratified Combustion with a Solenoid Injector. *SAE Int. J. Engines* **2009**, *2*, 1359–1366. [CrossRef]

5. Kim, J.S. Effects of Injection Rate and Spray Characteristics of Piezo Injectors on Diesel Combustion. Master's Thesis, Soongsil University, Seoul, Korea, 2015.

6. Jo, I.; Jeong, M.; Kim, S.; Sung, G.; Lee, J. Experimental Investigation and Hydraulic Simulation of Dynamic Effects on Diesel Injection Characteristics in Indirect Acting Piezo-driven Injector with Bypass-Circuit System. *Int. J. Autom. Technol.* **2015**, *16*, 173–182. [CrossRef]

7. Hardy, M.; Tolliday, A.; Delphi Technologies, Inc. Improvements Relating to Fuel Injector Control. Patent EP 2136062 A1, 2009.
8. Karimi, K. Characterisation of Multiple-Injection Diesel Sprays at Elevated Pressures and Temperatures. Ph.D. Thesis, University of Brighton, Brighton, UK, 2007.
9. Raul, P.; Jaime, G.; Michele, B.; Plazas, A.H. Study liquid length penetration results obtained with a direct acting piezo electric injector. *Appl. Energy* **2013**, *106*, 152–162.
10. Detlev, S.; Stefan, Z.; Martin, H.; Derk, G.; Rainer, W.; Nigel, B. Delphi common rail system with direct acting injector. *MTZ Worldw.* **2008**, *69*, 32–36.

 *applied sciences*

Article

On the Link between Diesel Spray Asymmetry and Off-Axis Needle Displacement

Giancarlo Chiatti [†], Ornella Chiavola [†], Pierluigi Frezzolini [†] and Fulvio Palmieri [*,†]

Dipartimento di Ingegneria, Università degli Studi Roma TRE, 00146 Rome, Italy;
giancarlo.chiatti@uniroma3.it (G.C.); ornella.chiavola@uniroma3.it (O.C.); p.frezzolini@gmail.com (P.F.)
* Correspondence: fulvio.palmieri@uniroma3.it; Tel.: +39-0657-333-493
† All authors contributed equally to the work.

Academic Editor: Jose Ramon Serrano
Received: 31 January 2017; Accepted: 31 March 2017; Published: 11 April 2017

Abstract: Cutting edge experiments and thorough investigations have pointed out that radial components affect the needle lift of diesel nozzles. So far, the effects of such needle "off-axis" have been investigated within the nozzle and immediately downstream from the hole outlet. Here, the focus has been extended to the spray ambient, far outside a multi-hole VCO (Valve Covered Orifice) nozzle. A reference off-axis configuration of the needle has been defined and used to investigate its effects on the spray, in terms of hole-to-hole differences. Indeed, the spray alterations due to the needle position have been addressed for those factors, such as the velocity-coefficient C_V and the area-coefficient C_A, able to describe the nozzle flow behavior under needle off-axis. The investigation has been based on 3D-CFD (three-dimensional computational fluid dynamics) campaigns. The modeling of diesel nozzle flow has been interfaced to the Eulerian–Eulerian near-nozzle spray simulation, initializing the break-up model on the basis of the transient flow conditions at each hole outlet section. The dense spray simulation has been on-line coupled to the Eulerian–Lagrangian modeling of the dilute spray region. Quantitative results on each fuel spray have been provided (in terms of penetration and Sauter Mean Diameter). The range of variability within the spray characteristics are expected to vary has been found and reported, providing reference information for lumped parameter models and other related investigations.

Keywords: diesel spray; needle off-axis; hole-to-hole spray differences

---

1. Introduction

Tracing the evolution of the multi-hole nozzles used in direct injection diesel engines, VCO (Valve Covered Orifice) layout was introduced with the advent of the first generation of common rail injectors, replacing the mini-sac layout. This step allowed the substantial reduction of unburned hydrocarbons emission. In fact, the fuel remaining inside the injector tip is released after the end of injection [1]; very recent studies are better focusing the causes of the so called "injector dribble", whose impact on emission is becoming crucial [2]. The implications of in-nozzle phenomena at the early beginning and after the end of injection have been related to pollutant emission and to deposit formation within the holes. In such a scenario, where low lifts and short injection shots are usual, some recent contributions have concerned the phenomenology of the very early phases of injection [3], of the primary atomization of the liquid [4] and of the phases in which fuel is trapped in the holes and in the sac at the beginning and at the end of injection [5].

Significant efforts have been devoted at comparing sac and VCO nozzles, focusing the attention on the internal nozzle flow, on the rate of injection and on the spray behavior [6–8]. Besides some advantages, the major drawback of the VCO layout is represented by the irregularities of the spray that occur at low needle lift; these abnormalities assume relevance because they are related to the formation of particulate matter [1]. At the time of the introduction of the VCO layout, this aspect had a relatively little importance, since low needle lift was the only operating condition of pilot injections.

The introduction of multiple-injection technique was complemented by the wide adoption of micro-sac geometries. These geometries resulted in being more suited to the characteristics of the second generation injectors (in the compromise between the reduction of unburned hydrocarbons and particulate matter). Indeed, the multiple-injection strategies have required a significant improvement of the injector dynamics, which was also translated into a substantial reduction of the maximum needle lift. Since the needle was located to work almost always at low lift (ballistic displacement), the micro-sac layout represented a better compromise than the VCO layout in producing regular sprays, with low formation of particulate matter [9] and low release of unburned fuel, in agreement with the emission regulations at that time.

As highlighted by the latest research activities [10], the release level of unburned fuel typical of micro-sac nozzles is no longer tolerable, according to the trend of unburned hydrocarbons emission regulation. From this viewpoint, the VCO design will predictably be reassessed and developed again [11,12], in order to minimize the content of liquid fuel that is trapped when the needle is closed. In the light of this, it is believed that investigations on behavior of VCO nozzles under real operating conditions represent an interesting and significant topic.

The abnormal behavior of the spray produced by the VCO nozzle has been observed by several investigators. The VCO-spray features reported in the literature show differences in the macroscopic characteristics among the fuel jets. The differences relate to the spray penetration and to the spray angle [13,14]. If the presence of significant defects of the nozzle is excluded, this behavior suggests that the fuel flows through each hole in a particular way. After the first experiments carried out on sprays from VCO injectors, investigators had suggested that the spray abnormalities were due to the misalignment of the needle during injection; the misalignment had been considered capable of significantly influencing the fuel flow pattern at the hole inlet section; indeed, in VCO nozzles, the needle remains in the immediate proximity of the inlet section of the holes for the whole injection event [14,15]. The super-accurate design and manufacturing of the nozzles, provided with expensive double guides for the needle in some cases, has been the main technical approach that enabled VCO-injector manufacturers to stabilize the behavior of the spray as much as possible.

Some experimental investigations based on X-ray techniques have confirmed that needle displacement is affected by major radial oscillations [16]. The experiments have later also allowed for detecting the amount of needle movement in the radial direction [17].

In agreement with these findings, further contributions and simulation campaigns were carried out to quantify the influence of the needle misalignment on the distribution of the fuel flow out of each hole. The results reported in the literature have highlighted that the flow field upstream of the holes is directly affected by the needle position during injection. In more detail, the position of the needle determines the flow characteristics at each hole inlet, influencing the downstream fuel flow [18–20].

The current experimental techniques allow for making a very fine analysis both on the needle displacement during injection and the internal multiphase nozzle flow [21]. Equally advanced are the techniques developed for the diagnosis of the spray [22] features.

In principle, the link between the needle alignment and the characteristics of each spray could be simultaneously studied by combining nozzle and spray diagnostics. Unfortunately, research activities that combine the two types of analysis are not yet reported in the literature.

The 3D-CFD (three-dimensional computational fluid dynamics) is the only practical way to consider, step by step in time, the link between what happens inside the nozzle and what happens outside; this work is oriented in this direction, and the aim is to observe in what measures the off-axis position of the needle is reflected by the spray anomalies, quantifying and delimiting the differences that could be expected due to the off-axis movement of the needle; according to the aforementioned motivations, the study considers a VCO layout.

In the first phase of the work, the reference nozzle geometry and the reference injection event have been identified. Subsequently, the 3D-CFD simulation approach has been defined, relying on the modeling of the internal nozzle flow; once characterized, it has been interfaced to the spray simulation, initializing the primary break-up model on the basis of the transient flow conditions at each outlet section of the multi-hole VCO nozzle. Among the possibilities, the spray modeling has been based on a coupled Euler–Lagrangian approach, whose features have recently been reported in the literature [23]. The simulations have been carried out in *FIRE*™ environment [24] (by AVL List GmbH, Graz, Austria). The following section describes the relevant assumptions and the details of the proposed approach. The results section reports the main points of investigation; and the conclusions are discussed in the last paragraph.

2. Materials and Methods

2.1. Reference Nozzle and Off-Axis Configuration of the Needle

The reference VCO layout for the investigation has been identified among the real applications, yet preserving the general validity of the study. The choice fell on a commercial six hole nozzle that equips the second generation electro-injectors (injection pressure up to 1600 bar) in the automotive field (cylinder displacement in the range of 500 cm^3). Once defined, the real nozzle has been sectioned and optically investigated (Figure 1); thus, the internal geometry has been digitalized to allow the construction of the computational grids. The relevant nozzle features are reported in Table 1. The simulations have been extended to a half-sector of the whole geometry, decreasing the computational costs, thanks to the plane-symmetry of the implemented off-axis configuration (as described in the next paragraph).

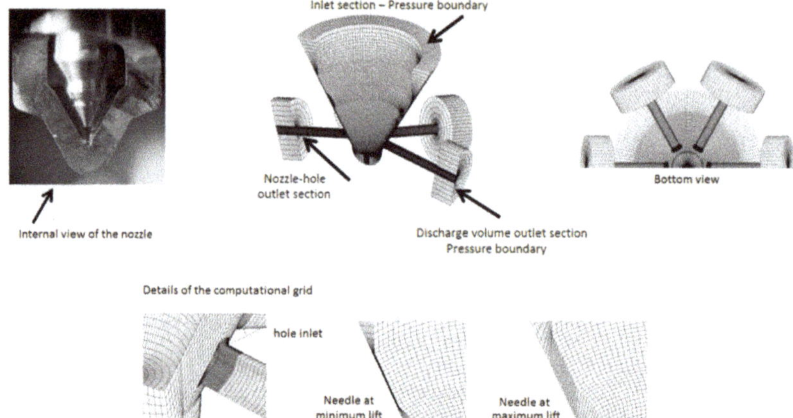

Figure 1. Internal view of the investigated nozzle and computational nozzle grid.

Table 1. Specifications of nozzle layout.

Nozzle Layout	
Nozzle type	VCO [1]
Number of nozzle holes	6
Hole diameter (mm)	0.175
Length to diameter ratio	5.7
Hole plane angle (deg)	156

[1] VCO: Valve Covered Orifice.

The nozzle flow simulation considers the operation of the needle under ballistic conditions; the adopted operating condition of the needle reflects the multiple-injection strategies, where low lifts and short injection shots are usual. The purely axial needle lift law is obtained by means of a lumped-parameter mechanical-hydraulic model of the complete electro-injector. The model has been built in the *LMS Amesim* environment [25] (by Siemens AG Digital Factory, Nurnberg, Germany), and it has been experimentally validated in terms of Rate of Injection (ROI), transient pressure signal at injector inlet and cumulative injected mass. ROI measure is based on the well-known Bosch Tube system [26], in which the pressure wave is measured by means of a piezo-resistive pressure sensor (*4067-type* by Kistler, Winterthur, Switzerland). The same sensor is used to measure the pressure signal at injector inlet, and the cumulative mass of 1000 injection shots has been measured by analytical balance. The complete information about the model and its validation is reported in [27]. The adopted needle lift curve and the related test conditions are reported in Figure 2.

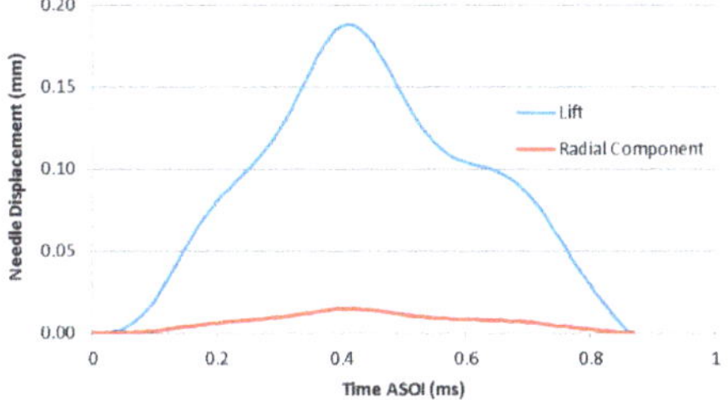

Injection pressure	1400 bar
Injection duration	0.88 ms
Spray chamber pressure	7 bar
Spray chamber temperature	393.15 K
Fuel type	ISO 4113

Figure 2. Needle displacement time traces (**left**); test conditions (**right**).

The grid of the nozzle body is structured and made of hexahedral elements. The hole grids are structured as well, and they are connected to the nozzle body through arbitrary grid interfaces. Such a meshing approach allows for adopting identical grids for the holes. Different meshing approaches are possible, such as the Cartesian cut cell method [17], but the same mesh topology for each hole is not guaranteed. Therefore, the adopted meshing approach is viewed as a good practice in the attempt to avoid hole-to-hole grid dependencies. The modeling of needle displacement in the current CFD analysis is based on "mesh deformation", which consists of the use of a mesh-set to reproduce the needle displacement step-by-step. In such an approach, some cell layers undergo a deformation according to the needle displacement at each simulation time-step and never change in number. To prevent the adoption of extremely small cells at a near-zero needle lift, the approach proposed in [17] has been adopted, so that a minimal gap has been used (4 µm, which correspond to about 2.2% of the maximum lift). It has to be noted that the cell layers in the gap at minimum lift, made of hexahedral elements, have orthogonal faces; during the needle lift, the cells of the layer are deformed, with a well-tolerated skewness level at maximum lift. The nozzle computational grid and the relevant details are visible in Figure 1.

2.2. Off-Axis Displacement of the Needle

As reported in [17], the off-axis displacement is due to needle oscillations during the injection; these have been identified as cantilever-type vibrations of the needle with respect to the needle guide, which is located upstream of the tip. Separate analysis of x and y motion components (assuming the needle is lifting along z) revealed that these lateral oscillations are quite often in-phase. Moreover, it has been evidenced that the oscillation occurs mainly on one side of the nozzle and the needle tip does not move back to the opposite side, crossing the centerline. In [17], the attention is focused on relatively long injections (2 ms), in which a specific needle wobble profile has been considered, with the needle performing three oscillation peaks in the x-plane. Since these lateral oscillations are purely mechanical in nature, the oscillation period is independent from injection duration [28]. In the current investigation, as the considered injection shot is shorter, the occurrence of just one oscillation peak is assumed. According to [17], it is assumed that the needle performs a pure two-dimensional translation, starting from the closing position; the radial component affecting the needle displacement has been set as a percent (8%) of the maximum vertical displacement (Figure 2, left).

As shown in the scheme of Figure 3, the needle displacement is along a straight direction, represented here by the A-B line. The A-B line and the axes of two opposite holes ($N1$ and $N4$) lie in the same plane. As visible, hole $N1$ is the most approached by the needle, whereas the other $N4$ is in the opposite condition. Thus, in the case of $N1$ and $N4$, the off-axis displacement allows for evaluating the influence of the needle proximity to the hole during the injection. In the case of $N2$ and $N3$, the off-axis displacement alters the proximity and also the symmetry of the needle towards the axes of the holes ($N2$ and $N3$). In summary, the current off-axis needle displacement has been chosen to realize a different configuration between the needle and each of the four holes, as evidenced by Figure 3 (right).

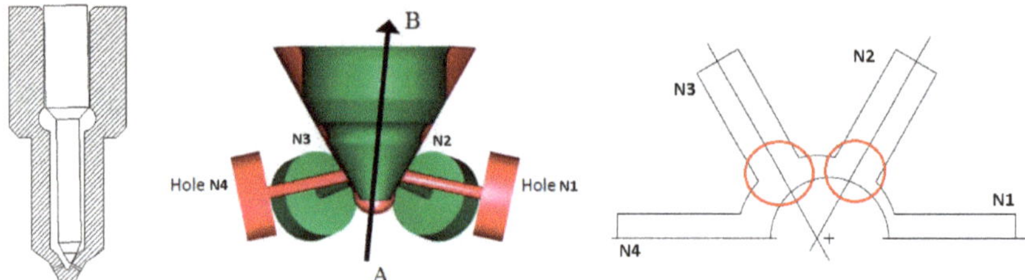

Figure 3. Off-axis needle displacement (cantilever-type) (**left**); adopted scheme for off-axis displacement (translation along A-B line) (**center**); displacement influence on the hole inlets (**right**).

2.3. Modeling Approach

The adopted concept of the coupling among the nozzle flow, the Eulerian and the Lagrangian spray models is shown in Figure 4 (left). The first, independent, simulation step is the transient nozzle flow simulation, taking into account the three-dimensional needle movement and cavitation effects; for certain discrete time steps (360 in the current analysis), the flow data of the bulk liquid and the vapor phase at each orifice outlet are stored on a data file providing the boundary conditions for the dense spray simulation, in agreement with the conclusions reported in [29], where this approach is outlined as appropriate when evaluating the effects of asymmetries in multihole nozzles. The dense spray region is simulated with the Eulerian spray approach, whereas the dilute spray is based on the DDM (Discrete Droplet Method)–Lagrangian approach.

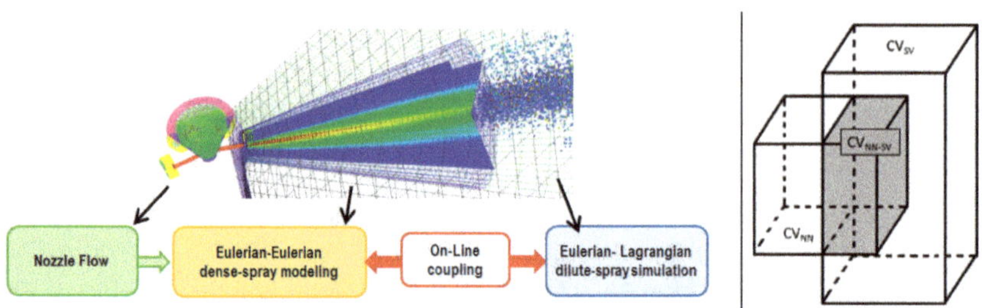

Figure 4. Simulation approach based on coupling among models (**left**); scheme of Control Volume intersection between elements belonging to the Near-Nozzle mesh and to the Spray Volume mesh (**right**).

2.3.1. Coupled Eulerian–Lagrangian Spray Simulation

On the modeling of spray process, the most of the strategies falls into the aforesaid two basic formulations: the Eulerian–Lagrangian (EL) and the Eulerian–Eulerian (EE) methods. Compared to the EL method, the EE method is suitable for calculation of flows with higher droplet concentration [30], whereas the other approach is appropriate in the simulation of the diluted spray region. Here, the EL and EE methods are used in combination. In the light of the contributions reported in the literature, the combination of EE and EL methods is placed in the frame of two coupling approaches; these are the "ELSA model" (Euler Lagrange Spray Atomisation) [31–35] and the "ACCI server" (AVL Coupling Code Interface) [36–39].

In the ELSA methodology, the Eulerian spray is treated as a single-phase flow represented by a liquid-gas mixture. By means of additional transport equations for the liquid mass fraction and the liquid surface density, the spray atomization is modeled; if the spray is dilute enough, Lagrangian spray parcels are initiated.

In the current work, the approach based on the ACCI server is adopted, where the coupling between Eulerian and Lagrangian spray is achieved as follows. The dense spray is calculated with the Eulerian spray approach in a separate simulation client, on a highly resolved computational grid representing the near nozzle region. In the second simulation client (here, the spray-volume ambient), the Lagrangian approach is used for the dilute spray modeling. Indeed, the computational grid of the spray-volume client covers the entire simulation domain, including also the Eulerian spray region. Thus, an overlapping region is defined, in which the simulation is performed on both clients. The interactions between the gaseous and the liquid phases in the Eulerian spray simulation are transferred as source terms to the spray-volume client simulation. The data transfer between the simulation clients is managed via the ACCI server. The coupled simulation starts with the computational initialization of both simulation clients; the fluid properties are determined and the flow field of the Lagrangian client is initialized. The first exchange event taking place is the initialization, when the initial gas flow field of the Eulerian spray client is fully determined by the initial flow field of the Lagrangian client. The three velocity components—pressure, enthalpy, species mass fractions and turbulence transport quantities—are mapped and transferred via the server. The other exchange events are determined by the simulation time steps of the clients. Data exchange from the Lagrangian to the Eulerian spray client is performed to obtain the flow field boundaries of the Eulerian spray simulation; at every exchange event, the 3D flow field of the Lagrangian client is mapped onto the boundary surface of the Eulerian spray client. Data exchange in the other direction occurs due to the interaction between droplets and gas. Drag forces introduce source terms in the momentum conservation equation; mass exchange from evaporation causes sources in the continuity and species transport equations. In the overlapping domain, the gas phase flow field is calculated on both simulation clients. Thus, the source terms from the phase interactions in momentum, mass and energy conservation equations as well as the sources from the species transport equations are transferred from the Eulerian spray to the Lagrangian client. This means that, although there is no liquid phase in the overlapping domain of the Lagrangian client, the gas phase is fully encountered by the interactions with the droplet phases.

The liquid droplets leaving the Eulerian spray client initiate the spray parcels at the Lagrangian client. Conservation of liquid mass, momentum, energy and number of droplets are required criteria for the interface. The algorithm for creating new parcels by fulfilling the conservation criteria is based on the idea that a new parcel is created, if the accumulated liquid mass of a droplet phase exceeds a certain mass threshold. The preparation of new spray parcels is fully calculated by the Eulerian spray client; the interface server sends the parcel initialization data to the Lagrangian client where the new parcels are introduced. A loop over all droplet phases and over all open boundary faces is performed for every time step to check the droplet mass leaving the domain.

Since different meshes are used for the near-nozzle and spray ambient simulations (according to the scheme of Figure 4 (right)), the source terms mapping is performed in a conservative manner using weighting factors. For extensive attributes, such as mass or momentum sources, weighting factors wf_{ex}

are calculated from the intersection volumes between the two client domains, as shown in Equation (1). For example, the mass source term S in a control volume of the spray mesh is calculated from all values in the control volumes in near nozzle mesh, as shown in Equation (2). In the case of intensive attributes such as mass fraction, velocity, pressure or temperature, where attributes do not depend on the size of control volumes, and the weighting factors wf_{in} are defined as shown in Equation(3):

$$wf_{ex} = \frac{CV_{NN-SV}}{CV_{NN}}, \tag{1}$$

$$S_{SV} = \sum_s wf_{ex}S_{NN}, \tag{2}$$

$$wf_{in} = \frac{CV_{NN-SV}}{CV_{SV}}. \tag{3}$$

2.3.2. Near Nozzle Eulerian Spray Modeling

The Eulerian spray modeling is based on a multiphase method [40]. The gas and the liquid are treated as interpenetrating continuous phases, characterized by their volume fraction in the control volume. A number n of phases is considered. The first ($n = 1$) is the gaseous one, and it represents the gas and the fuel vapor in the spray ambient. The other phases ($n = 2, \ldots, n-1$) represent the droplet size classes, whereas the phase n is the bulk liquid phase preceding the break-up process. For each phase, the set of conservation equations is solved separately. Mass—Equation (4), momentum—Equation (5) and enthalpy—Equation (6) conservation equations for the phase k are reported in the following. On the right hand side of each equation, there are the exchange-terms Γ_{kl}, M_{kl}, H_{kl} (mass, momentum, enthalpy) between the phases k and l; these are the terms that contain the physics of the spray. The left-hand side of each equation is made of two terms, the rate of change and the convective transport of the phase flow property. The volume fraction of each phase must fulfill the compatibility condition, Equation (7):

$$\frac{\partial \alpha_k \rho_k}{\partial t} + \nabla \cdot (\alpha_k \rho_k v_k) = \sum_{l=1,\, l \neq k}^{n} \Gamma_{kl}, \tag{4}$$

$$\frac{\partial \alpha_k \rho_k v_k}{\partial t} + \nabla \cdot (\alpha_k \rho_k v_k v_k) = -\alpha_k \nabla p + \nabla \cdot \alpha_k \left(\tau_k + \tau_k^t\right) + \alpha_k \rho_k f + \sum_{l=1,\, l \neq k}^{n} M_{kl} + v_k \sum_{l=1,\, l \neq k}^{n} \Gamma_{kl}, \tag{5}$$

$$\frac{\partial \alpha_k \rho_k h_k}{\partial t} + \nabla \cdot (\alpha_k \rho_k v_k h_k) = \nabla \cdot \alpha_k \left(q_k + q_k^t\right) + \alpha_k \rho_k f \cdot v_k + \alpha_k \rho_k \theta_k + \alpha_k \tau_k : \nabla v_k + \alpha_k \frac{dp}{dt} \sum_{l=1,\, l \neq k}^{n} H_{kl} + h_k \sum_{l=1,\, l \neq k}^{n} \Gamma_{kl}, \tag{6}$$

$$\sum_{k=1}^{n} \alpha_k = 1. \tag{7}$$

The Eulerian spray model is framed in the RANS (Reynolds-averaged Navier-Stokes) approach, with k-epsilon closure. The Eulerian–Eulerian spray model here adopted has been extensively validated against experimental data, as reported in [39], concerning atomization and evaporation process, and in [23] concerning reactive cases.

2.4. Primary Break-Up

2.4.1. Core Injection Approach under Nozzle Flow Local Information

The primary break-up rate in the spray domain considers two independent mechanisms and is determined by the liquid turbulence in the nozzle orifice and by the aerodynamic conditions in the spray region. According to this approach, the turbulent fluctuations in the liquid jet create the initial perturbations on the surface. These grow under the action of aerodynamic pressure forces until they detach as atomized droplets. The coherent liquid core region at the nozzle exit, where primary break-up occurs, is calculated from a mass balance at the volume elements of the liquid core. The liquid core itself is modeled through a blob-injection scheme, in which the blob diameter is constant and the blob number varies according to the mass loss due to primary break-up. The determination of mass loss from this region is based on the rate-approach dR/dt, where dR/dt is viewed as the artificial-radius change of the assumed blob injection. The rate dR/dt is an artificial dimension for the mass loss because the number of blobs decreases while the blob radius stays constant. In the current scheme, such a concept is applied locally, by means of the separate multi-phase nozzle flow simulation (Figure 5, left). For each "*j*" face of the nozzle orifice, the rate dR_j/dt is calculated from the two-phase nozzle flow simulation; the atomization length scale $L_{A,j}$ is equal to the turbulent length scale $L_{T,j}$ as reported in Equation (8). The turbulent length-scale $L_{T,j}$ and the atomization time scale $\tau_{A,j}$ are determined locally for each j face. The expression for $L_{T,j}$ is given by Equation (9), whereas the expression for the local atomization time scale $\tau_{A,j}$ is given by Equation (10):

$$\frac{dR_j}{dt} = \frac{L_{A,j}}{\tau_{A,j}}, \tag{8}$$

$$L_{T,j} = C_2 C_\mu \frac{k_j^{1.5}}{\varepsilon_j}, \tag{9}$$

$$\tau_{A,j} = C_1 \tau_{T,j} + C_3 \tau_{W,j}. \tag{10}$$

The local turbulent time scale $\tau_{T,j}$ is calculated from Equation (11) and the local aerodynamic time and length scale, $\tau_{W,j}$ and $L_{W,j}$, are calculated according to Equation (12):

$$\tau_{T,j} = C_\mu \frac{k_j}{\varepsilon_j}, \tag{11}$$

$$\tau_{W,j} = \frac{L_{W,j}}{\sqrt{\frac{\rho_1 \rho_N |V_N - V_1|^2}{(\rho_N + \rho_1)^2} - \frac{\sigma}{(\rho_N + \rho_1) L_{W,j}}}} \quad \text{and} \quad L_{W,j} = 2 L_{T,j}. \tag{12}$$

The local diameter of the product droplets resulting from this model is proportional to the turbulent length scale, Equation (13):

$$D_{d,j} = 2 L_{T,j}. \tag{13}$$

The adopted model gives different break-up rates dR_j/dt and different product droplet diameter for each face of the nozzle orifice. Each face of the orifice determines the break-up rate on an idealized blob injection surface; therefore, the local rate of change dR_j/dt is mapped onto the corresponding blob surface and represents a certain fraction of the mass loss (Figure 5, right).

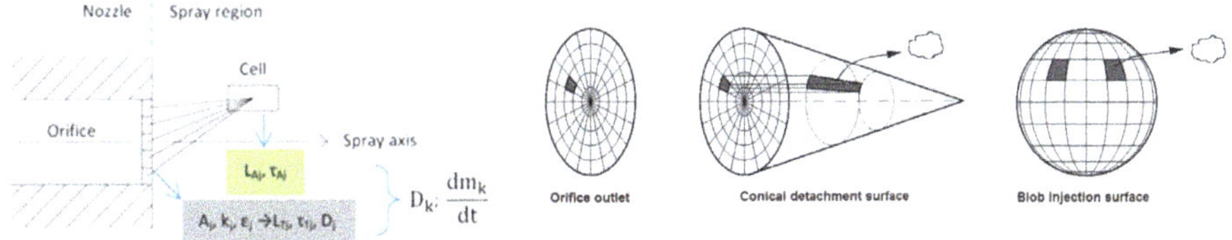

Figure 5. Assignment of orifice cells to primary break-up rate of continuous liquid phase (**left**); local nozzle flow conditions influence on primary break-up rate at the blob (**right**).

The target phase of the liquid droplet phases is determined by the local product droplet diameter $D_{d,j}$. These droplet diameters are sorted into five different predetermined droplet size classes [39].

The primary break-up mass rate per unit volume from the bulk liquid phase (*N*-phase) into the droplet phase k, $\Gamma_{P,Nk}$ is computed according to Equation (14), where α_N is the volume fraction of *N*-phase, A_{nozz} is the surface area of the entire nozzle orifice and A_j is the local surface area of each face in the nozzle orifice. $\dot{M}_{N,j}$ denotes the local primary break-up mass transfer rate per blob surface and is determined by Equation (15). The function $\delta_k\,(D_{d,i})$ results from the sorting process of the local primary break-up rate values, and it is either one or zero. If the local droplet diameter $D_{d,i}$ belongs to size class k, and the function $\delta_k\,(D_{d,i})$ is one; otherwise, it is zero:

$$\Gamma_{P,Nk} = \frac{6\alpha_N}{D_N} \sum_{j=1}^{n_{orifice}} \frac{A_j}{A_{nozz}} \dot{M}_{N,j}\delta_k\left(D_{d,j}\right) = -\Gamma_{P,kN}, \tag{14}$$

$$\dot{M}_{N,j} = \rho_N \frac{dR_j}{dt} = \rho_N \frac{L_{A,j}}{\tau_{A,j}}. \tag{15}$$

2.4.2. Lagrangian Spray Modeling

The injection process in the current investigation has been modeled within a quiescent spray-chamber (7 bar pressure) in non-evaporative conditions (393.15 K). The domain is the same in all simulation cases and it is built on polar symmetry, where the nozzle axis is in the central position. In the near tip region, in all the tested cases, the mesh is built with the same resolution of the nozzle hole, and it gradually widens as the distance from the hole increases. The spray calculation is based on the Discrete Droplet Method (DDM); the k-ε turbulence closure for the RANS-equations solution has been adopted. The WAVE model [41] has been used to simulate the droplet secondary break-up process, assuming that the growth of an initial perturbation on a liquid surface is linked to its wavelength and to other physical and dynamic parameters of the injected fuel and the fluid domain. The particle interaction model based on the statistical approach proposed by O' Rourke [42] has been adopted.

2.5. Multiphase Nozzle Flow Modeling

The internal nozzle flow is simulated through the Eulerian multi-fluid method, Equations (4)–(7). In the current application, the flow is supposed to be isothermal and comprises $n = 3$ phases, namely, the liquid fuel, its vapor and air. The flow model is based on the RANS approach with k–ζ–f turbulence closure model, with a standard wall function. The turbulent stresses in the continuous phases are computed by adding to the standard turbulent viscosity a bubble-induced viscosity term, according to the Sato model [43]. The interfacial exchange terms, relevant for cavitation between liquid and fuel vapor, are represented by the mass- and momentum-exchange terms, which take into account the microscopic effects at the interface between the phases. Concerning the mass exchange term between liquid fuel and vapor, the non-linear form of the Rayleigh–Plesset equation is used to describe bubble growth dynamics. The interfacial momentum exchange (between liquid fuel and vapor) is modeled taking drag and turbulent dispersion effects into account, while neglecting inertia and lift effects. The drag model on the bubbles uses drag coefficient C_D, based on spherical shape and is dependent on Reynolds number. The turbulent mixing process between phases relies on the momentum interaction at the interface, which induces turbulence production on the liquid phase through the Sato model [43]. The state of bubble diameter is a function of position and time (poly-dispersed diameter); coalescence due to turbulent random collisions, breakup (induced by turbulent impact) and bubble generation due to cavitation are the mechanisms taken into account. The complete mathematical description of this approach in the modeling of multiphase nozzle flow has been published and can be found in [44,45]; these authors also report the model validation against several reference flows.

2.6. Cavitation Model Assessment

The cavitation model adopted for the current investigation and the related settings have been assessed simulating two reference cases, i.e., the experiment proposed by Winklhofer et al. [46] and the more recent experiment proposed by Sou et al. [47]. As recently reported in [19], many other contributors [48,49] have considered the experiment of Winklhofer as a significant reference case. This experiment is based on the investigation of an optically accessible channel (inlet section 0.300 × 0.301 mm; outlet section 0.284 × 0.301 mm; 1 mm length, 0.02 mm inlet curvature radius).

During the experimental research, the inlet pressure was fixed (10 MPa), at a uniform fuel temperature of 300 K, whereas the outlet pressure was adjusted in order to obtain the required flow rate. The same technique has been applied in the simulations. Figure 6 (left) shows the computational grid used in the cavitation model assessment; it represents 1/4th of the real channel geometry, thanks to symmetry of domain. The flow behavior of the throttle has been positively tested in several operating points. On one side, the hydraulic prediction has been found to be good agreement with experiments, as visible in Figure 6 (right) and in Table 2, which reports the details of mass flow rate in the conditions evidenced by [46] (cavitation incipience—Point A, critical cavitation—Point B and super cavitation—Point C). On the other side, the cavitation patterns related to the three fundamental conditions, as evidenced in [46], have been consistently reproduced, as reported in Figure 7; it has to be explicitly mentioned that the cavitation patterns observed by Winklhofer were fluctuating, due to the highly turbulent flow environment and they have been obtained by an ensemble averaging process of a set of twenty images taken under the same operating conditions.

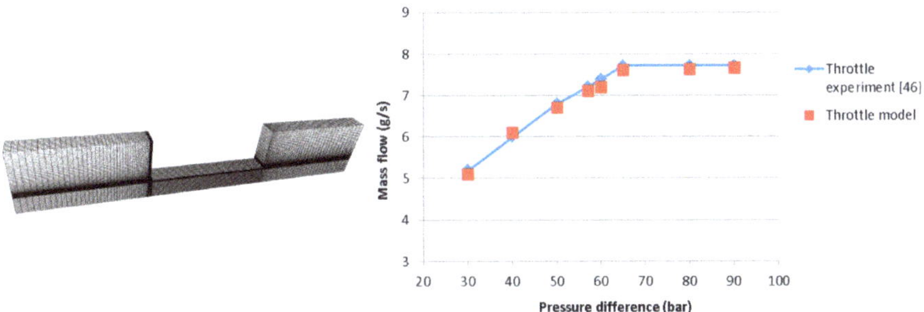

Figure 6. Computational grid used to model the experiment of Winklhofer et al. [46] (**left**); mass flow within the throttle channel (**right**).

Table 2. Mass flow conditions in throttle channel.

Conditions	Cavitation Start (A)		Critical Cavitation (B)		Super Cavitation (C)	
	Experiment	Model	Experiment	Model	Experiment	Model
Pressure difference (bar)	57.0	57.0	65.0	65.0	80	80
Mass flow (g/s)	7.21	7.1	7.72	7.60	7.72	7.63

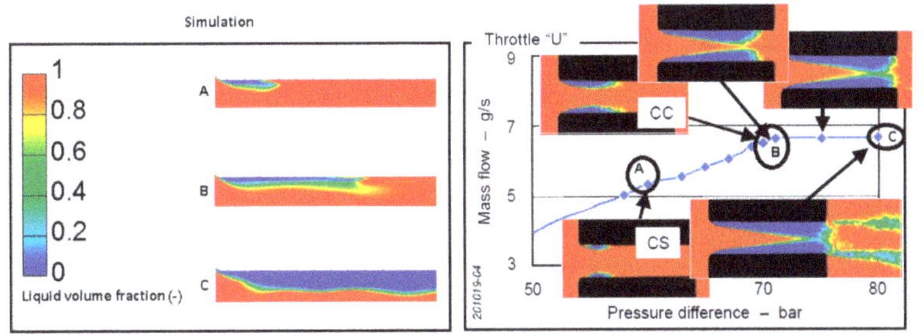

Figure 7. Cavitation at throttle flow conditions (**A, B, C**); simulated cavitation regions (**left**); experiment of Winklhofer et al. [46] (**right**); CS: cavitation start; CC: critical cavitation.

Moving to the second assessment phase, the experiment of Sou [47] has been reproduced by simulation. The experiment consists of a flow through a transparent channel allowing the visualization (based on Laser Doppler Velocimetry) of cavitation incipience and development. In the experiments, a water flow is produced. The discharge ambient is kept at a constant pressure (ambient pressure) and the inflow condition is varied to realize different flow regimes, represented by the mean liquid velocity (V_n) in the channel. The incipience and the development of cavitation is well reproduced by the simulation, as indicated by the contours of liquid mass fraction shown in Figure 8. The quantitative comparison between the experiment and the simulation is reported in Figure 9, where the available data on velocity profile provided by Sou [47] are compared with the simulations. The trends are referred to three different axial locations along the throttle axis, as evidenced by the sketch of throttle layout of Figure 9. Good agreement has been found in all of the cases.

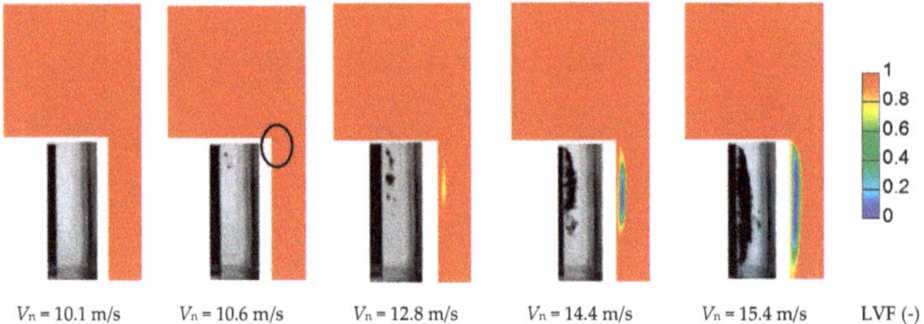

Figure 8. Cavitation at throttle flow conditions; simulation LVF (-) (liquid volume fraction) (**right**); Experiment of Sou et al. [47] (**left**), cavitation inception at black circle.

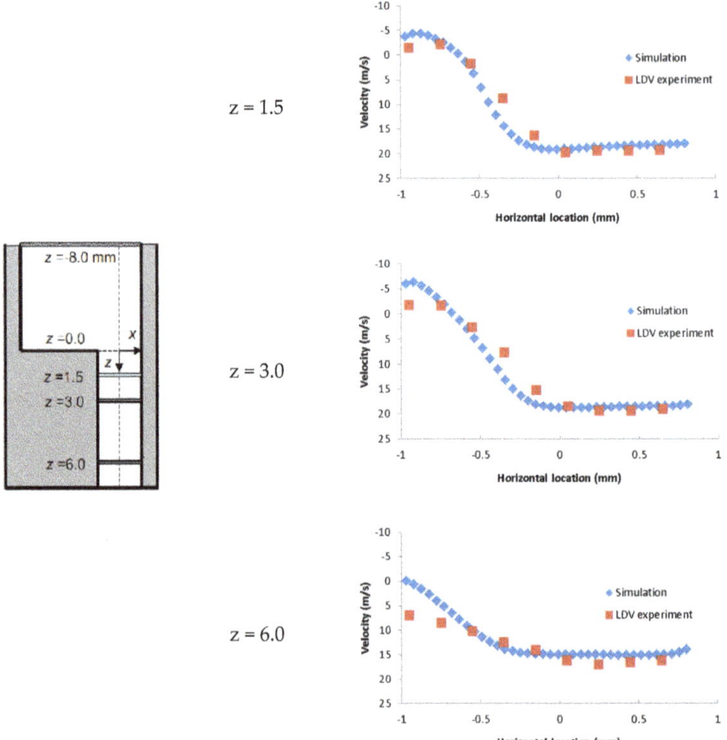

Figure 9. Velocity profiles in the nozzles at V_n = 12.8 m/s [47].

2.7. Grid Sensitivity Tests and Spray Model Assessment

Once the cavitation model is validated, a thorough pre-computation procedure has been performed to find the best computation settings on the used grids; the effect of local refinements has been evaluated as well. The tested grid types and the adopted cell number are listed in Table 3. In the same table, the mass flow rate differences among the nozzle flow tests are reported. Figure 10 reports the graphical results in terms of the velocity profile and cavitation development, when adopted, and the most refined nozzle grids are compared.

Table 3. Grid properties and adopted refinements.

Injector Nozzle			
mesh type Hexahedral-structured	min cell number 195,500	max cell number 3,800,000	adopted cell number 490,000
Mass flow rate % difference during refinement tests	7.4% more	1.6% less	reference case
Near nozzle region			
mesh type Hexahedral-structured	min cell number 11,760	max cell number 687,000	adopted cell number 94,080
Spray volume			
mesh type Hexahedral-structured	min cell number 54,230	max cell number 372,000	adopted cell number 105,984

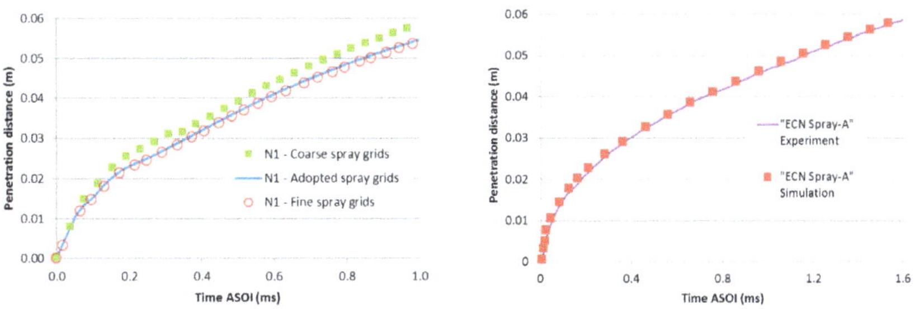

Figure 10. Velocity field contours and cavitation patterns in the case of the adopted and finest nozzle grid.

Since the spray model is used to chase the hole-to-hole spray differences, a crucial check in the current investigation is to assess the penetration dependence on computational grid refinement. Figure 11 (left) reports the trends obtained for the hole $N1$, depending on the refinement of spray domains.

Figure 11. Penetration dependence on spray-domain refinement (**left**); spray model assessment against "ECN Spray-A" reference case (**right**).

The spray model capability has also been assessed against experimental data; Figure 11 (right) reports the comparison between experimental and simulated penetration lengths, referring to "ECN Spray-A" case [50], Table 4.

Table 4. Spray-A conditions.

Rail Pressure (MPa)	Ambient Pressure (MPa)	Ambient Temperature (K)	Nozzle Number	Fluid
150	6	900	210677	N-C12-H26

2.8. Non-Dimensional Coefficients

To allow the comparison among the holes, the results of the nozzle CFD model have been lumped in non-dimensional parameters that qualify the flow at each outlet section [45]. In the adopted scheme, Equation (16), the liquid-phase mass flow rate (m_r) passes through an effective area (A_e) with a uniform velocity (v_e); thus, vena contracta and cavitation affect the flow reducing the geometrical area (A_g) to the effective one (A_e). Considering an ideal flow, the theoretical velocity v_t, Equation (17), is defined by means of a Bernoulli equation, assuming negligible velocity at the inlet section. The theoretical mass flow passing through the geometrical nozzle outlet section (A_g) is given by Equation (18). The first non-dimensional parameter is the discharge coefficient, C_D defined as follows (Equation (19)):

$$m_r = \rho_l \, A_e v_e, \tag{16}$$

$$v_t = \sqrt{\frac{2\left(P_{inj} - P_{back}\right)}{\rho_l}}, \tag{17}$$

$$m_t = \rho_l \, A_g v_t, \tag{18}$$

$$C_D = \frac{m_r}{m_t}. \tag{19}$$

In order to isolate the area from the velocity contribute, the discharge coefficient is viewed as the product of two further non-dimensional parameters C_V and C_A, Equation (20):

$$C_D = C_A \cdot C_V, \tag{20}$$

$$C_V = \frac{v_{eff}}{v_t}, \tag{21}$$

$$C_A = \frac{A_{eff}}{A_g}. \tag{22}$$

3. Results

3.1. Hole-to-Hole Difference on Spray Penetration

As a first step in analyzing the results, it has been checked whether the off-axis of the needle is tied to significant differences on the spray; accordingly, the penetration trends over time for each nozzle hole have been observed. The penetration distance is defined as the distance along the spray axis to the boundary of the spray, according to [51]. The curves of Figure 6 originate at time zero ASOI (After Start Of Injection) and reach the penetration of 0.01 m at different times. Incidentally, up to 0.01 m, the spray is in the Eulerian–Eulerian domain; when droplets exceed this limit, they are taken over by the Eulerian–Lagrangian simulation. The spray emerging from the hole N2 is the slowest, both in the Eulerian tract for the half part of the Lagrangian tract. Referring to the off-axis configuration of Figure 3, the comparisons were made in couples, first N4 and N1, then N2 and N3 and finally the pair N1–N3; the related trends, in terms of percentage differences, are shown in Figure 12 (right).

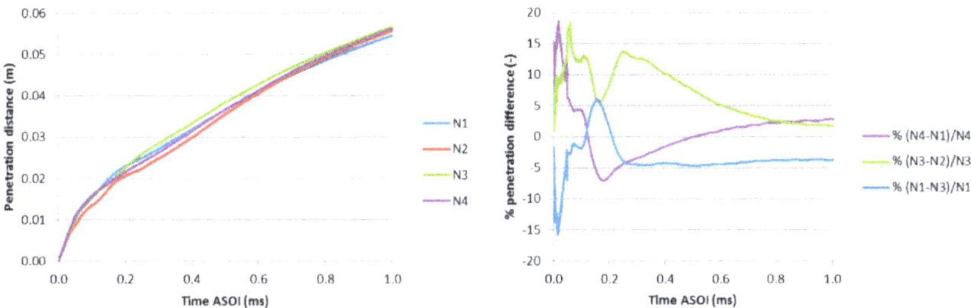

Figure 12. Spray penetration of simulated sprays (**left**); % penetration difference among the sprays (**right**).

Since the opening of the nozzle up to 0.1 ms ASOI, spray $N4$ prevails over spray N1; this difference decreases rapidly, vanishes in 0.11 ms ASOI and reverses back, up to more than half of the injection process. In the closing phase, the spray N4 returns to prevail on the spray N1 but with a more limited deviation to what occurs at the opening (3% versus 18%).

During the opening, the pair $N3$–$N2$ behaves likewise to the pair $N4$–$N1$, with the prevalence of $N3$ on $N2$, but, thereafter, the difference, while decreasing, does not fade and 0 is not reversed.

The trend of the pair $N1$–$N3$ shows the spray $N1$ below $N3$, but not in the period between 0.11 and 0.2 ms ASOI.

A graphical evolution of the spray shape, accompanied by the cut view of the nozzle flow, is reported in Figure 13; even if the thorough analysis of the flow structures within the nozzle is beyond the aim of the current contribution, it is desirable to focus on how the penetration dispersion among the holes is reflected by the shape of the spray plumes; the asymmetry of velocity and cavitation patterns is clearly evident.

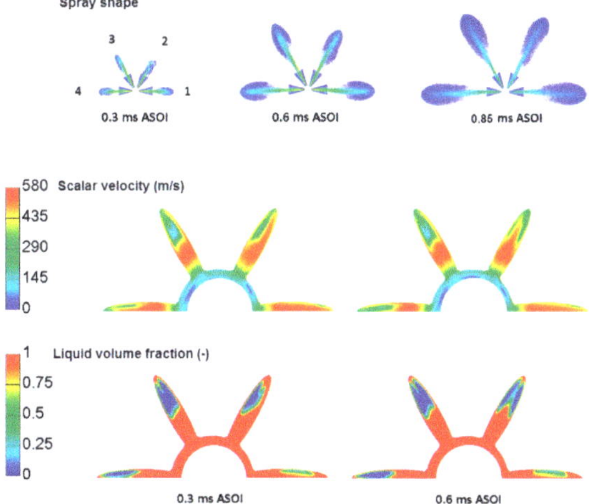

Figure 13. Shape of the four sprays (numbered according to Figure 3) and cut views of the internal nozzle flow.

3.2. Hole-to-Hole Differences for Spray Sauter Mean Diameter (SMD)

Time trends of the global SMD (Sauter Mean Diameter) for different sprays are shown in Figure 14, while maintaining the same comparison method based on couples seen in the previous paragraph.

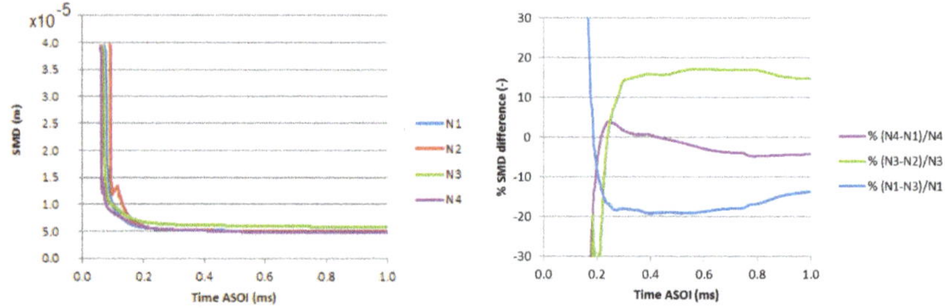

Figure 14. SMD (Sauter Mean Diameter) of the spray versus time (**left**); % difference on SMD among the sprays (**right**).

With reference to the pair $N4$–$N1$, from the opening up to 0.21 ms ASOI, spray $N1$ shows significantly higher SMD values. The behavior is reversed for a short period, up to 0.42 ms ASOI, after which the spray $N4$ returns to exhibit lower SMD, with percentage differences within 5%.

The trend of the pair $N3$–$N2$ sees much lower initial values for the spray $N3$, but the behavior is reversed in 0.24 ms ASOI. The SMD of spray $N2$ assumes the values of the other cases throughout the rest of the injection, about 15% below $N3$.

The pair $N1$–$N3$ still highlights the high SMD value of $N3$, which exceeds $N1$ as low as 0.21 ms ASOI with differences very close to 20%.

3.3. Flow Features at the Outlet of the Holes

In the interpretation of the spray behavior seen previously, the trends of discharge-coefficient C_D, velocity-coefficient C_V and area-coefficient C_A have been obtained for each hole, based on the 3D-CFD nozzle flow modeling. Maximum value of C_D is encountered when the needle is at maximum lift, and it is limited slightly below 0.3. According to [6], such a value could be expected for the same conditions in terms of pressure but at lower needle lift. Here, the obtained value reflects the geometrical features of the considered nozzle; indeed, the configuration of the needle closing passage in the modeled nozzle is able to influence the flow even at relatively high lift values, affecting the flow rate significantly.

Up to 0.2 ms ASOI, the hole $N2$ is characterized by the lesser fuel delivery, Figure 15. In this time interval, the flow rate is penalized by both coefficients C_V and C_A, shown in the graphs in Figure 16. This indicates that, in the initial stages of injection, the fluid reaches the output section of hole $N2$ with relatively low velocity and that the section of the hole is moderately active. The trends of Figure 13 reflect this scenario; the intensity of cavitation at the outlet section is relatively moderate (the liquid volume fraction is relatively high) in addition to the turbulence kinetic energy. In this initial period of the injection process, the mass flow rate is affected by an evident irregularity, which is slightly reproduced in the closing phase of injection. Such a behavior is in agreement with what was found in [17], and it is typically due to the flow perturbations induced by the needle eccentricity.

After 0.2 ms ASOI, the behavior of hole $N2$ changes. The trend of discharge coefficient C_D is modified and from 0.4 ms ASOI becomes similar to that of $N1$ and $N4$ cases. The trend of the liquid fraction is similar to those of the other holes, except in the closing phase, where an increase is visible, accompanied by similar deviations of turbulence kinetic energy and C_A coefficient.

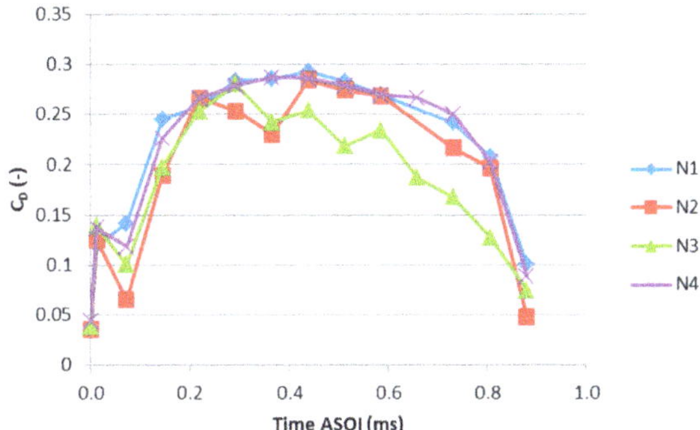

Figure 15. Nozzle hole discharge coefficient Cd and mass flow of injected liquid phase.

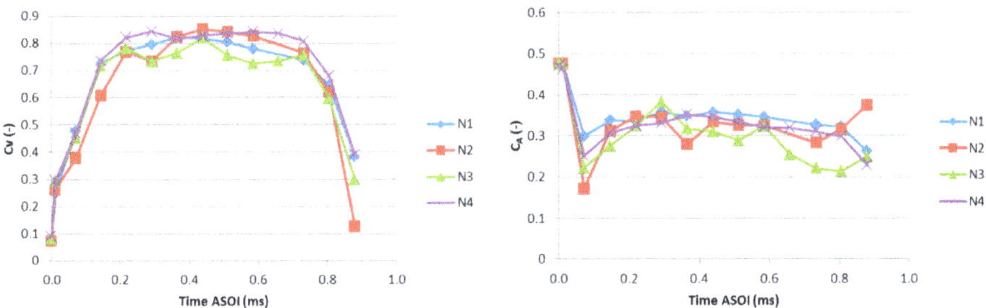

Figure 16. Nozzle hole velocity coefficient C_V (**left**), nozzle hole Area coefficient C_A (**right**).

The discharge coefficient C_D of hole $N3$ is relatively high during the initial stage of the injection, in contrast to what was seen in the $N2$ case. The velocity coefficient C_V is aligned to the values of the other cases and the area coefficient C_A indicates a relatively good utilization of the hole area. The liquid fraction reaches the lowest value at about 0.1 ms ASOI (Figure 17), and the trend of turbulence kinetic energy is relatively high with downward concavity. This scenario changes at 0.3 ms ASOI. The discharge coefficient C_D starts decreasing significantly (Figure 15), due to the drop of C_V and C_A (Figure 16).

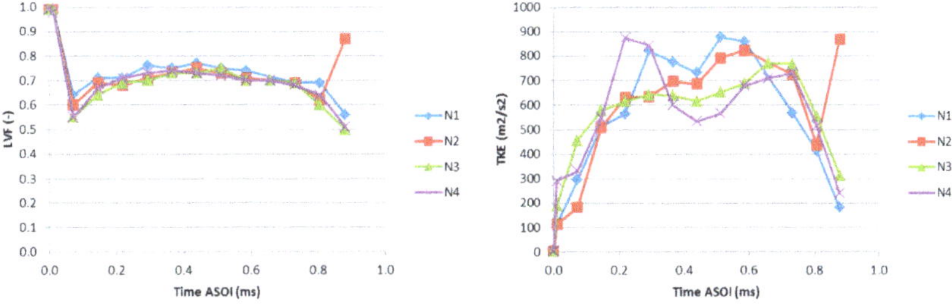

Figure 17. Liquid volume fraction at nozzle hole outlet section (-) (**left**); turbulent kinetic energy at nozzle hole outlet section (m^2/s^2) (**right**).

The $N1$ and $N4$ holes show slight differences in mass flow rate (Figure 15) and the needle off-axis effects are better highlighted by the trends of coefficients C_A and C_V (Figure 16). The coefficients indicate that the outlet velocity at hole $N4$ is higher; this factor does not significantly increase the mass flow since it is balanced by the reduced utilization of the geometric section of the hole, due to stronger cavitation. Indeed, the trends of Figure 17 show the lower liquid fraction at the exit section of hole $N4$, compared to the case $N1$.

4. Conclusions

The current study showed that the lack of homogeneity in the mass-flow distribution is certainly a key factor behind the spray irregularities, but also that it is not the only one. In fact, even when the same discharge coefficient is found among the holes, irregularities are observed among the far sprays. The irregularities have been evidenced by the dispersion of penetration among the different orifices, and they have been addressed to the nozzle configuration by analyzing the flow features in terms of non-dimensional parameters. The velocity-coefficient C_V and the area-coefficient C_A have been found to provide valuable information to completely represent the flow conditions at nozzle outlets when needle off-axis is encountered.

The 3D-CFD modeling retains major relevance in the determination of these coefficients, due to the difficulties that would arise by the experimental techniques. In this scenario, the coupling with the spray models is the crucial step to quantify the differences that cannot be assessed only through the analysis of mass flow rate distribution.

From the industrial viewpoint, the current study made it possible to identify the range of variability that can be expected from the VCO nozzles, in the typical operating conditions of common rail systems (ballistic needle displacement and reduced lift), and the range of variability within the spray characteristics, expected to vary, has been defined. In the current case, these ranges have been found on the order of 15% for penetration and in the order of 20% for global SMD; following the same approach, these ranges can be identified for other specific injectors and operating conditions.

The values of C_V and C_A coefficients can be taken as a reference in further investigations (three-dimensional, multi-zone, or lumped-parameter based), in order to explore the effects of spray non-idealities on the engine performance, on the combustion behavior and, more generally, on those cases affected by sensitivity to the spray characteristics.

Acknowledgments: The authors acknowledge the fundamental contribution of AVL List–Graz providing the *FIRE*™ code and giving the necessary support during the research activities.

Author Contributions: Giancarlo Chiatti conceived and organized the work; Ornella Chiavola and Fulvio Palmieri designed the investigations, analyzed the data and wrote the article; Pierluigi Frezzolini performed numerical simulations.

Conflicts of Interest: The authors declare no conflict of interest.

Nomenclature

Roman	*Description (Unit)*
k,l	Eulerian class index
∇p	pressure gradient (Pa/m)
f	body force vector (N/m^3)
h	specific enthalpy (J/kg)
$M_{k,l}$	momentum exchange term between phase k and l (N/m^3)
$H_{k,l}$	heat flux vector (W/m^2)
t	time (s)
dt	calculation time step
WF	weighting factor
v	velocity vector (m/s)
Greek	*Description (Unit)*
α	volume fraction (–)

θ	enthalpy volumetric source (W/kg)
ρ	density (kg/m^3)
τ	shear stress tensor (N/m^2)
$\Gamma_{k,l}$	mass exchange term between phase k and l (kg/(m^3 s))
ε	turbulence dissipation rate (m^2/s^3)
Subscripts	*Description*
k	phase index
ex	extensive property
in	intensive property
NN	near nozzle
SV	spray volume
Superscripts	*Description*
t	turbulent index
Abbreviations	*Description*
ASOI	after start of injection
3D-CFD	three-dimensional computational fluid dynamics
CV	control volume
DDM	discrete droplet method
RANS	Reynolds-averaged Navier-Stokes
SMD	Sauter mean diameter
VCO	valve covered orifice
LVF	liquid volume fraction

1. Stan, C. *Direct Injection Systems for Spark-Ignition and Compression-Ignition Engines*; Society of Automotive Engineers (SAE): Troy, MI, USA, 2000.
2. Eagle, W.E.; Musculus, M.P.B. Cinema-Stereo Imaging of Fuel Dribble after the End of Injection in an Optical Heavy-Duty Diesel Engine. In Proceedings of the Thiesel Conference on Thermo and Fluid Dynamic Processes in Direct Injection Engines, Valencia, Spain, 9–12 September 2014.
3. Crua, C.; Heikal, M.R.; Gold, M.R. Microscopic imaging of the initial stage of diesel spray formation. *Fuel* **2015**, *157*, 140–150. [CrossRef]
4. Ghiji, M.; Goldsworthy, L.; Brandner, P.A.; Garaniya, V.; Hield, P. Analysis of diesel spray dynamics using a compressible Eulerian/VOF/LES model and microscopic shadowgraphy. *Fuel* **2017**, *188*, 352–366. [CrossRef]
5. Papadopoulos, N.; Aleiferis, P. Numerical Modelling of the in-Nozzle Flow of a Diesel Injector with Moving Needle during and after the End of a Full Injection Event. *SAE Int. J. Eng.* **2015**, *8*, 2285–2302. [CrossRef]
6. Bermúdez, V.; Payri, R.; Salvador, F.J.; Plazas, A.H. Study of the influence of nozzle seat type on injection rate and spray behavior. *Proc. Inst. Mech. Eng. Part D J. Automob. Eng.* **2005**, *219*, 677–689.
7. Salvador, F.J.; Carreres, M.; Jaramillo, D.; Martínez-López, J. Comparison of microsac and VCO diesel injector nozzles in terms of internal nozzle flow characteristics. *Energy Convers. Manag.* **2015**, *103*, 284–299. [CrossRef]
8. Moro, A.; Zhou, Q.; Xue, F.; Luo, F. Comparative study of flow characteristics within asymmetric multi hole VCO and SAC nozzles. *Energy Convers. Manag.* **2017**, *132*, 482–493. [CrossRef]
9. Chiatti, G.; Chiavola, O.; Recco, E.; Palmieri, F. Soot Particles Experimental Characterization during Cold Start of a Micro Car Engine. *Energy Proced.* **2016**, *101*, 662–669. [CrossRef]
10. Eagle, W.E.; Musculus, M.P.B. Image-Based Correlation of Engine Operating Parameters with Occurrence and Duration of Diesel Fuel Injector Dribble. In Proceedings of the Oral Communication at SAE World Congress 2015, Detroit, MI, USA, 21–23 April 2015.
11. Mitroglou, N.; Gavaises, M.; Arcoumanis, D. *Spray Stability from VCO and a New Diesel Nozzle Design Concept*; Fuel Systems for IC Engines; IMechE: London, UK, 2012.
12. Chiatti, G.; Chiavola, O.; Palmieri, F. Diesel Nozzle Flow Investigation in Non-Radial Multi Hole Geometry, ASME Paper 5556. In Proceedings of the 2014 Internal Combustion Engine Division Fall Technical Conference, Columbus, IN, USA, 19–22 October 2014.
13. Bae, C.; Kang, J. Diesel Spray Development of VCO Nozzles for High Pressure Direct-Injection. *SAE Tech. Pap.* **2000**. [CrossRef]

14. De Risi, A.; Colangelo, G.; Laforgia, D. An Experimental Study of High-Pressure Nozzles in Consideration of Hole-To-Hole Spray Abnormalities. *SAE Tech. Pap.* **2000**. [CrossRef]
15. Oda, T.; Hiratsuka, M.; Goda, Y.; Kanaike, S.; Ohsawa, K. Experimental and Numerical Investigation about Internal Cavitating Flow and Primary Atomization of a Large-scaled VCO Diesel Injector with Eccentric Needle. In Proceedings of the ILASS-Europe, Brno, Czech Republic, 6–9 September 2010.
16. Fezzaa, K.; Lee, W.K.; Cheong, S.; Powell, C.F.; Wang, J.; Li, M.; Lai, M.C. *High Pressure Diesel Injection Studied by Time-Resolved X-ray Phase Contrast Imaging*; ICES2006 ASME: New York, NY, USA, 2006.
17. Battistoni, M.; Xue, Q.; Som, S.; Pomraning, E. Effect of Off-Axis Needle Motion on Internal Nozzle and Near Exit Flow in a Multi-Hole Diesel Injector. *SAE Int. J. Fuels Lubr.* **2014**, *7*, 167–182. [CrossRef]
18. Palmieri, F. The Influence of Actual Layout and Off-Axis Needle Stroke on Diesel Nozzle Flow under Ballistic Needle Displacement. *J. Eng. Gas Turbines Power* **2013**, *135*, 101502. [CrossRef]
19. Xue, F.; Luo, F.; Cui, H.; Moro, A.; Zhou, L. Numerical analyses of transient flow characteristics within each nozzle hole of an asymmetric diesel injector. *Int. J. Heat Mass Transf.* **2017**, *104*, 18–27. [CrossRef]
20. Salvador, F.J.; Martínez-López, J.; Romero, J.-V.; Roselló, M.-D. Study of the influence of the needle eccentricity on the internal flow in diesel injector nozzles by computational fluid dynamics calculations. *Int. J. Comput. Math.* **2014**, *91*, 24–31. [CrossRef]
21. Xue, Q.; Battistoni, M.; Powell, C.F.; Longman, D.E.; Quan, S.P.; Pomraning, E.; Senecal, P.K.; Schmidt, D.P.; Som, S. An Eulerian CFD model and X-ray radiography for coupled nozzle flow and spray in internal combustion engines. *Int. J. Multiph. Flow* **2015**, *70*, 77–88. [CrossRef]
22. Payri, R.; Viera, J.P.; Gopalakrishnan, V.; Szymkowicz, P.G. The effect of nozzle geometry over the evaporative spray formation for three different fuels. *Fuel* **2017**, *188*, 645–660. [CrossRef]
23. Petranović, Z.; Edelbauer, W.; Vujanović, M.; Duić, N. Modelling of spray and combustion processes by using the Eulerian multiphase approach and detailed chemical kinetics. *Fuel* **2017**, *191*, 25–35. [CrossRef]
24. FIRE Rev.14.2. *User's Guide, Solver Manual. Eulerian Multiphase Manual, Spray Manual*; AVL List: Graz, Austria, 2016.
25. LMS Imagine. Lab. AMESim Manuals. In *Technical Bulletins and Libraries*; Release 14; Steris Life Sciences: Mentor, OH, USA, 2014.
26. Bosch, W. The Fuel Rate Indicator: A New Measuring Instrument for Display of the Characteristics of Individual Injection. *SAE Tech. Pap.* **1966**. [CrossRef]
27. Marini, J. Studio e Modellazione del Comportamento Meccanico-Idraulico di Polverizzatori di Elettroiniettori Diesel Common-Rail. Master's Thesis, Roma TRE University, Rome, Italy, 2016.
28. Manin, J.; Kastengren, A.; Payri, R. Understanding the acoustic oscillations observed in the injection rate of a common-rail direct injection diesel injector. *J. Energy Power Eng.* **2012**, *134*, 122801. [CrossRef]
29. Desantes, J.M.; García-Oliver, J.M.; Pastor, J.M.; Pandal, A.; Baldwin, E.; Schmidt, D.P. Coupled/decoupled spray simulation comparison of the ECN spray a condition with the σ-Y Eulerian atomization model. *Int. J. Multiph. Flow* **2016**, *80*, 89–99. [CrossRef]
30. Pandal, A.; Pastor, J.M.; García-Oliver, J.M.; Baldwin, E.; Schmidt, D.P. A consistent, scalable model for Eulerian spray modeling. *Int. J. Multiph. Flow* **2016**, *83*, 162–171. [CrossRef]
31. Vallet, A.; Burluka, A.A.; Borghi, R. Development of an Eulerian model for the "atomization" of a liquid jet. *At. Sprays* **2001**, *11*, 619–642. [CrossRef]
32. Blokkeel, G.; Barbeau, B.; Borghi, R. A 3D Eulerian model to improve the primary breakup of atomizing jet. *SAE* **2003**. [CrossRef]
33. Demoulin, F.X. Coupling vaporization model with the Eulerian–Lagrangian Spray Atomization (ELSA) model in diesel engine conditions. *SAE Tech. Pap.* **2005**. [CrossRef]
34. Ning, W.; Reitz, R.D.; Diwakar, R.; Lippert, A.M. An Eulerian–Lagrangian spray and atomization model with improved turbulence modeling. *At. Sprays* **2009**, *19*, 727–739. [CrossRef]
35. Hoyas, S.; Gil, A.; Margot, X.; Khuong-Anh, D.; Ravet, F. Evaluation of the Eulerian–Lagrangian Spray Atomization (ELSA) model in spray simulations: 2D cases. *Math. Comput. Model.* **2013**, *57*, 1686–1693. [CrossRef]
36. Krueger, C. Validation of a 1D spray model for simulation of mixture formation in direct injection Diesel engines. Ph.D. Thesis, RWTH Aachen, Aachen, Germany, 2001.

37. Edelbauer, W.; Suzzi, D.; Sampl, P.; Tatschl, R.; Krueger, C.; Weigand, B. New concept for on-line coupling of 3D Eulerian and Lagrangian spray approaches in engine simulations. In Proceedings of the 10th International Conference on Liquid Atomisation and Spray Systems, Kyoto, Japan, 27 August–1 September 2006.

38. Edelbauer, W. Coupling of 3D Eulerian and Lagrangian spray approaches in industrial combustion engine simulations. *J. Energy Power Eng.* **2014**, *8*, 190–200.

39. Vujanović, M.; Petranović, Z.; Edelbauer, W.; Duić, N. Modelling spray and combustion processes in diesel engine by using the coupled Eulerian–Eulerian and Eulerian–Lagrangian method. *Energy Convers. Manag.* **2016**, *125*, 15–25. [CrossRef]

40. Drew, D.A.; Passman, S.L. *Theory of Multicomponent Fluids*; Springer: Berlin, Germany, 1998.

41. Patterson, M.A.; Reitz, R.D. Modeling the Effects of Fuel Spray Characteristics on Diesel Engine Combustion and Emissions. *SAE Tech. Pap.* **1998**. [CrossRef]

42. O'Rourke, P.J.; Bracco, F.V. *Modelling of Drop Interactions in Thick Sprays and a Comparison with Experiments*; IMechE: London, UK, 1980.

43. Sato, Y.; Sekoguchi, K. Liquid Velocity Distribution in Two-Phase Bubble Flow. *Int. J. Multiph. Flow* **1975**, *2*, 79–95. [CrossRef]

44. Wang, D.M.; Greif, D. Progress in modeling injector cavitating flows with a multi-fluid method. In Proceedings of the ASME 2006 2nd Joint US-European Fluids Engineering Summer Meeting Collocated with the 14th International Conference on Nuclear Engineering, Miami, FL, USA, 17–20 July 2006; American Society of Mechanical Engineers: New York, NY, USA, 2006; pp. 153–162.

45. Von Berg, E.; Edelbauer, W.; Alajbegovic, A.; Tatschl, R.; Volmajer, M.; Kegl, B.; Ganippa, L.C. Coupled simulations of nozzle flow, primary fuel jet breakup, and spray formation. *J. Eng. Gas Turbines Power* **2005**, *127*, 897–908. [CrossRef]

46. Winklhofer, E.; Kull, E.; Kelz, E.; Morozov, A. Comprehensive hydraulic and flow field documentation in model throttle experiments under cavitation conditions. In Proceedings of the ILASS-Europe Conference, Zurich, Switzerland, 2–6 September 2001; pp. 574–579.

47. Sou, A.; Biçer, B.; Tomiyama, A. Numerical simulation of incipient cavitation flow in a nozzle of fuel injector. *Comput. Fluids* **2014**, *103*, 42–48. [CrossRef]

48. Salvador, F.J.; Romero, J.V.; Roselló, M.D.; Martínez-López, J. Validation of a code for modeling cavitation phenomena in Diesel injector nozzles. *Math. Comput. Model.* **2010**, *52*, 1123–1132. [CrossRef]

49. Som, S.; Aggarwal, S.K.; El-Hannouny, E.M.; Longman, D.E. Investigation of nozzle flow and cavitation characteristics in a diesel injector. *J. Eng. Gas Turbines Power* **2010**, *132*, 042802. [CrossRef]

50. Spray-A Characterization Data, Engine Combustion Network. Available online: https://ecn.sandia.gov/ecn-data-search/ (accessed on 1 January 2017).

51. Naber, J.; Siebers, D. Effects of Gas Density and Vaporization on Penetration and Dispersion of Diesel Sprays. *SAE Tech. Pap.* **1996**. [CrossRef]

 *applied sciences*

Article

An Experimental Investigation into Combustion Fitting in a Direct Injection Marine Diesel Engine

Yu Ding *, Congbiao Sui and Jincheng Li

College of Power and Energy Engineering, Harbin Engineering University, Harbin 150001, China; suicongbiao@hrbeu.edu.cn (C.S.); heuljc@163.com (J.L.)
* Correspondence: dingyu@hrbeu.edu.cn; Tel.: +86-451-8258-9370

Received: 16 November 2018; Accepted: 30 November 2018; Published: 4 December 2018

Abstract: The marine diesel engine combustion process is discontinuous and unsteady, resulting in complicated simulations and applications. When the diesel engine is used in the system integration simulation and investigation, a suitable combustion model has to be developed due to compatibility to the other components in the system. The Seiliger process model uses finite combustion stages to perform the main engine combustion characteristics and using the cycle time scale instead of the crank angle shortens the simulation time. Obtaining the defined Seiliger parameters used to calculate the engine performance such as peak pressure, temperature and work is significant and fitting process has to be carried out to get the parameters based on experimental investigation. During the combustion fitting, an appropriate mathematics approach is selected for root finding of non-linear multi-variable functions since there is a large amount of used experimental data. A direct injection marine engine test bed is applied for the experimental investigation based on the combustion fitting approach. The results of each cylinder and four-cylinder averaged pressure signals are fitted with the Seiliger process that is shown separately to obtain the Seiliger parameters, and are varied together with these parameters and with engine operating conditions to provide the basis for engine combustion modeling.

Keywords: seiliger process model; combustion fitting; marine diesel engine; pressure signal; experimental investigation

---

1. Introduction

Diesel engine combustion is an unsteady and discontinuous process, in which the fuel chemical energy is transferred to the work medium internal energy. Due to its complicated energy conversion process, modeling the diesel engine combustion is quite challenging to both engine designers and users [1]. When the engine is applied in an integrated system such as the diesel engine control system, after-treatment system, ship propulsion system, etc., the combustion details are not the main concern, but the main performance parameters of the engine, such as engine work, peak pressure, peak temperature, are strongly affected by the engine combustion process. A suitable approach on fitting diesel engine combustion process with a few parameters is useful for combustion modeling, where the mathematics analysis is essentially required to solve the rooting finding problems in particular for the experimental investigation, because plenty of experimental data needs to be handled and used in the multi-dimensional system of equations [2,3].

Diesel engine combustion process is varied in comparison with the other combustion machines, resulting in the specific theoretical investigation method. Recently there has been lots of diesel engine combustion research using both theoretical and experimental approaches. With the rapid development of three-dimension numerical calculation technology, the diesel engine combustion can be simulated in quite detailed ways to obtain the detailed combustion information [4,5]. A three-dimension simulation of diesel engine combustion requires the turbulence flow model, spray model and combustion model, among which the turbulence flow model plays an important role. Normally there are three ways to model the turbulence flow of the engine: Large Eddy Simulation (LES), Direct Numerical Simulation (DNS) and Reynolds-Averaged Navier-Stokes (RANS) [6]. Kahila et al. developed the LES together with a finite-rate chemistry model to investigate the engine dual-fuel ignition process [7]. Knudsen et al. proposed the compressible Eulerian numerical model to describe how liquid fuel injector nozzle geometry and operation strategies influence gas phase fuel distribution [8]. An et al. used the renormalization group K-epsilon model to describe the turbulent flow field and predict the soot particles evolution [9]. When modeling the turbulent flow, it is important to calculate the average chemical reaction rate and the turbulent combustion rate of fuel [10,11]. Yu et al. used the characteristic time combustion model coupled with the chemical kinetic mechanism to study the mixture formation's influence on the smoke [12]. Cheng et al. studied the methyl esters of soybean and coconuts formed by spray combustion and related emissions of the diesel engine and further to the effect of EGR on these biodiesel fuel combustion [13]. The diesel engine spray models are mainly divided into homogeneous gaseous jet models and gas-liquid two phase models, and the latter use the gas and liquid two phases in modeling engine spray process [14–16].

Besides the three-dimension approaches in modeling the engine combustion process, the multi-zone models are also efficient tools. Compared with the commonly used single-zone models, the multi-zone models take into account the instantaneous details of combustion process, such as the formation and development of fuel spray, the relative motion of fuel droplets and air. A popularly used multi-zone combustion model is Hiroyasu's fuel droplet evaporation combustion model [17], which includes the fuel injection sub-model, combustion thermodynamic calculation sub-model and emission generation sub-model. Zhang et al. used the in-cylinder steam injection method to establish a two-zone combustion model to research the waste heat recovery and NOx emission control [18]. Fang et al. built a two-dimension combustion model to study the effect of electric fields on the combustion characteristics of a lean-burning methane air mixture [19]. Xiang et al. proposed a two-zone combustion model in nature gas engine application and used this model for engine knocking predictions [20].

The diesel engine combustion measurement techniques include visualization, PIV (particle image velocimetry), LIF (laser induced fluorescence) and PDPA (phase Doppler particle anemometry). The PIV velocity measurement is a non-contact, transient and full-flow velocity measurement method, which has the advantages of not interfering with the quantitative information of the measured field and fast dynamic response [21,22]. LIF method produces relatively strong light signals with high spatial resolution and the fluorescence intensity is proportional to the incident light intensity so that it can be used to measure the concentration of substances [23]. The PDPA method depends on the frequency difference between the scattered light and the irradiated light of the moving particles, whose size is determined by analyzing the phase shift of the scattered light reflected or refracted by the spherical particles passing through the laser measuring body [24]. The diesel engine combustion experimental research supplemented with the theoretical investigation improve the engine combustion performance and reduces emissions.

Although the detailed engine combustion models are developed rapidly with the progress of simulation tool and experimental facilities, when the engine is used as a component of systems such as the control system, after treatment system and propulsion system, the engine combustion process has to meet the requirement of the system real-time characteristics. Mean value model is a time-domination model with cyclic time scale instead of the crank angle time scale, having the characteristics of fast calculation to get the engine main performance parameters, and it is commonly used in the control system and real-time dynamic simulation. Yin et al. introduced a turbocharged diesel engine model for hardware-in-the-loop simulation, which has the capability of observing engine state parameters and capturing engine transient response [25]. Sui et al. introduced a two finite stages Seiliger process model and simulates the engine combustion process by several parameters, which were applied in the ship propulsion system simulation [26]. Baldi et al. combined the zero-dimension and mean value models to calculate the engine performance parameters including the in-cylinder ones in relatively short simulation time [27]. Theotokatos et al. developed an extended engine mean value model to predict the thermodynamic parameters of two-stroke marine engine at different injection times [28]. Meanwhile, plenty of "black-box" methods such as multi-level hierarchical, neural networks, fuzzy logic, and wavelet networks are also applied in the modeling engine combustion process [29–31].

During the modeling engine combustion process, the experimental methods are frequently used together with theoretical analysis to parameterize the combustion process, in which the mathematics analysis is necessary to find the roots in the systems of equations deriving from the balance in the engine combustion process. Heat release rate calculation based on the measured in-cylinder pressure, energy conservation equation and empirical heat transfer formula is a fundamental method to obtain the engine combustion phenomena. On the other hand, when the heat release is known, the statistical analysis of a large amount of actual heat release can be used to build the heat release models based on the empirical formula or curve fitting [32]. The general semi-empirical formula for calculating heat release rate is Vibe formulas and the isosceles triangle combustion model [33,34]. In addition, if one only needs to know the engine in-cylinder main operating parameters under various conditions instead of the engine combustion details, the Seiliger process is the appropriate method for engine combustion modeling [26,35,36].

Nevertheless, it is critical to the mean value simulation model research that the derivation of Seiliger parameters such as the combustion parameters to give a global description of the combustion process obtains lots of experimental data or a deep theoretical investigation. Most of the researchers merely focused on obtaining the Seiliger parameters according to the engine experiments and ignoring the mathematics applications during the fitting process, resulting in the low accuracy. On the other hand, the Seiliger parameters of only one engine operating point, normally the nominal operating point, were used in these literatures, and the others were obtained based on the mathematical interpolation methods.

In the case of getting Seiliger parameters based on the combustion fitting process, the multiple variables non-linear system of equations has to be set up with equivalence criteria. When solving non-linear equations in engineering problems, the Bisection method, Secant method, Newton-Raphson method, etc. are usually used [37,38]. Sciniaka et al. introduced several iterative algorithms for solving equations and proposed an adaptive method based on the Newton-Raphson rooting method, which can better control the algorithm dynamics [39]. Furthermore, when more variables are chosen, the system of equivalence criteria becomes a multi-variable function and the Newton-Raphson method is a suitable way to find the roots of the system of equations. The Newton-Raphson method has a prominent advantage in that it has a square convergence around a single root of the equation $f(x) = 0$. It can also be used to find the multiple roots and the complex roots of the equation [40,41]. Meanwhile, the method converges linearly but can be super linear convergence by some conversions.

In this paper, an experimental investigation into combustion fitting is carried out based on the in-cylinder pressure signals measurement. With the processed measurement data, the Seiliger process model is introduced, and after determining the Seiliger parameters and equivalence criteria, the diesel engine measurement is fitted with Seilgier process model for engine combustion analysis. The fitting results of engine running at nominal point are shown in both single cylinder and four-cylinder average signals, in which the discrepancy of cylinder working condition can be obtained. Last but not least, the results of engine running at constant speed are discussed for further research on combustion fitting models on the systematic simulation of marine diesel engine applications.

2. Experiments

Diesel engine in-cylinder pressure measurement is the fundamental method to investigate the combustion process and an effective way to get insight into the combustion phenomena. In order to obtain the heat release, in-cylinder temperature, etc., a plenty of other engine measurements should be carried out, such as fuel consumption, inlet pressure, and engine speed.

2.1. Engine Test Bed Installation

The engine test bed is a MAN4L 20/27 diesel engine, which is a commonly used medium speed, direct injection marine diesel engine. The general data of the engine are listed in Table 1.

Table 1. The dimensions of engine used in the model.

Parameter	
Model	MAN4L 20/27
Cylinder Number	4
Bore	0.20 m
Stroke	0.27 m
Connection Rod Length	0.52 m
Nominal Engine Speed	1000 rpm
Maximum Effective Power	340 kW
Compression Ratio	13.4 [-]
Fuel Injection	Plunger pump Direct injection
Fuel Injection Pressure	80 MPa
SOI	4° before TDC
IC	20° after BDC
EO	300° after BDC

Figure 1 illustrates the engine test bed layout, which is installed in the laboratory with a hydraulic dynamometer for applying the engine load. The pressure sensors with a water cooling system are arranged in the cylinder head of each cylinder. Due to the two in-cylinder pressure sensors using one cooling tubule, in the two red circles, there are two pressure sensors mounted in each circle. There are also inlet receiver pressure and temperature, pressure and temperature before turbine, fuel consumption measurements, etc. Together with these measurements, the engine main performance parameters can be obtained to investigate the engine combustion process.

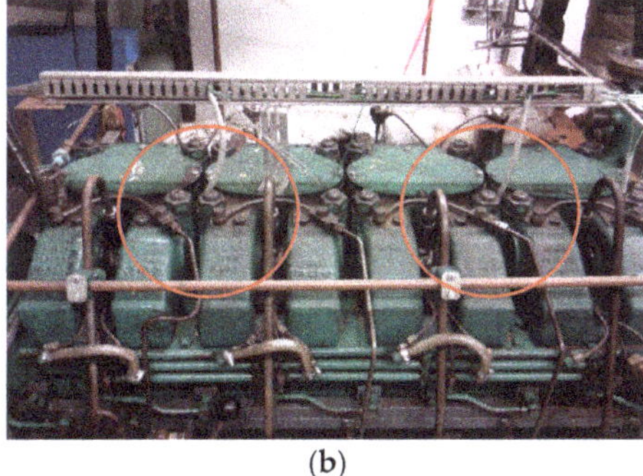

(**a**)   (**b**)

Figure 1. Engine test bed: (**a**) Overview of the test bed; (**b**) the in-cylinder pressure sensor installation in the test bed (the red circle).

2.2. Measurements Procedures

Figure 2 illustrates the engine test bed layout. The sensors arranged on the test bed are: in-cylinder pressure sensors, inlet pressure and temperature sensors, exhaust pressure and temperature sensors, air flow meters, hydraulic dynamometer, etc. The fuel flow measurement in the test bed uses the weighing scale and clock with a two-position three-way valve instead of a flow meter, which is a traditional method but ensures the accuracy of the fuel flow measurement.

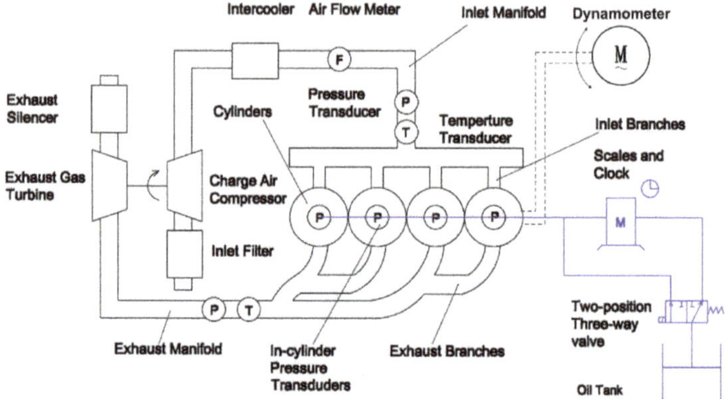

Figure 2. Schematic diagram of engine test bed layout.

2.3. Measurements samples Preparations

The marine diesel engine investigated in this paper is frequently used to drive a generator, therefore, two engine speed 900 rpm and 1000 rpm are selected in the measurement, and four operating points for each engine speed are measured (see Figure 3). In order to get the engine running points to be as thermodynamically stable as possible, the operating points at constant speeds were traversed alternatively up and down, as indicated by the arrows.

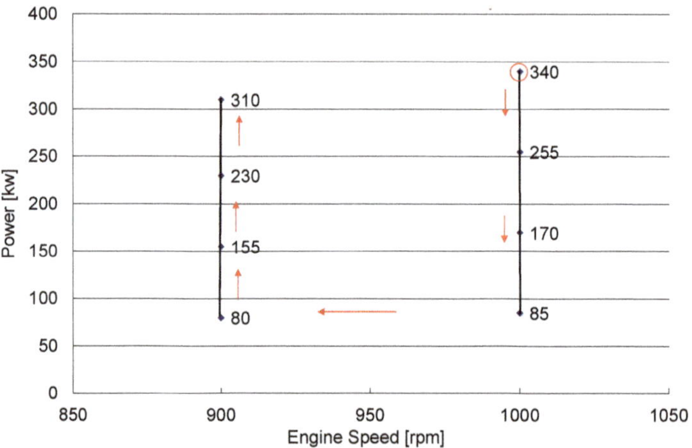

Figure 3. Map of measured points.

When the diesel engine is used for power generation, the engine speed has to be kept constant in order to ensure the constant frequency of current. Due to fixed current frequency f used in the marine power generation such as 50 Hz in China and 60 Hz in US, together with the number of poles P, the engine speed can be determined by Equation (1).

$$n = \frac{60 \cdot f}{P} \tag{1}$$

Therefore, the 50 Hz and 60 Hz are corresponding to 1000 r/min and 900 r/min when the poles numbers are selected to be 3 and 4 separately.

3. Methodology

3.1. In-Cylidner Measured Pressure Signals Processing

The measured in-cylinder pressure signals fluctuate especially in the peak pressure region, which will easily bring calculation errors during the combustion fitting process. Therefore, the measured pressure signals have to be processed firstly based on mathematics methods. On the other hand, some useful and effective methods can be used to improve the accuracy such as a pressure signals smooth approach based on the heat release calculation proposed by the author to ensure the accuracy of the signals' application to some extent [32].

3.2. Seiliger Process and Seliger Process Parameters Obtain

The Seiliger process is an efficient way to characterize the diesel engine combustion process, which divides the combustion process into finite stages to describe main phenomena of combustion characteristics. The definition and interpretation of Seiliger process and Seiliger parameters were described in author's previous articles. Based on the stages number definition, there are 5-point Seiliger and 6-point Seiliger cycles [42]. Figure 4 shows the six-point Seiliger process model with both the basic and the advanced Seiliger process. The stages can be described as follows:

1–2: polytropic compression;
2–3: isochoric combustion;
3–4: isobaric combustion and expansion;
4–5: isothermal combustion and expansion;
5–6: polytropic expansion indicating a net heat loss, used when there is no combustion in this stage (basic);
5–6′: polytropic expansion indicating a net heat input caused by late combustion during expansion (advanced).

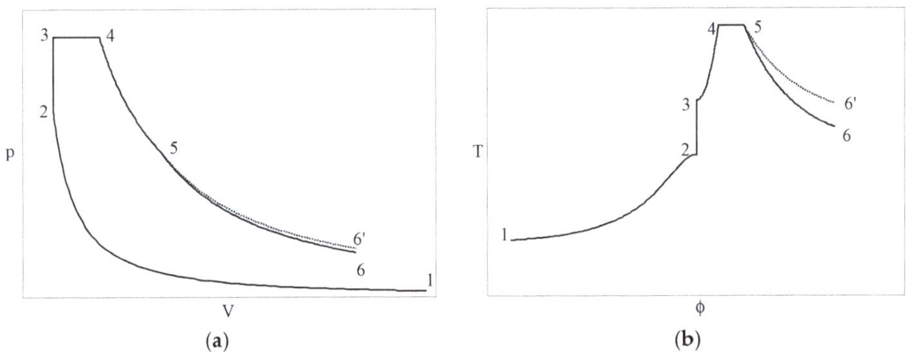

Figure 4. Six-point Seiliger process definition: (**a**) *p-V* diagram; (**b**) *T-φ* diagram.

The Seiliger process can be described by a finite number of parameters that fully specify the process together with the initial (trapped) condition and the working medium properties. The definition of the Seiliger stages and the Seiliger parameters are given in Table 2. Among these Seiliger parameters a, b and c are the combustion parameters indicating the isochoric combustion stage, isobaric combustion stage and isothermal combustion stage respectively. The polytropic compression exponent n_{comp} and effective compression ratio r_c model the polytropic compression process while the polytropic expansion exponent n_{exp} and expansion ratio r_e model the polytropic expansion process. In the case that all the Seiliger parameters are known, the pressures, temperatures, work and heat in the various stages of the Seiliger cycle can be calculated.

Table 2. Seiliger process definition and parameters.

Seiliger Stage	Seiliger Definition	Parameter Definition	Seiliger Parameters
1–2	$\frac{p_2}{p_1} = r_c^{n_{comp}}$	$\frac{V_1}{V_2} = r_c$	$r_c,\ n_{comp}$
2–3	$\frac{V_3}{V_2} = 1$	$\frac{p_3}{p_2} = a$	a
3–4	$\frac{p_4}{p_3} = 1$	$\frac{V_4}{V_3} = b$	b
4–5	$\frac{T_5}{T_4} = 1$	$\frac{V_5}{V_4} = c$	c
5–6 (5–6′)	$\frac{p_5}{p_6} = r_e^{n_{exp}}$	$\frac{V_6}{V_5} = r_e$	$r_c,\ n_{exp}$

Figure 5 shows how to obtain the Seiliger parameters based on measures in-cylinder pressure signals. The first line is the fit procedure to smooth the in-cylinder pressure signals, after which the Seiliger parameters are determined according to the combination of equivalence criteria between Seiliger process and in-cylinder process of the real engine. The Seiliger parameters are obtained from the 'Seiliger fit model' (model (4)), in which the fitting algorithms have to be investigated to improve the accuracy and speed the simulation time. Finally, the in-cylinder performance based on the Seiliger process characterizes the cylinder process and in particular the 'combustion shape' in order to compare the fitting results to the real engine (smoothed signals).

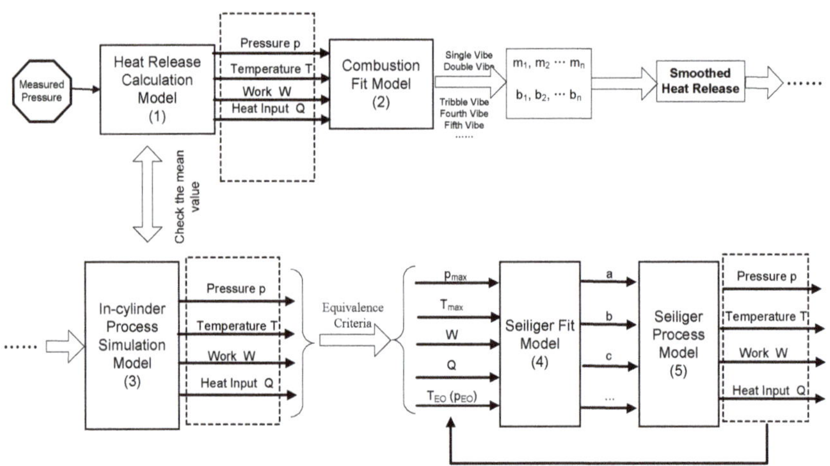

Figure 5. Flow chart of the overall simulation procedure to calculate Seiliger parameters.

3.3. Combustion Fitting Approach

The Newton-Raphson method applied for a single variable function is briefly introduced. If the initial estimate of the root is x_n, a tangent can be extended from the point $[x_n, f(x_n)]$. The point where this tangent crosses the x-axis represents an improved estimate of the root. The first derivative $f'(x_n)$ is equivalent to the slope of the tangent and can be derived on the basis of a geometrical interpretation (Figure 6) and an iteration scheme is constructed:

$$f'(x_n) = \frac{f(x_n) - 0}{x_n - x_{n+1}} \tag{2}$$

Then

$$x_{n+1} = x_n - \frac{f(x_n)}{f'(x_n)} \tag{3}$$

While $|x_{n+1} - x_{n+1}| \leq tolerance$, the x_{n+1} can be considered to be the root of function $f(x)$.

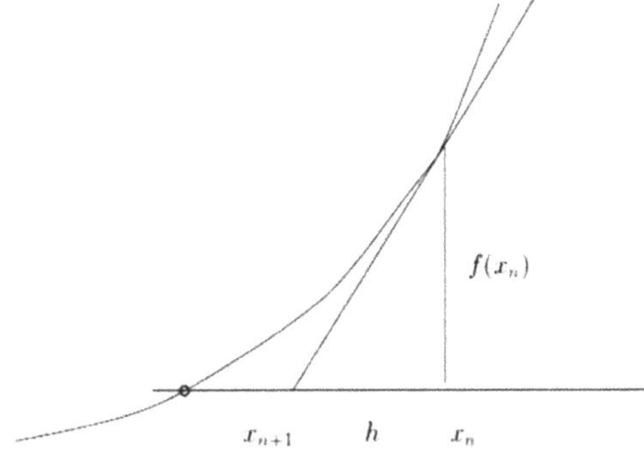

Figure 6. Graphical description of Newton-Raphson method.

Furthermore, the Newton-Raphson method can be used to acquire the solution for a multi-variable function $\mathbf{f(x)}$. If the zero was considered to occur at $\mathbf{x} = \mathbf{x}^*$, where $\mathbf{x}^*$ is a vector, then the Taylor series for multi-variable function is applied. If $\mathbf{x_n}$ is assumed to be the current estimation of the functions and $\mathbf{x_{n+1}} = \mathbf{x_n} + \mathbf{h}$, due to the demand that $\mathbf{f(x^*)} = \mathbf{0}$, the Taylor series is acquired:

$$\mathbf{f(x)}|_{\mathbf{x=x_{n+1}}} = \mathbf{f(x_n + h)} = \mathbf{f(x_n)} + \nabla\mathbf{f}|_{\mathbf{x_n}} \cdot \mathbf{h} + \mathbf{o}\left(|\mathbf{h}|^2\right) = \mathbf{0} \tag{4}$$

If $\mathbf{o}\left(|\mathbf{h}|^2\right)$ is assumed small enough to neglect, Equation (4) becomes:

$$\mathbf{f(x_n)} = -\nabla\mathbf{f}|_{\mathbf{x_n}} \cdot \mathbf{h} \tag{5}$$

When $-\nabla\mathbf{f}|_{\mathbf{x_g}}$ is replaced by matrix $\mathbf{A}$:

$$\mathbf{h} = -\mathbf{A}^{-1} \cdot \mathbf{f(x_n)} \tag{6}$$

$$\mathbf{A} = \begin{bmatrix} \frac{\partial f_1}{\partial x_1} & \frac{\partial f_1}{\partial x_2} & \cdots & \frac{\partial f_1}{\partial x_n} \\ \frac{\partial f_2}{\partial x_1} & \frac{\partial f_2}{\partial x_2} & \cdots & \frac{\partial f_2}{\partial x_n} \\ \vdots & \vdots & \ddots & \vdots \\ \frac{\partial f_n}{\partial x_1} & \frac{\partial f_n}{\partial x_2} & \cdots & \frac{\partial f_n}{\partial x_n} \end{bmatrix} \text{ and } \mathbf{f(x)} = \begin{bmatrix} f_1(x_1, x_2, \cdots x_n) \\ f_2(x_1, x_2, \cdots x_n) \\ \vdots \\ f_n(x_1, x_2, \cdots x_n) \end{bmatrix} \tag{7}$$

Finally, the next estimate of $\mathbf{x_{n+1}}$ is obtained and used as the initial value of the next iteration and the iteration is terminated when the stop criteria are met.

3.4. Implementation in the Combustion Fitting Based on Seiliger Process

The Newton-Raphson multi-variable root finding method is applied to find the solution for the fitting of the (real) engine equivalence values by means of varying Seiliger process parameters. The Seiliger process is an engine combustion model to divide the engine working cycle into finite stages based on different Seiliger parameter numbers choosing. The case with four equivalence criteria as functions of four variables (Seiliger parameters) is taken as an example.

Seiliger parameters a, b, c and n_{exp} are the variables in these equations and the differences between the calculated result from the equivalence criteria functions of the Seiliger process (p_{max}, T_{max}, $q_{in,Seiliger}$ and $w_{i,Seiliger}$) and the measured engine cycle(p_3, T_4, $q_{in,measured}$ and $w_{i,measured}$) are the functions for which the zero has to be found. The other parameters in Equations (8)–(11) are considered to be constant but changed with engine working conditions. Therefore, the simulation model is important to determine them at different iteration steps rather than analytical method.

In summary, the Equations (8)–(11) are expressed by standard equation form

$$f_1(x_1, x_2, x_3, x_4) = p_3 - p_{max} \tag{8}$$

$$f_2(x_1, x_2, x_3, x_4) = T_4 - T_{max} \tag{9}$$

$$f_3(x_1, x_2, x_3, x_4) = q_{in,Seiliger} - q_{in,measured} \tag{10}$$

$$f_4(x_1, x_2, x_3, x_4) = w_{i,Seiliger} - w_{i,measured} \tag{11}$$

The five functions are set to a column vector as **func**:

$$\mathbf{func} = [f_1, \ f_2, \ f_3, \ f_4] \tag{12}$$

A matrix **PDE** then is set to solve the derivatives:

$$\mathbf{PDE} = \begin{bmatrix} \frac{\partial f_1}{\partial x_1} & \cdots & \frac{\partial f_1}{\partial x_4} \\ \vdots & \ddots & \vdots \\ \frac{\partial f_4}{\partial x_1} & \cdots & \frac{\partial f_4}{\partial x_4} \end{bmatrix} \tag{13}$$

$$\mathbf{Vect} = -\mathbf{PDE}/\mathbf{func} \tag{14}$$

Vect is the increment or decrement of the root that is obtained from the previous step:

$$x_i = x_i + \mathbf{Vect}\,(i) \quad 1, 2, 3, 4 \tag{15}$$

The partial derivatives of the elements in matrix **PDE** are obtained with numerical analysis rather than with an analytical method. The latter is impossible because some parameters in equivalence criteria functions are interdependent and a numerical method is used to avoid this problem.

Variable x_i is changed one percent and then the partial derivatives are obtained relying on the local linearity.

$$\Delta x_i = 0.01 x_i \quad i = 1, 2, 3, 4, 5 \tag{16}$$

$$\frac{\partial f_i}{\partial x_j} = \frac{f(x_j + \Delta x_j, \ldots) - f(x_j, \ldots)}{\Delta x_j} \quad i = 1, 2, 3, 4, 5; \quad j = 1, 2, 3, 4, 5 \tag{17}$$

4. Results and Analysis

4.1. The Process Pressure Signals of Four Cylinders

The frequency of logging the in-cylinder pressure signals is 0.02 ms meaning that there are around 6000 samples per cycle for this engine operating at 1000 rpm. Following the in-cylinder pressure measurements, the signals have to be processed first. In this engine test, 25 cycles were recorded for one operating point measurement selecting from the total measured cycle (around 200 cycles) and the in-cylinder pressure signals are averaged cycle by cycle to eliminate the fluctuation. Figure 7 shows the pressure signals before and after the cycle's average in the p-φ and p-V diagram respectively, in which it is obvious to see that the in-cylinder pressure signals after 25 cycle's average are smoother than before but the fluctuation still remains, especially in the peak pressure region.

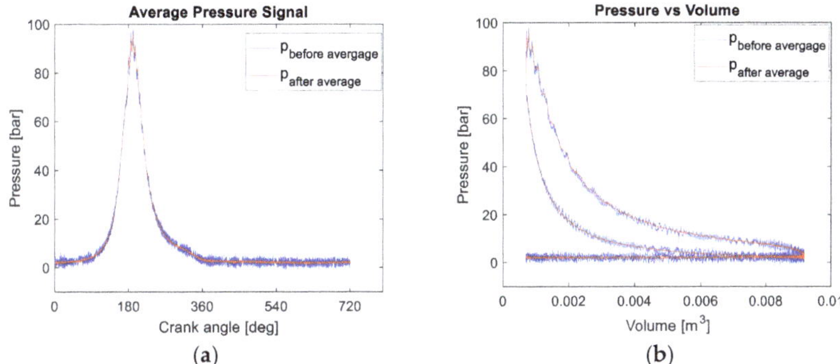

Figure 7. In-cylinder pressure before and after cyclic average: (**a**) *p-φ* diagram; (**b**) *p-V* diagram.

Figure 8 shows the in-cylinder pressure signals of 4 cylinders. Due to the fire order in the multi-cylinder to keep the engine operating stable, this 4-cylinder diesel engine has 90 °CA differences between two cylinders and the fire order is 1-3-4-2 as shown in Figure 8a. In Figure 8b, the 4-cylinder pressure signals are moved together along with them being averaged on point by point in the overall cycle (the red curve). Figure 8c shows the zoom-in curves in the peak pressure region. Due to the pressure sensors mounted position, there are regular-looked waves in the peak pressure region. The tendency of the four cylinders is same in regards to the wave crest position and frequency, from which it seems that the four in-cylinder pressure signals average is necessary to eliminate the fluctuation effect in order to get the signals close to the reality.

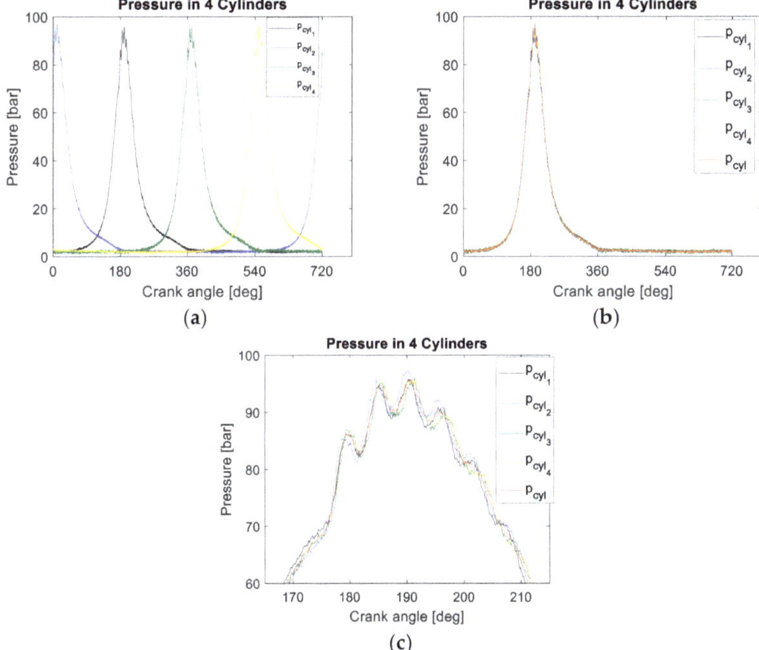

Figure 8. In-cylinder pressure of 4 cylinders: (**a**) pressure of 4 cylinders with fire order; (**b**) pressure of 4 cylinders after moving; (**c**) zoom in of peak pressure area.

Although the in-cylinder pressure signals are smoothed according to cycle by cycle average and the four-cylinder average, the fluctuations still exist in particular in the peak pressure region. Since the waves are caused by the channel effect, the averaged method cannot solve these wave problems resulting in the difficulties for combustion fitting research. Therefore, the smoothing procedure has to continue to eliminate the channel effect's influence as much as possible. Figure 9 compares the final smoothed pressure signals with measurement based on the new smooth method presented in reference [32]. According to Figure 9b, there is only one wave of the peak in-cylinder pressure region and for the other parts the pressure curve is smoothed sufficiently for the following combustion fitting investigation.

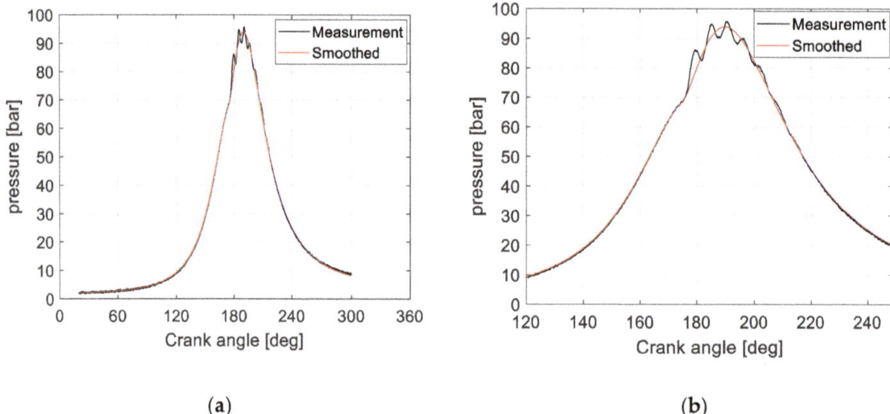

(a) (b)

Figure 9. In-cylinder pressure of 4 cylinders: (**a**) The smoothed pressure signals; (**b**) Zoom in of peak pressure region.

Regarding to the smoothing accuracy, as shown in Table 3, in the aspect of mathematics, there are 'sum of squares due to error (SSE)', 'R-square', 'Adjusted R-square', 'Degrees of freedom in the error (DFE)' and 'Root mean squared error (RMSE)' to evaluate the quality of fit. Also, the raw data and the smoothed results of the engine main performance parameters, such as P_{max}, T_{max}, P_{EO}, T_{EO} and P_i, are compared with each other to verify the signals processing accuracy.

Table 3. The goodness of fit of the pressure signals.

Fit Indicator	SSE	R-Square	Adjusted R-Square	DFE	RMSE
	0.4306	0.9921	0.9921	1025	0.0205
Engine parameters	P_{max} (bar)	T_{max} (K)	P_{EO} (bar)	T_{EO} (K)	P_i (kW)
Raw data	95.87	1569.70	8.95	1159.98	83.59
Smoothed	93.19	1513.58	8.61	1119.60	83.32

4.2. The Combustion Fit Results of Engine Nominal Operating Point

With smoothed in-cylinder pressure signals and the definition of Seiliger process model, the engine combustion fitting can be carried out using the Newton-Raphson root finding method. The combustion fitting results of the nominal operating point are discussed firstly to validate and verify the approaches, after which the fitting results of engine running at generator conditions are presented for the further application of marine diesel engine combustion modelling.

4.2.1. Four-Cylinder Averaged Pressure Signals

According to definition of the basic and advanced Seiliger process presented in Section 3, the combustion fitting results of these two Seiliger process types are compared and analyzed in terms of the four-cylinder averaged pressure signals at engine nominal operating point. Table 4 shows the Seiliger parameters and engine performance parameters chosen in the system of equations as variables and equivalence criteria for basic and advanced Seiliger process models, which actually are the same as each other, although the engine performance parameter q_{in} includes a different heat input definition as explained in Section 3.

Table 4. Variables and equivalence criteria in combustion fit functions.

	Seiliger Parameters (Variables)	Engine Performance (Equivalence Criteria)
basic Seiliger process	a, b, c, n_{exp}	$p_{max}, T_{max}, q_{in}, w_i$
advanced Seiliger process	a, b, c, n_{exp}	$p_{max}, T_{max}, q_{in}, w_i$

Before combustion fitting numerical calculation, the Seiliger parameters n_{comp}, r_c and ΔEO, which are not taken into account as the variables in the fitting process, have to be set as constants. Considering the engine operating conditions in reality, the n_{comp} is set to be 1.36 to indicate proper heat loss during engine compression process; the r_c 13.107 means constant volumetric process occurring at top dead centre and ΔEO is 0 without regarding the effect of exhaust valve open. Then based on Equations (12)–(15), the system of equations of combustion fitting is set up. During the system of equations root finding procedure, the mathematic setting should be thought over, including initial values, iterations, termination criteria, etc.

The termination criteria for the iteration in this paper, which affect the accuracy and the calculation speed, in both basic and advanced Seiliger process models fitting is set as follows:

$$|p_3 - p_{max}| \leq 1 \text{ bar,}$$

$$|T_4 - T_{max}| \leq 1 \text{ K,}$$

$$|q_{in,Seiliger} - q_{in,measured}| \leq 10 \text{ J/kg,}$$

$$|w_{i,Seiliger} - w_{i,measured}| \leq 10 \text{ J/kg.}$$

The combustion fitting results of both basic and advanced Seiliger process models are shown in Table 5, in which the Seiliger parameters values and heat input during the Seiliger parameters effect stages are compared. The values of a and b are the same of these two types, the latter's small difference is caused by the numerical calculation errors. The value of c is quite different between these two Seiliger process models, together with heat input ratio in Seiliger stage 4–5 (isothermal process) large differences. In the basic Seiliger process, besides in stage 1–2 (isochoric process), the heat input occurs in isothermal process and the ratio is 36.02%, but in the advanced Seiliger process model it is only 16.53% with the others at 19.51% in stage 5–6 (expansion process), which means there is a very late combustion during operation of the engine.

Table 5. Results of combustion fit of basic and advanced Seiliger process models.

		Basic Seiliger Process		Advanced Seiliger Process	
Constant	n_{comp}	1.36			
	r_c	13.107			
	ΔEO	0			
		Value	**Heat Input Ratio**	**Value**	**Heat Input Ratio**
Seiliger Variables	a	1.331	22.11%	1.331	22.11%
	b	1.339	41.87%	1.338	41.85%
	c	2.428	36.02%	1.505	16.53%
	n_{exp}	1.367	0	1.197	19.51%

Figure 10 illustrates the basic and advanced Seiliger process models of combustion fitting results compared with the measurements in diversified diagrams. Figure 10a,b presents the comparison between pressure signals, due to relatively small scale, the difference among the three curves are not distinct. However, in the in-cylinder temperature diagram shown in Figure 10c, it is more likely to observe the trend and comparison of these curves, in which the basic and advanced Seiliger process before peak temperature completely coincide with each other and the duration of the isothermal process are different, and have a longer crank angle time with a larger c value in the basic Seiliger process model. The temperature at Exhaust Open (EO) in the basic Seiliger process is lower than that in the advanced Seiliger process model (around 150 K), which is caused by the late combustion occurring during expansion process in advanced Seiliger process model.

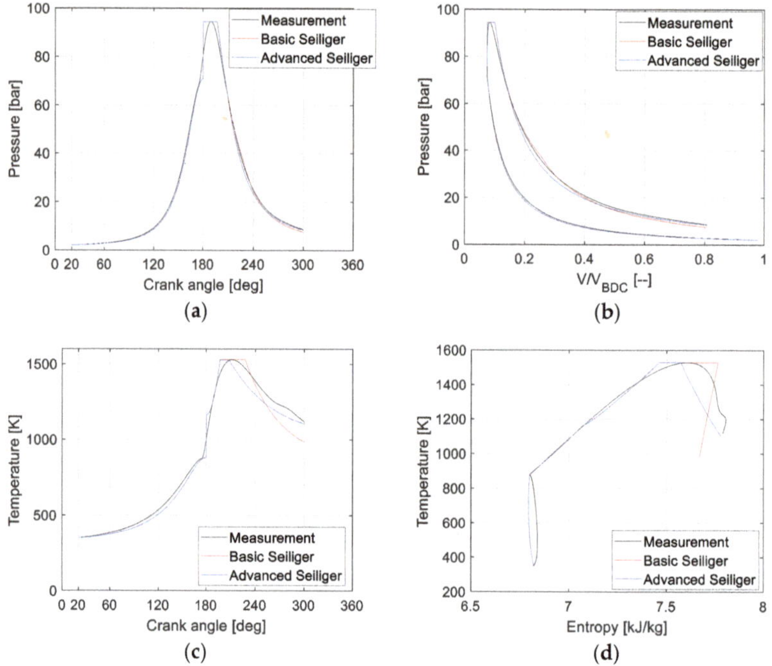

Figure 10. Comparison of basic and advanced Seiliger process models: (**a**) p-φ diagram; (**b**) p-V diagram; (**c**) T-φ diagram; (**d**) T-s diagram.

Figure 10d shows the fitting results compared with measurement in temperature-entropy diagram, which is even more clear to reveal the differences between the two Seiliger process models. The same trend as the in-cylinder temperature in Figure 10c before peak temperature and the two fitting curves are slightly different from the measurement. Nevertheless, the discrepancy among them in the expansion process is especially evident. The vertical curve in the T-S diagram means adiabatic expansion and the negative slope indicated heat input to the work medium and vice versa. As a result, the advanced Seiliger process model fitted the same case as the measurement with heat input in the expansion process and in the basic Seiliger process model, the result is totally different, i.e., heat loss during expansion.

Based on the fitting definition, when the basic Seiliger process model is used, the basic Seiliger process fitting shares the same equivalence criteria as the advanced Seiliger process model, but some of the heat enters the in-cylinder work medium in the isothermal process instead of the expansion process of advanced Seiliger process. From the fitting diagram, the advanced Seiliger process model fitting seems closer to the measurements, and when the ancient engine is operating with very late combustion, which is not expected, the advanced Seiliger process model is closer to the reality to describe the detailed phenomena of engine combustion. If the Seiliger process model fitting is used in the modern engine without late combustion, the basic process model fitting is preferred to avoid making errors in the heat input analysis with relatively simple modelling.

4.2.2. Each Cylinder Pressure Signals

The combustion fitting results discussed in the previous sections are based on the four-cylinder average pressure signals measurement in order to eliminate the measurement fluctuation effect. In fact, for the 4-cylinder diesel engine, four cylinders behave different performance to some extent in particular for the pressure signals but not too much due to engine operating balance demand. Table 6 shows the fitting results of the four cylinders respectively. The Seiliger parameters a and b are slightly different for the four cylinders, referring to the heat input ratio, there is maximum 2.22% difference for Seiliger parameter a (cylinder 1 and cylinder 2) and maximum 1.24% difference for Seiliger parameter b (cylinder 1 and cylinder 4).

Table 6. Results of advanced Seiliger process fit in 4 cylinders.

Seiliger Variables	Cylinder 1		Cylinder 2		Cylinder 3		Cylinder 4		Cylinder (average)	
	Value	Heat Input Ratio	Value	Heat Input Ratio	Value	Heat Input Ratio	Value	Heat Input Ratio	Value	Heat Input Ratio
a	1.311	20.83%	1.349	23.05%	1.329	21.04%	1.338	22.44%	1.331	22.11%
b	1.344	41.92%	1.330	40.92%	1.347	41.07%	1.330	40.68%	1.338	41.85%
c	1.375	12.77%	1.516	16.74%	**1.127**	**4.64%**	1.427	14.24%	1.505	16.53%
n_{exp}	1.180	24.49%	1.198	19.29%	**1.151**	**33.25%**	1.185	22.65%	1.197	19.51%

As to Seiliger parameter c and n_{exp}, there is much discrepancy such as 3.97% of Seiliger parameter c and 5.2% of Seiliger parameter n_{exp}. This does not apply to cylinder 3, which has irregular cylinder fitting results. However, the fitting results of averaged in-cylinder pressure signals is not much affected by that of cylinder 3 and closed to the result of cylinders 1, 2 and 4. Although there are irregular cylinder fitting results, which are probably caused by the engine operating in reality or a numerical calculation procedure, the averaged in-cylinder pressure signals are relatively reliable to obtain the fitting results in comparison with reality.

4.3. The Combustion Fit Results of Engine Running with Generator Conditions

After obtaining the Seiliger parameters fitting results of one engine operating point, the overall measurements are calculated. Figure 11 shows the trend of Seiliger parameters a, b, c and n_{exp} as functions of effective power P_e when engine running at 900 r/min and 1000 r/min respectively.

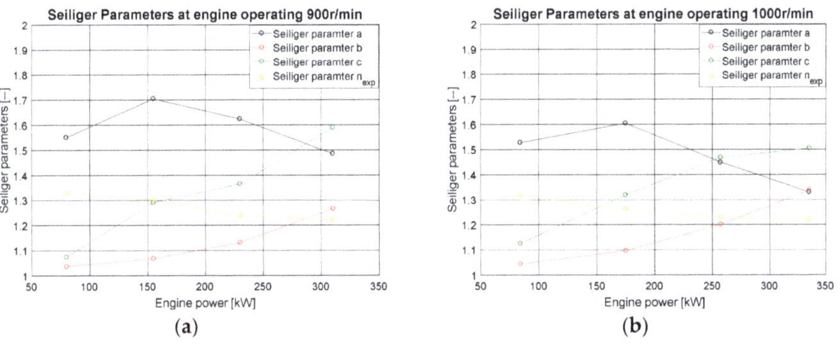

Figure 11. Seiliger parameters at engine constant speed: (**a**) 900 r/min; (**b**) 1000 r/min.

Seiliger parameter a indicates the premixed combustion phase and a large value of a is associated with more premixed combustion. At each engine speed, the value of a increases at low load and reaches a maximum, then decreases when going to higher load. As to the effect of engine speed on the

Seiliger parameter a, it can be observed that for a certain load point parameter a is lower for a higher engine speed. At a certain engine speed, b goes up with increasing load. The engine speed seems to have less effect on b, i.e., there is hardly any differences for different engine speeds at the same load.

Seiliger parameter c represents the isothermal combustion stage, to some extent, the large value causing late engine combustion. The value of c goes up with the engine power increasing, which means, for this engine, when the engine is running at higher load, the later combustion occurs more frequently. Seiliger parameter n_{exp} indicates the very late combustion phase during expansion. Due to the polytrophic expansion exponent, the value of n_{exp} varies in a relatively small range (around 1.2 to 1.35) and decreased with engine power going up, representing less heat input during the very late combustion stage.

5. Conclusions

This paper has proposed an approach on engine combustion fitting using the Seiliger process model, in which the in-cylinder pressure data measured from the engine test bed is applied for the experimental investigation. The combustion fitting results of each cylinder and average four-cylinder pressure signals obtaining the Seiliger parameters are discussed to verify that the combustion fitting is suitable to parameterize the engine combustion process and then for investigating the tendency of Seiliger parameters along with engine operating conditions. The conclusions can be drawn as follows:

(1) The Seiliger process provides an efficient way on parameterizing engine combustion process in particular for the engine in system integration simulation. The advanced Seiliger process model takes into account the engine late combustion and extends the basic Seiliger process application.

(2) According to the fitting results on comparison of in-cylinder pressure measurements and Seiliger parameters mathematics solutions, it is verified that Newton-Raphson method is an efficient way to solve multi-variable differential equations, especially for engineering applications.

(3) There is some cylinder performance discrepancy in multi-cylinder diesel engine, where after averaging the overall cylinders, the single cylinder ambiguous data could be eliminated. Therefore, the averaged pressure signals are preferred to be used in the combustion fitting process.

The combustion fitting approach investigated in this paper is applied in diesel engines, but it can be also used in the substitute fueled engine due to the first principle used in the research. The method provides a way to develop a new engine combustion model with fast response when used in the integration system simulations.

Author Contributions: Conceptualization, Y.D.; Data curation, C.S.; Formal analysis, J.L.; Investigation, C.S.; Methodology, Y.D. and J.L.; Project administration, Y.D.; Software, Y.D.; Validation, Y.D. and C.S.; Visualization, C.S.; Writing—original draft, Y.D. and J.L.; Writing—review & editing, C.S.

Funding: This research was funded by 'International Science & Technology Cooperation Program of China', 2014DFA71700; Marine Low-Speed Engine Project-Phase I.

Abbreviations

BDC	bottom dead center
DFE	degrees of freedom in the error
EGR	exhaust gas recirculation
EO	exhaust valve open angle
HCCI	homogeneous charge compression ignition
IC	intake valve close angle
NO	nitric oxide
PDE	partial differential equation
RMSE	root mean squared error
SOI	start of injection
SSE	sum of squares due to error
Vect	vector quantity

Symbols

f	frequency
a	Seiliger parameter
b	Seiliger parameter
c	Seiliger parameter
func	functions
n	speed
n_{comp}	compression exponent
n_{exp}	expansion exponent
P	number of poles
P_e	effective power
P_{EO}	pressure of exhaust valve open angle
P_i	indication pressure
p_{max}	maximum pressure
q_{in}	inlet quantity of heat
r_c	compression ratio
T_{max}	maximum temperature
w_i	indicator work

References

1. Kumar, S.; Kumar Chauhan, M. Varun Numerical modeling of compression ignition engine: A review. *Renew. Sustain. Energy Rev.* **2013**, *19*, 517–530. [CrossRef]
2. Carbot-Rojas, D.A.; Escobar-Jiménez, R.F.; Gómez-Aguilar, J.F.; Téllez-Anguiano, A.C. A survey on modeling, biofuels, control and supervision systems applied in internal combustion engines. *Renew. Sustain. Energy Rev.* **2017**, *73*, 1070–1085. [CrossRef]
3. Nahim, H.M.; Younes, R.; Shraim, H.; Ouladsine, M. Oriented review to potential simulator for faults modeling in diesel engine. *J. Mar. Sci. Technol. Jpn.* **2016**, *21*, 533–551. [CrossRef]
4. Huang, M.; Gowdagiri, S.; Cesari, X.M.; Oehlschlaeger, M.A. Diesel engine CFD simulations: Influence of fuel variability on ignition delay. *Fuel* **2016**, *181*, 170–177. [CrossRef]
5. Benajes, J.; García, A.; Monsalve-Serrano, J.; Boronat, V. Achieving clean and efficient engine operation up to full load by combining optimized RCCI and dual-fuel diesel-gasoline combustion strategies. *Energy Convers. Manag.* **2017**, *136*, 142–151. [CrossRef]
6. Oefelein, J.C.; Drozda, T.G.; Sankaran, V. Large eddy simulation of turbulence-chemistry interactions in reacting flows. *J. Phys. Conf. Ser.* **2006**, *46*, 16. [CrossRef]
7. Kahila, H.; Wehrfritz, A.; Kaario, O.; Vuorinen, V. Large-eddy simulation of dual-fuel ignition: Diesel spray injection into a lean methane-air mixture. *Combust. Flame* **2019**, *199*, 131–151. [CrossRef]
8. Knudsen, E.; Doran, E.M.; Mittal, V.; Meng, J.; Spurlock, W. Compressible Eulerian needle-to-target large eddy simulations of a diesel fuel injector. *Proc. Combust. Inst.* **2017**, *36*, 2459–2466. [CrossRef]

9. An, Y.; Jaasim, M.; Raman, V.; Im, H.G.; Johansson, B. In-Cylinder Combustion and Soot Evolution in the Transition from Conventional Compression Ignition (CI) Mode to Partially Premixed Combustion (PPC) Mode. *Energy Fuel* **2018**, *32*, 2306–2320. [CrossRef]

10. Tay, K.L.; Yang, W.; Mohan, B.; Zhou, D.; Yu, W.; Zhao, F. Development of a reduced kerosene–diesel reaction mechanism with embedded soot chemistry for diesel engines. *Fuel* **2016**, *181*, 926–934. [CrossRef]

11. Petranović, Z.; Sjerić, M.; Taritaš, I.; Vujanović, M.; Kozarac, D. Study of advanced engine operating strategies on a turbocharged diesel engine by using coupled numerical approaches. *Energy Convers. Manag.* **2018**, *171*, 1–11. [CrossRef]

12. Yu, Y.; Song, G. Numerical Study of Diesel Lift-Off Flame and Soot Formation under Low-Temperature Combustion. *Energy Fuel* **2015**, *30*, 2035–2042. [CrossRef]

13. Cheng, X.; Ng, H.K.; Gan, S.; Ho, J.H.; Pang, K.M. Numerical Analysis of the Effects of Biodiesel Unsaturation Levels on Combustion and Emission Characteristics under Conventional and Diluted Air Conditions. *Energy Fuel* **2018**, *32*, 8392–8410. [CrossRef]

14. Lackmann, T.; Kerstein, A.R.; Oevermann, M. A representative linear eddy model for simulating spray combustion in engines (RILEM). *Combust. Flame* **2018**, *193*, 1–15. [CrossRef]

15. Dahms, R.N.; Manin, J.; Pickett, L.M.; Oefelein, J.C. Understanding high-pressure gas-liquid interface phenomena in Diesel engines. *Proc. Combust. Inst.* **2013**, *34*, 1667–1675. [CrossRef]

16. Cung, K.; Moiz, A.; Johnson, J.; Lee, S.; Kweon, C.; Montanaro, A. Spray–combustion interaction mechanism of multiple-injection under diesel engine conditions. *Proc. Combust. Inst.* **2015**, *35*, 3061–3068. [CrossRef]

17. Hiroyasu, H.; Kadota, T.; Arai, M. Development and Use of a Spray Combustion Modeling to Predict Diesel Engine Efficiency and Pollutant Emissions: Part 1 Combustion Modeling. *Bull. JSME* **1983**, *26*, 569–575. [CrossRef]

18. Zhang, Z.; Li, L. Investigation of In-Cylinder Steam Injection in a Turbocharged Diesel Engine for Waste Heat Recovery and NOx Emission Control. *Energies* **2018**, *11*, 936. [CrossRef]

19. Fang, J.; Wu, X.; Duan, H.; Li, C.; Gao, Z. Effects of Electric Fields on the Combustion Characteristics of Lean Burn Methane-Air Mixtures. *Energies* **2015**, *8*, 2587–2605. [CrossRef]

20. Xiang, L.; Song, E.; Ding, Y. A Two-Zone Combustion Model for Knocking Prediction of Marine Natural Gas SI Engines. *Energies* **2018**, *11*, 561. [CrossRef]

21. Zegers, R.P.C.; Luijten, C.C.M.; Dam, N.J.; de Goey, L.P.H. Pre- and post-injection flow characterization in a heavy-duty diesel engine using high-speed PIV. *Exp. Fluids* **2012**, *53*, 731–746. [CrossRef]

22. Rabault, J.; Vernet, J.A.; Lindgren, B.; Alfredsson, P.H. A study using PIV of the intake flow in a diesel engine cylinder. *Int. J. Heat Fluid Flow* **2016**, *62*, 56–67. [CrossRef]

23. Leermakers, C.A.J.; Musculus, M.P.B. In-cylinder soot precursor growth in a low-temperature combustion diesel engine: Laser-induced fluorescence of polycyclic aromatic hydrocarbons. *Proc. Combust. Inst.* **2015**, *35*, 3079–3086. [CrossRef]

24. Payri, R.; Gimeno, J.; Martí-Aldaraví, P.; Giraldo, J.S. Methodology for Phase Doppler Anemometry Measurements on a Multi-Hole Diesel Injector. *Exp. Tech.* **2017**, *41*, 95–102. [CrossRef]

25. Yin, J.; Su, T.; Guan, Z.; Chu, Q.; Meng, C.; Jia, L.; Wang, J.; Zhang, Y. Modeling and Validation of a Diesel Engine with Turbocharger for Hardware-in-the-Loop Applications. *Energies* **2017**, *10*, 685. [CrossRef]

26. Sui, C.; Song, E.; Stapersma, D.; Ding, Y. Mean value modelling of diesel engine combustion based on parameterized finite stage cylinder process. *Ocean Eng.* **2017**, *136*, 218–232. [CrossRef]

27. Baldi, F.; Theotokatos, G.; Andersson, K. Development of a combined mean value–zero dimensional model and application for a large marine four-stroke Diesel engine simulation. *Appl. Energy* **2015**, *154*, 402–415. [CrossRef]

28. Theotokatos, G.; Guan, C.; Chen, H.; Lazakis, I. Development of an extended mean value engine model for predicting the marine two-stroke engine operation at varying settings. *Energy* **2018**, *143*, 533–545. [CrossRef]

29. Shamekhi, A.; Shamekhi, A.H. A new approach in improvement of mean value models for spark ignition engines using neural networks. *Expert Syst. Appl.* **2015**, *42*, 5192–5218. [CrossRef]

30. Nikzadfar, K.; Shamekhi, A.H. An extended mean value model (EMVM) for control-oriented modeling of diesel engines transient performance and emissions. *Fuel* **2015**, *154*, 275–292. [CrossRef]

31. Geertsmaab, R.D.; Negenborna, R.R.; Visserab, K.; Loonstijna, M.A.; Hopman, J.J. Pitch control for ships with diesel mechanical and hybrid propulsion: Modelling, validation and performance quantification. *Appl. Energy* **2017**, *206*, 1609–1631. [CrossRef]

32. Ding, Y.; Stapersma, D.; Knoll, H.; Grimmelius, H.T. A new method to smooth the in-cylinder pressure signal for combustion analysis in diesel engines. *Proc. Inst. Mech. Eng. Part A J. Power Energy* **2011**, *225*, 309–318. [CrossRef]

33. Wang, Z.; Shi, S.; Huang, S.; Huang, R.; Tang, J.; Du, T.; Cheng, X.; Chen, J. Effects of water content on evaporation and combustion characteristics of water emulsified diesel spray. *Appl. Energy* **2018**, *226*, 397–407. [CrossRef]

34. Pagán Rubio, J.A.; Vera-García, F.; Hernandez Grau, J.; Muñoz Cámara, J.; Albaladejo Hernandez, D. Marine diesel engine failure simulator based on thermodynamic model. *Appl. Therm. Eng.* **2018**, *144*, 982–995. [CrossRef]

35. Ding, Y.; Stapersma, D.; Grimmelius, H. Using Parametrized Finite Combustion Stage Models to Characterize Combustion in Diesel Engines. *Energy Fuel* **2012**, *26*, 7099–7106. [CrossRef]

36. Uchida, N.; Okamoto, T.; Watanabe, H. A new concept of actively controlled rate of diesel combustion for improving brake thermal efficiency of diesel engines: Part 1—Verification of the concept. *Int. J. Eng. Res.* **2017**, *19*, 474–487. [CrossRef]

37. Hartmann, S. A remark on the application of the Newton-Raphson method in non-linear finite element analysis. *Comput. Mech.* **2005**, *36*, 100–116. [CrossRef]

38. Smith, M.R.; Lin, Y. Analysis of the convergence properties for a non-linear implicit Equilibrium Flux Method using Quasi Newton–Raphson and BiCGStab techniques. *Comput. Math. Appl.* **2016**, *72*, 2008–2019. [CrossRef]

39. Gościniak, I.; Gdawiec, K. Control of Dynamics of the Modified Newton-Raphson. *Commun. Nonlinear Sci.* **2018**, *67*, 76–99. [CrossRef]

40. Pakdemirli, M.; Boyacı, H. Generation of root finding algorithms via perturbation theory and some formulas. *Appl. Math. Comput.* **2007**, *184*, 783–788. [CrossRef]

41. Abbasbandy, S. Improving Newton–Raphson method for nonlinear equations by modified Adomian decomposition method. *Appl. Math. Comput.* **2003**, *145*, 887–893. [CrossRef]

42. Ding, Y. Characterising Combustion in Diesel Engines: Using Parameterised Finite Stage Cylinder Process Models. Ph.D. Thesis, Delft University of Technology, Delft, The Netherlands, 2011.

 *applied sciences*

Article

Selected Issues of the Indicating Measurements in a Spark Ignition Engine with an Additional Expansion Process

Marcin Noga

Institute of Automobiles and Internal Combustion Engines, Division of Mechatronics,
Cracow University of Technology, Al. Jana Pawla II 37, 31-864 Kraków, Poland; noga@pk.edu.pl;
Tel.: +48-12-628-3688

Academic Editor: Jose Ramon Serrano
Received: 1 February 2017; Accepted: 14 March 2017; Published: 17 March 2017

Abstract: The paper presents the results of research on the turbocharged spark ignition engine with additional exhaust expansion in a separate cylinder, which is commonly known as the five-stroke engine. The research engine has been constructed based on the four cylinder engine in which two outer cylinders work as the fired cylinders, while two internally connected inner cylinders constitute the volume of the additional expansion process. The engine represents a powertrain realizing an ultra-expansion cycle. The purpose of the study was to find an effective additional expansion process in the five-stroke engine. Cylinder-pressure indicating measurements were carried out for one of the fired cylinders and the additional expansion cylinder. The study was performed for over 20 different points on the engine operation map. This allowed us to determine a dependence between the pressure indicated in the fired cylinders and in the additional expansion cylinders. A function of the mean pressure indicated in the additional expansion cylinder versus a brake mean effective pressure was also presented. This showed a load threshold from which the work of the cylinders of additional expansion produced benefits for the output of the experimental engine. The issues of mechanical efficiency and effective efficiency of this engine were also discussed.

Keywords: spark ignition; efficiency; Miller cycle; Atkinson cycle; five-stroke engine; additional expansion process

---

1. Introduction

Spark ignition engines, which have been used in motor vehicles for over 140 years, still have a relatively low effective efficiency of conversion of fuel combustion energy into mechanical energy of rotation, despite significant developments. The best spark ignition engines used in motor vehicles currently obtain a peak effective efficiency of about 40% [1], what means that in the best case scenario, 60% of the energy delivered with the fuel is lost. In the case of low engine load, the loss of energy may exceed 80% of the value resulting from fuel combustion. To solve this and other problems, the European Commission has initiatives aimed at obtaining automotive engines with significantly increased efficiency. A peak thermal conversion efficiency of more than 50% is expected for engines developed in the Horizon 2020 Program [2]. However, at the moment this is only one of the targets for R&D projects which will be finished in the years 2020 to 2021, so an implementation of the developed engines for mass production will undoubtedly take place in the next few years. There are many different methods to improve this situation. It is worth mentioning here the most important methods, such as direct injection, thermal insulation of the combustion chamber, turbocharging, downsizing, downspeeding, the implementation of exhaust energy recovery using thermoelectric generators [3], or the advanced mechatronic systems, such as variable valve timing or continuously variable valve lift.

Among the ways to improve the effective efficiency (fuel conversion efficiency) of an internal combustion engine, an interesting method is to increase the expansion ratio to be significantly higher than the compression ratio of the engine. This leads to a significant increase in the thermal efficiency of the theoretical comparative cycle of the engine [4], but also gives tangible benefits in the effective efficiency of the engine performing such a cycle [5]. Figure 1 presents the concept of using an increased expansion ratio to improve the efficiency of an internal combustion engine. In a particular case, the working medium pressure can be decreased to match the ambient pressure.

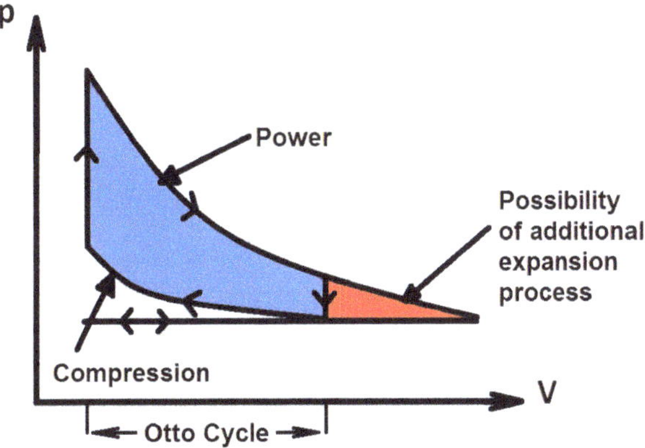

Figure 1. Potential of exhaust energy recovery using additional expansion.

Differentiation of the compression and expansion ratios of the engine may be accomplished in several ways. The first well-known solution was developed by the English engineer James Atkinson in the second half of the nineteenth century [6]. The engine concept by Atkinson was characterized by shorter intake and compression strokes than the power and exhaust strokes. This solution, however, was a significant disadvantage due to the sophisticated crank mechanism with respect to classical four-stroke engine. The increased interest in construction solutions of engines with increased expansion occurred in the second half of the twentieth century, when the concept by Ralph Miller [7] was developed. This concept was based on the classic four-stroke engine, but with suitably modified valve timing so as to get a reduced effective length of the compression stroke—Late Intake Valve Closing—or by reducing the degree of filling of the cylinder in a supercharged engine with an additional charge cooling by expansion in the intake stroke as a result of an Early Intake Valve Closing strategy. Engines performing an Atkinson/Miller cycle, but based on the classic four-stroke engine, have been used in motor vehicles since the mid-1990s [8]. Due to the specific characteristics of this type of solution, they often occur in vehicles with hybrid drive systems, diesel locomotives, and in industrial applications [9], where the engine operates with an average high load. The Atkinson cycle is also used in modern engines for conventional propulsion systems in order to reduce pumping losses in the low load region [10]. Research and development of engines performing Atkinson/Miller cycles are also carried out in Poland at several universities [11,12].

An entirely different approach to the problem of an internal combustion engine with an expansion ratio greater than the compression ratio was applied in the five-stroke engine developed according to the concept by Gerhard Schmitz [13,14]. In this engine, a significantly increased expansion is done after the exhaust process by a second expansion of the working medium in a separate cylinder. This cylinder has a displacement volume that is about twice as high in comparison to the cylinder where the combustion occurs. The engine provides practically unlimited possibilities in terms of establishing the compression and expansion ratios, without losing a significant part of the displacement volume of a working cylinder, as it is a classic engine implementing a Atkinson/Miller cycle.

Five-Stroke Engine Developed at Cracow University of Technology

In the years 2012–2014 at the Cracow University of Technology (CUT), an engine which worked according to the concept of Gerhard Schmitz [13] was developed. The main difference of this engine is the fact that it was not designed from scratch as a new engine, but its design was based on the classic four-stroke engine [15]. The tested engine was based on a mass-produced four-cylinder turbocharged spark ignition engine and a displacement volume of 2.0 L [16]. The concept of retrofitting the classic engine to five-stroke is shown in Figure 2.

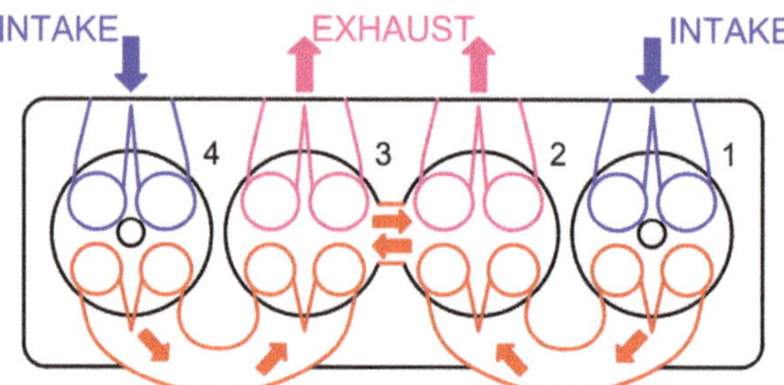

Figure 2. Scheme of the engine with an additional expansion process developed at Cracow University of Technology.

As may be seen in the figure above, the engine after modification has two fired cylinders operating in a classical four-stroke mode with a phase shift of 360 Crank Angle Degrees (CAD), while cylinders No. 2 and 3 are permanently connected by the channel in the cylinder head where the process of additional expansion of the working medium occurs. These cylinders work as one in the two-stroke mode and are filled by exhaust gasses alternately from cylinder No. 1 and cylinder No. 4. The research engine has 4 valves per cylinder.

During the development of the engine with the additional expansion of exhaust gas, a thermodynamic cycle taking into account the specific characteristics of the tested engine was proposed [17]. Extensive simulation studies of a similar engine, but with a single-cylinder of additional expansion, was also carried out by Li et al. at Shanghai Jiao Tong University [18,19]. Initial simulation studies on a five-stroke engine were also performed by Palanivendhan et al. [20].

2. Materials and Methods

2.1. Background

The described engine has a high overall expansion ratio (ε_{dcp}) of the cylinder charge equal to 21. This value results from the fact that the compression and expansion ratios of the fired cylinders of the engine are equal to 10.5. The volume of the additional expansion cylinders at Bottom Dead Center (BDC), at the end of the second expansion, is twice as high as the volume of the gas at the end of the first expansion. The mentioned value of ε_{dcp} allows us to obtain a relatively high degree of energy recovery of the exhaust gas in the additional expansion cylinders at wide open throttle (WOT). In one study [5], the results of the theoretical calculations of the increase of the thermal efficiency of an engine cycle with the additional expansion were presented. To standardize the analysis of the engine cycle, processes are described as though the additional expansion was carried out in the same working volume. In the Otto-cycle, in which the compression and expansion ratios are the same, the thermal efficiency of the cycle is independent of the amount of heat added to the cycle. In the cycle in which the expansion ratio is higher than the compression ratio (e.g., five-stroke engine cycle), the increase in thermal efficiency is higher with more heat is added to the cycle. Under some simplification, the heat added to the theoretical cycle may be identified as the load of a real engine, so the higher the load, the higher the increase in the efficiency can be expected. On the other hand, in the real engine at partial load, when the cylinder pressure does not reach high values, the additional expansion process may give a charge pressure lower than the ambient pressure, which causes negative work and energy losses. In this situation, you should take into account that the cylinders of additional expansion, instead of giving additional power to the engine shaft, will be required to bring the power from the fired cylinders, which is a highly undesirable phenomenon for obvious reasons. The idea of an additional expansion process carried out in a separate cylinder together with the undesirable phenomenon of expansion below the ambient pressure in some cylinders is shown in Figure 3. As seen in the figure above, when the maximum pressure of the cycle is too low, a vacuum occurs in the additional expansion cylinders. This will produce an area of negative work of the cycle. When it becomes dominant over the rest of the cycle of the second expansion, it turns out that the overall balance of the additional expansion cylinder is negative—instead of giving power to the engine shaft, it will require its delivery and will act as an additional load of the engine, decreasing its efficiency.

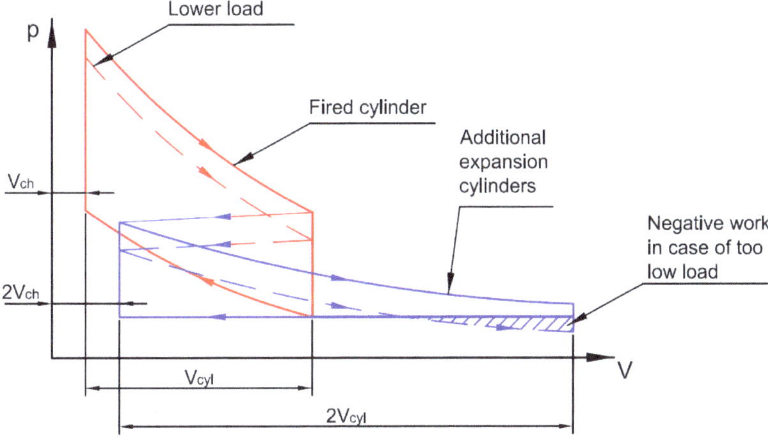

Figure 3. The concept of realizing the additional expansion process in a separate cylinder; V_{cyl}—displacement of one cylinder of the engine; V_{ch}—volume of the combustion chamber.

In addition, in the five-stroke engine, the volume of the transfer port connecting the fired cylinder with the cylinder of an additional expansion is a parasitic volume, because expansion of exhaust gases occurring there causes the loss of part of the energy. In an ideal situation, this volume should be equal to 0. In practice, this is impossible due to the constraints of the conventional engine arrangement. Similarly, it would be preferable that the volume of the chamber above the piston in the cylinder of the additional expansion would be as small as possible. Unfortunately, in the tested engine, improvement of these parameters was not feasible, because during the construction of the test engine, a modified cylinder head of the original mass-produced engine was used.

2.2. Purpose of the Work

The purpose of this work was to evaluate the relationship between the indicated mean effective pressure of the fired cylinders ($IMEP_{frd}$) and the indicated mean effective pressure of the additional expansion cylinders ($IMEP_{add}$) of the tested engine, depending on the load and rotational speed of the engine. This approach allows us to estimate the load level below which the operation of the engine with the additional expansion of exhaust gases is not fully consistent with the main assumptions of such an engine, i.e., when an additional expansion process does not bring a positive effect. The results of this research also allow us to assess the mechanical efficiency of the tested engine and to show how the effective efficiency of the engine changes in various test conditions.

2.3. Specific Features of the Test Engine

As mentioned above, the basis for the development of the test engine was a mass-produced four-cylinder turbocharged spark-ignition engine. This engine is equipped with a gasoline direct injection system. The engine adaptation for this research was mainly focused on the modification of the cylinder head and camshaft. A different turbocharger dedicated to the engine with a smaller displacement was used. The control of the boost pressure was done through the wastegate valve integrated in the turbine housing. For the test engine, the exhaust and intake manifolds were designed and manufactured. The test engine is fully autonomous, which means that the power to drive the high-pressure fuel pump, alternator, water, and oil pumps is derived from the engine rather than from external sources. Basic technical data of the engine are summarized in Table 1.

Table 1. Basic technical data of the test engine.

Parameter	Value
Bore	82.5 mm
Stroke	92.8 mm
Compression ratio, ε_{cp}	10.5
No. of cylinders	4 in-line, 2 fired and 2 add. expansion
Displacement of a fired cylinder, V_{frd}	2 cylinders, 496 cm^3 each
Displacement of both add. expansion cylinders, V_{add}	992 cm^3
Overall expansion ratio ε_{dcp}	21
No. of valves	4 per cylinder
Turbocharger type	KP39
Control system	AEM EMS 30-1010
High-pressure injector driver	Denso 131000-1041
Control method of high-pressure of fuel	pressure relief valve

An engine management system (EMS) was built based on the stand alone AEM controller cooperating with the wide-band oxygen sensor. This EMS allowed us to change the operating parameters of the engine in real time during operation. High-pressure fuel injectors supplied with increased voltage were controlled using the Denso injector driver, because the EMS only allowed us to control high-resistance injectors.

Implementation of the engine concept presented in the scheme in Figure 2 required us to also develop modified camshafts, which provided valve timing as shown in Figure 4.

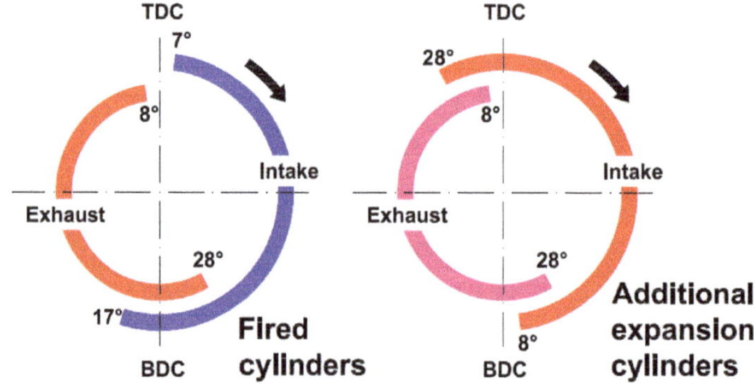

Figure 4. Valve timing of the test engine; Top Dead Center (TDC).

2.4. The Test Stand

The engine was mounted on a test stand with an Eddy-current brake dynamometer with a nominal power of 100 kW. A brake controller has a module for measuring and recording the temperature measured at selected positions. Fuel consumption measurement was conducted using a gravimetric method with electronic registration of instantaneous values of this parameter. Figure 5 shows a general view of the test bench.

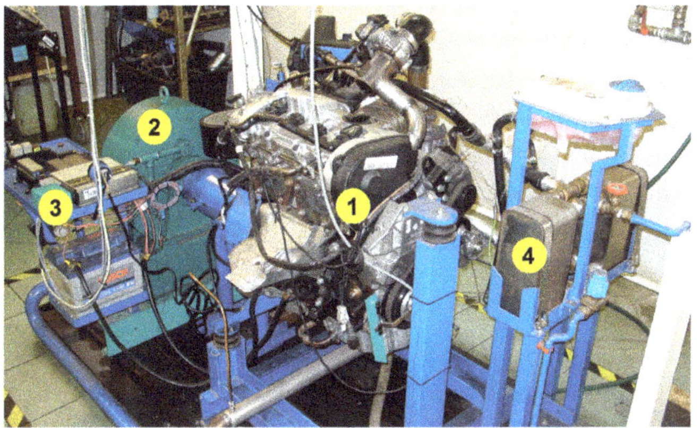

Figure 5. General view of the test stand of the engine with the additional expansion of exhaust gases; 1—Tested engine, 2—Eddy-current brake dynamometer, 3—Engine management system, 4—Engine coolant heat exchangers.

2.5. Methodology for Measurement of the In-Cylinder Pressure

Pressure measurements in the fired cylinder and the additional expansion cylinder were carried out by the use of an optical sensor type Optrand AutoPSI-TC (D822D6-SP) with temperature-compensated output signal [21]. The pressure sensor with a threaded adapter was fitted to the combustion chamber of the fired cylinder (No. 4) through the drilled hole in its side surface. The hole was machined parallel to the bottom plane of the cylinder head.

Figure 6 shows a photo of the combustion chamber of the fired cylinder with a visible hole for in-cylinder pressure measurement and the adapter for the pressure sensor.

Figure 6. Combustion chamber of the fired cylinder (No. 4); 1—Adapter of the combustion pressure sensor, 2—Hole in the combustion chamber wall, 3—Intake Valves, 4—Exhaust valves.

The view of the pressure sensor installed in the fired cylinder of the test engine is shown in Figure 7.

Figure 7. View of the combustion pressure sensor installed in the fired cylinder of the research engine (location of the sensor is indicated by the yellow arrow).

The in-cylinder pressure sensor for the cylinders of additional expansion was installed in cylinder No. 3 to the adapter mounted in the place of the spark plug. The sensors were able to measure the combustion pressure up to 10 MPa (with a possible short-term overload of 50%) and sensitivity equal to 385.8 mV/MPa. In order to perform a correlation of the recorded waveform of the in-cylinder pressure with the momentary position of the crankshaft, an incremental encoder was mounted to the test engine. The encoder had a resolution of 360 pulses per 1 revolution with a separate output phase for the Top Dead Center (TDC) marker.

The registration of waveforms of the in-cylinder pressure of the fired and additional expansion cylinders using Optrand sensors was performed using a PC equipped with a multifunctional data acquisition card with a 14-bit resolution of Analog-to-Digital Converter (ADC). The data acquisition process was carried out using a specialized application developed for this purpose in the LabView environment. The results presented in the following parts of this work were obtained by averaging the data from dozens of engine cycles for each measurement point. The study was preceded by an experimental procedure of the encoder settings to TDC of the cylinder No. 1.

3. Results

3.1. Range of the Implemented Research

Pressure measurements in the fired cylinder and additional expansion cylinder were carried out for the selected points of the engine operating map. The maximum boost pressure (p_{bst}) was limited in the study to 0.9 bar. Figure 8 shows the operating map of the engine with the additional expansion of exhaust gas, with the highlighted points in which the in-cylinder pressure waveforms were recorded.

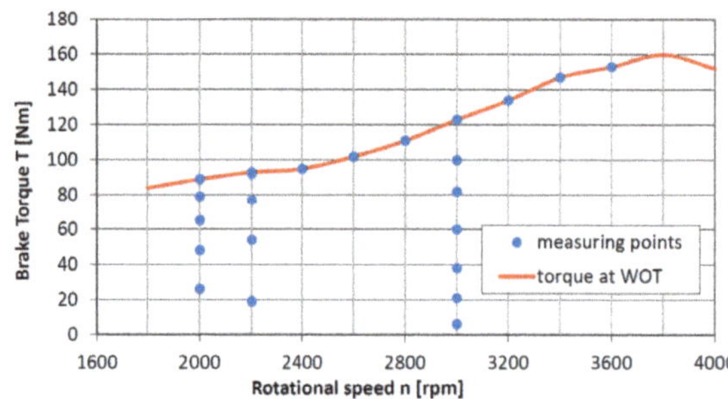

Figure 8. Measuring points for indicating measurements of the engine; Brake Torque (T).

As shown in Figure 8, the in-cylinder pressure of the tested engine were measured at various values of the engine brake torque in the rotational speed (n) range from 2000 to 3600 rpm. During these tests, the engine throttle was fully open (WOT), and the wastegate valve of the exhaust gases remained closed. In addition, measurements of pressure in the cylinders at partial load for rotational speeds of 2000, 2200, and 3000 rpm were carried out. For the rotational speed of 3000 rpm of the engine, a minimum brake specific fuel consumption was achieved. During the course of the tests, a stoichiometric composition of the air-fuel mixture was maintained—relative air-to-fuel ratio was equal to 1 (rel. AFR). The engine was fueled with gasoline during the tests.

3.2. Measurements of the In-Cylinder Pressure of the Tested Engine

After testing the engine the results saved to the files were calculated in a popular spreadsheet software, but using a specialized template in order to shorten the calculation time. This allowed us to obtain the average waveforms of pressure in the fired cylinder and in the cylinders of additional expansion as a function of the crank angle (CA). Furthermore, after introducing the geometrical data of the engine into the calculation program, p-V diagrams were developed for the fired cylinders, as well as for the cylinders of the additional expansion. Development of the p-V indicator diagrams allowed the calculation of the indicated mean effective pressure and the indicated power for fired cylinders and additional expansion cylinders for each of the measurement points.

The following are examples of the work carried out on the recorded waveforms of in-cylinder pressure for the fired and additional expansion cylinders.

In Figure 9 the p-CA diagram of the engine with the additional-expansion registered for a full load 123 Nm at 3000 rpm rotational speed is presented. A turbocharger wastegate valve remained closed, and the boost pressure was 0.5 bar. The exhaust gas temperature measured at the outlet of the turbine had a value of 437 °C.

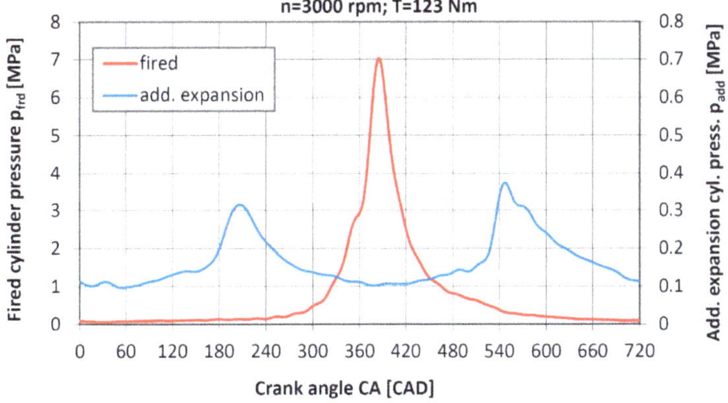

Figure 9. Chart in a p-CA for the fired cylinder and additional expansion cylinder recorded at 3000 rpm and a torque equal to 123 Nm.

The peak pressure in the fired cylinder reached 7 MPa for the case above, for the position of the crankshaft at 380 CAD ATDC (Crank Angle Degrees) (After Top Dead Center). The difference in pressure peaks recorded in the cylinders of additional expansion in the case of filling by exhaust cylinder No. 3 from fired cylinder No. 4 and No. 1 (through cylinder No. 2 and the passage) was approximately 0.06 MPa.

The aforementioned difference in the value of the maximum pressure in the additional expansion cylinder for the crank angle of approximately 180 CAD and approximately 540 CAD are caused by throttling the flow in the connecting passage between the expansion cylinders 2 and 3. The pressure value obtained for 540 CAD occurs when the exhaust goes to the additional expansion cylinder No. 3 from the adjacent fired cylinder No. 4, in which the in-cylinder pressure is also measured. The pressure registered at 180 CAD resulted from filling the assembly of the cylinders of additional expansion by exhaust gases from cylinder No. 1. In this case, the exhaust gases were first delivered into cylinder 2, then through the connecting channel in the cylinder head to cylinder No. 3, where the measurement was carried out. During the calculation of the indicated work of the additional expansion cylinders, its value was determined in consideration of the fact that in the cylinders of additional expansion, the pressure is variable in a cyclic manner as described above.

In Figure 10, the in-cylinder pressure curves are shown as a function of cylinder volume (V_c) for the fired and additional expansion cylinders registered for a full load of 123 Nm at 3000 rpm rotational speed.

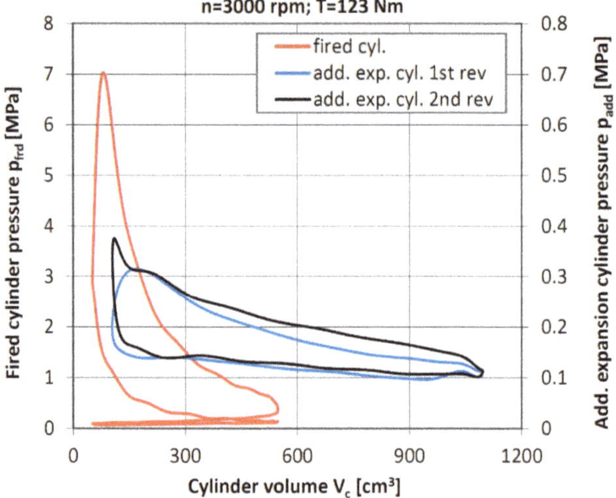

Figure 10. p-V diagrams for the fired cylinder and additional expansion cylinder recorded at 3000 rpm and a torque equal to 123 Nm.

The assembly of additional expansion cylinders of the engine works in a two-stroke mode, therefore the in-cylinder pressure waveform shown in the above chart includes two complete processes of additional expansion and the exhaust process, as shown in the p-CA diagram. The values of the indicated mean effective pressure in the fired cylinder ($IMEP_{frd}$) and in the additional expansion cylinders ($IMEP_{add}$) were calculated by numerical integration of relevant areas of the p-V chart. These values were: $IMEP_{frd}$ was equal to 1.72 MPa, and $IMEP_{add}$ was equal to 0.08 MPa. The indicated power in the working cylinders (P_{i_frd}) was then calculated using Formula (1):

$$P_{i\_frd} = \frac{2 \cdot IMEP_{frd} \cdot V_{frd} \cdot n}{120000}, \text{ kW} \tag{1}$$

The formula presented above takes into account the used units, including the speed and four-stroke nature of the work of the two working cylinders. Similarly, the formula for calculating the indicated power of the cylinder of additional expansion (P_{i_add}) is presented as Formula (2):

$$P_{i\_add} = \frac{IMEP_{add} \cdot V_{add} \cdot n}{60000}, \text{ kW} \tag{2}$$

This formula takes into account that both cylinders of additional expansion work in a two-stroke mode as a single working volume. The calculated value of the indicated power of the fired cylinders at this operating point is 42.77 kW, and the value of the indicated power in the cylinders of additional expansion is approximately 3.98 kW, which is more than 9% of the P_{i_frd} value.

A similar analysis of the waveforms of in-cylinder pressure for the fired and additional expansion cylinders was carried out for the results obtained in the other measuring points. This made it possible to determine the relationship between $IMEP_{frd}$ and $IMEP_{add}$. This is shown graphically in the form of a plot in Figure 11.

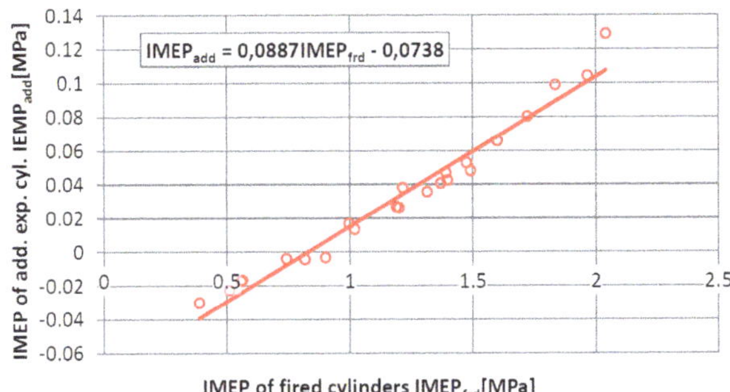

Figure 11. The Indicated Mean Effective Pressure in the cylinders of additional expansion as a function of the Indicated Mean Effective Pressure in the fired cylinders.

The course has been approximated by a straight line, and the equation is presented in the chart. The achieved value for the $IMEP_{add}/IMEP_{frd}$ ratio is greater when the $IMEP_{frd}$ is higher. The highest value (the last point on the top of the graph) is 0.063. With regards to the relations of the indicated power, the mentioned ratio will be about 12.6%, as the additional expansion cylinders operate in a two-stroke mode. This is an important energy recovery ratio in the additional expansion process of the working medium.

In order to determine the load of the five-stroke engine from which the cylinders of additional expansion start to give power to the output, the dependence of the Brake Mean Effective Pressure (BMEP) as a function of the Indicated Mean Effective Pressure of the additional expansion cylinders ($IMEP_{add}$) was defined. Figure 12 shows a graph of this function. Similarly, as in the previous case, the obtained waveform is approximated by a straight line whose equation is given in the chart.

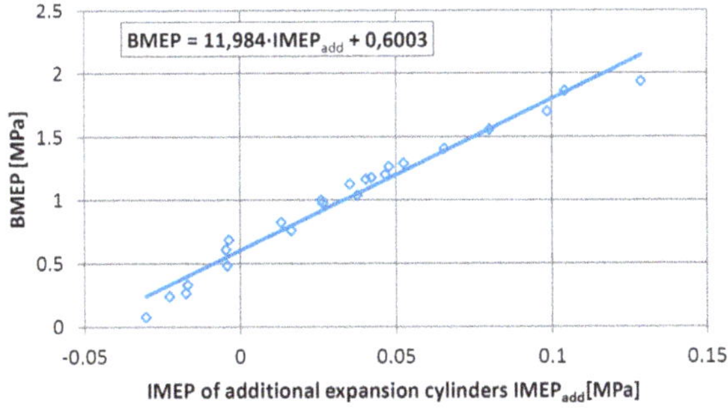

Figure 12. The Brake Mean Effective Pressure of the five-stroke engine as a function of the Indicated Mean Effective Pressure of the additional expansion cylinders.

BMEP values were determined from the value of the torque, taking into account the displacement volume of the fired cylinders ($2 \times V_{frd} = 992$ cm^3). The volume of the cylinders of additional expansion cannot be taken into account while calculating BMEP, because there is no combustion process in these cylinders. The determination of this function was aimed at finding a load limit before which the additional expansion cylinders take power from the engine instead of delivering power to the engine. The value of the intercept of the function from Figure 12 is equal to about 0.6. This means that if the BMEP is lower than 0.6 MPa, then additional expansion cylinders require power from the engine. Above this value, the additional expansion process becomes positive for the engine work. Taking into account the volume of the cylinder, the engine, and the mechanical efficiency, this means that the assembly of the cylinders of additional expansion transmit additional power to the engine output from the time when the measured torque exceeds 60 Nm. This fact means that the five-stroke engine would be most suitable for applications where it would work mainly in the field of medium and high loads.

3.3. Brake Specific Fuel Consumption and Effective Power

The results of this research also allow us to determine the effective power (P_e) and Brake Specific Fuel Consumption (BSFC) of the engine. In Figure 13, the curves of BSFC and effective power as a function of engine rotational speed at wide open throttle (WOT) are presented. A stoichiometric air-fuel mixture composition was maintained.

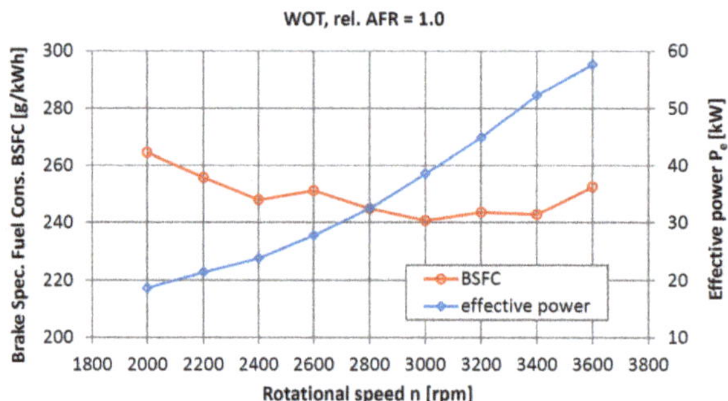

Figure 13. Effective power and BSFC as a function of engine rotational speed at WOT and with rel. AFR = 1.0; Brake Specific Fuel Consumption (BSFC); Wide Open Throttle (WOT); Relative Air-to-Fuel Ratio (rel. AFR).

The minimum value of the specific fuel consumption was 240 g/kW·h. This value was obtained at a speed of 3000 rpm and a boost pressure p_{bst} of 0.5 bar. For higher values of the rotational speed, especially for 3600 rpm, the obtained value of the brake specific fuel consumption tended to increase, which resulted in a discontinuation of the tests for higher values of the rotational speed.

3.4. Mechanical Efficiency of the Tested Engine

For conventional internal combustion engines, calculations of the mechanical efficiency (η_{m_cnv}) based on the known indicated mean effective pressure do not pose any problems. This is described by Formula (3):

$$\eta_{m\_cnv} = \frac{BMEP}{IMEP} = \frac{P_e}{P_i},$$

(3)

Brake Mean Effective Pressure at a certain rotational speed is determined by measuring the torque of the engine relative to a displacement volume of the engine and to the engine type, i.e., whether it is a two-stroke or four-stroke engine. Indicated Mean Effective Pressure is determined based on the measured in-cylinder pressure as a function of CA. For a multi-cylinder engine, the pressure is typically measured for one cylinder and it is assumed that all the other cylinders operate in the same way. For obvious reasons, for the classic engine, the BMEP/IMPEP ratio is exactly the same as the ratio of effective power (P_e) to the indicated power of the engine (P_i).

In the case of an engine with the additional expansion of exhaust gas in a separate cylinder, the situation becomes more complicated. Besides the cylinders, where a conventional four-stroke working cycle is carried out, the engine has a cylinder for the additional expansion of the exhaust, which should deliver work to the output of the engine. This cylinder is an integral part of the engine so its effect cannot be ignored in the analysis of the mechanical efficiency of the engine, because the calculated value of the mechanical efficiency would be artificially high and, in particular, the engine operating conditions could prove to be even higher than 1. Under the conditions of a sufficiently high load of the engine, where the cylinder of additional expansion provides additional power to the output, its indicated mean effective pressure is higher than zero. In the case of low engine load when the pressure of the exhaust gas going into the cylinder is low, the additional expansion cylinder requires additional propulsion from the engine, and the value of the indicated mean effective pressure becomes negative; of course, the same thing happens to the value of power indicated from the cylinder. Considering the case above, we conclude that the effect of the operation of the additional expansion cylinder has to be included in a certain way when determining the mechanical efficiency of the five-stroke engine. The values of IMEP for the fired and additional expansion cylinders are not additive. This happens because in a general case they relate the different displacement volumes, and it should be noted that the fired cylinders operate in four-stroke mode, while the cylinders of additional expansion operate in two-stroke mode. Avoiding this problem is possible by calculating the mechanical efficiency from the calculated values of the effective power and the indicated power of the engine. To determine the mechanical efficiency of the five-stroke engine, the author of this paper proposes a summation of the indicated power for the fired cylinders P_{i_frd} and indicated power of the cylinders of additional expansion P_{i_add} when the indicated power P_{i_add} is greater than zero—Formula (4). In contrast, if the indicated power of the cylinder of additional expansion P_{i_add} is zero or less than zero, it acts as a load (like the oil pump, water pump, and alternator), and in this situation the indicated power of the cylinder of additional expansion P_{i_add} should not be taken into account while calculating the mechanical efficiency of the engine with an additional expansion of exhaust gas in a separate cylinder (η_m)—Formula (5).

$$\text{for } P_{i\_add} > 0, \quad \eta_m = \frac{P_e}{P_{i\_frd} + P_{i\_add}}, \tag{4}$$

$$\text{for } P_{i\_add} \leq 0, \quad \eta_m = \frac{P_e}{P_{i\_frd}}, \tag{5}$$

Figure 14 shows curves of the mechanical efficiency of the engine with the additional expansion of the exhaust gases for various values of rotational speed and its dependence on the brake torque.

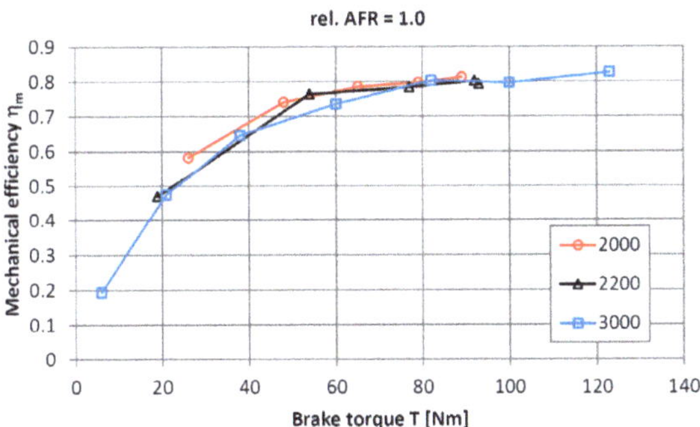

Figure 14. Mechanical efficiency as a function of the load for three different values of the rotational speed.

An analysis of the diagram indicates that the obtained value of the mechanical efficiency of the engine increases with engine load, reaching a value slightly larger than 0.8. This took place at the rotational speed of 3000 rpm, when the boost pressure p_{bst} was 0.5 bar. It is also seen that in the analyzed range of the rotational speed of the engine, the mechanical efficiency does not depend significantly on the rotational speed.

Figure 15 presents charts of the effective efficiency and boost pressure of the tested engine versus the rotational speed at full throttle (WOT) and with a stoichiometric air-fuel mixture composition.

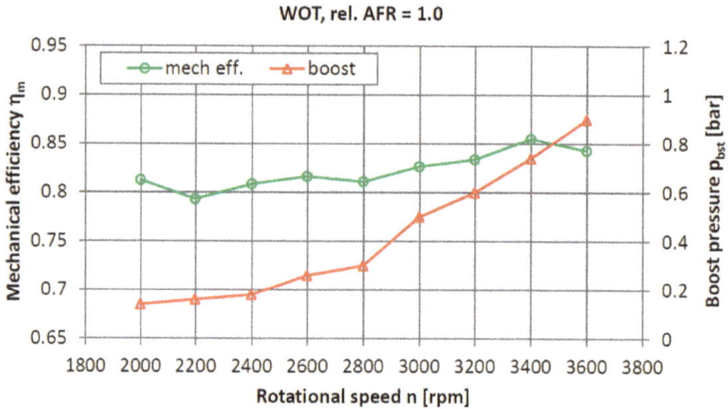

Figure 15. Mechanical efficiency and boost pressure as a function of the rotational speed at WOT.

The maximum value of the mechanical efficiency of the tested engine was 85.5% and was recorded at a rotational speed of 3400 rpm. Above this value, the boost pressure increased to 0.9 bar, and the mechanical efficiency began to decrease. The effect of a significant increase in BSFC was also demonstrated (shown in Figure 13). The resulting course of the boost pressure indicates that the used turbocharger starts to work effectively with the tested engine at rotational speeds of 3000–3200 rpm.

4. Discussion

The results of the experimental test of the engine with an additional expansion of exhaust gases are presented in this study. After the analysis, the author attempted to compare the obtained results mainly to that from [14], which shows the results obtained for a similar engine, but which was designed and built from scratch. Summary of the results of the comparison is as follows:

- During the tests, significant differences in the maximum pressure in the two cylinders of additional expansion were observed. These differences arise from the small area of the cross-section of the passage between the cylinders. This is a limitation of the base engine. In the developed engine, there are two smaller cylinders of additional expansion instead of one larger cylinder. Improvement of the existing motor can consist only of a larger cross-section of the channel, which connects the cylinders of additional expansion.
- The energy recovery ratio—the indicated power of the additional expansion cylinders—is up to 12.6% of the indicated power of the fired cylinders. Obstacles to achieving better results were: high volume of the transfer ports between the fired cylinders and additional expansion cylinder, as well as the unchanged volume of the former combustion chambers in the cylinders of additional expansion, and also the division of the additional expansion volume into two cylinders. Kéromnès et al. showed a significantly higher ratio of the indicated work of the additional expansion cylinder to the indicated work of the fired cylinder. They obtained a value of about 18%—[14] (p. 265).

- During the engine testing, a high value of load was obtained, from which the cylinders of additional expansion of the engine gave power to the output. This was about 60 Nm and was caused by the limitations of the above-mentioned structure of the base engine, mainly from the loss of energy in the parasitic volumes. Below 60 Nm of torque, the engine still operates at a relatively low BSFC, but is related to the effect of downsizing, instead of the additional effects of the second expansion of exhaust gas. This aspect of the five-stroke engine was not strongly emphasized in [14]. The problem of the selection of an appropriate displacement volume of the second expansion cylinder linked to this issue was analyzed by Li et al. in [18].

- This research revealed an additional problem, which was not described in the work content; the increased consumption of oil lubricating the engine when working with low load. The increased consumption of lubricating oil is caused by the occurrence of negative pressure in the inlet channel of the turbine at the opening of the exhaust valves of the engine. This significantly changes the operation of an oil seal of the turbocharger rotor shaft and the sucking of oil into the exhaust system occurs.

- The maximum values of the mechanical efficiency of the engine achieved by the author correspond to the results presented in [14]. This indicates a good level of technical modification made to the engine.

- The minimum brake specific fuel consumption amounted to 240 g/kWh. One should remember that the tested engine worked with all the equipment necessary for autonomous operation, i.e., with the alternator, oil pump, coolant pump, and the high pressure fuel pump. A low-pressure fuel pump was also supplied from the engine power grid. Kéromnès et al. obtained a BSFC of 226.4 g/kWh, but the engine was working without propelling an oil pump and a water pump, and probably also without the alternator. A drive of these devices by the engine undoubtedly caused an increase in the brake specific fuel consumption.

As a summary of this part of the work, Figure 16 shows the results of the above-described comparisons of these two engines, presented in graphical form.

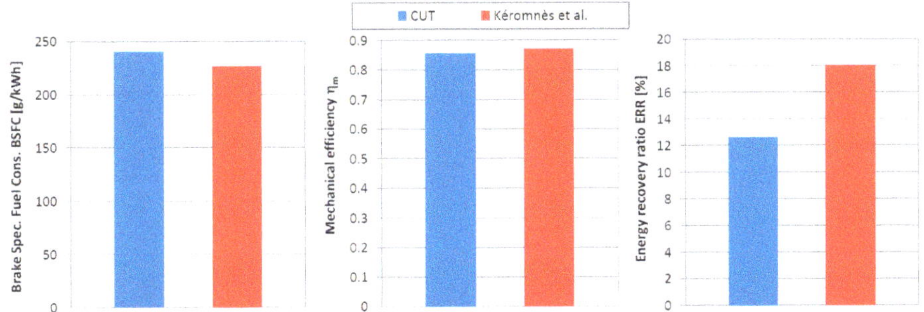

Figure 16. Comparison of the best results obtained for the engine developed at CUT and that developed by Kéromnès et al.

Considering the obtained results, but also the indicated imperfections and limitations of the developed engine, the author plans for further development of the design. These activities will mostly cover:

- Simulations to optimize the valve timing of the tested engine, or the development of new camshafts. The current valve timing is based on four-stroke engine settings and is not optimized for the new engine with a significantly different design and performance.

- An enlargement of the crossover passage between the additional expansion cylinders, and a reduction of the parasitic volume of the second expansion cylinders by the use of modified pistons, since it is the only modification to the combustion chambers of the cylinders of additional expansion that is possible to make in the developed engine.

- Selection of a turbocharger more matched to the specific parameters of the engine with the additional expansion of the exhaust gas. The plan is to apply a different turbocharger controlled by the wastegate valve. The initial analysis of this issue indicated that a VNT-turbocharger may not be suitable for the tested engine.
- Changing the method of control of the high fuel pressure—a pressure relief valve is currently used, which introduces a loss of power to the system. the plan is to develop a pressure control system by varying the flow rate of the high-pressure pump, similar to the base engine. This action requires the development of a separate electronic controller, because the used stand-alone engine management system does not offer such feasibility.
- After improving the engine, emissions testing will be carried out and selection of the aftertreatment system will be made, taking into consideration the specificity of the engine.

5. Conclusions

Basing on the results of the research described in the paper, the following conclusions were formulated:

- The results obtained by the author are so promising that despite the significant limitations in the feasibility of developing the design resulting from the adoption of a mass-produced four-cylinder engine as a base, he intends to lead its further development.
- The engine developed according to the concept of the five-stroke engine has many advantages, especially if it is designed from scratch, but it also has some drawbacks that make it not quite suitable for use in the classical drive system of cars. At low loads, there are some problems, such as the expansion of the exhaust gas below the ambient pressure, which results in energy loss, or problems with the operation of the turbocharger.
- Under heavy load, the engine achieves high efficiency while maintaining a high power-to-weight ratio. This makes this type of engine very well suited for applications where would it work with average high load, that is, for example, as a stationary electric generator or as an engine for the hybrid powertrain of a motor vehicle, in particular in series arrangement, but also under certain conditions for a power-split hybrid. A similar application field for such an engine is also indicated by the authors of study [14], who developed the five-stroke engine as the main power source for an extended-range electric vehicle or for a series-hybrid arrangement.

Acknowledgments: This research was conducted in the framework of task No. M-4/353/2016/DS, which was a subsidy for research granted by the Ministry of Science and Higher Education of the Republic of Poland. The author of the paper would like to thank Łukasz Rodak from Cracow University of Technology for support in the data acquisition process.

Conflicts of Interest: The author declares no conflict of interest. The founding sponsors had no role in the design of the study; in the collection, analyses, or interpretation of data; in the writing of the manuscript, and in the decision to publish the results.

1. Ota, Y.; Ito, Y.; Kawamura, A.; Nishiura, H.; Matsuo, S. Fuel Economy Improvement Technologies of the ESTEC 2ZR-FXE Engine. *Toyota Tech. Rev.* **2016**, *10*, 54–59.

2. European Commission. Horizon 2020 Work Programme 2016–2017 Smart, Green and Integrated Transport. Available online: https://ec.europa.eu/research/participants/data/ref/h2020/wp/2016_2017/main/h2020-wp1617-transport_en.pdf (accessed on 27 December 2016).

3. Merkisz, J.; Fuć, P.; Lijewski, P.; Ziółkowski, A.; Galant, M.; Siedlecki, M. Analysis of an Increase in the Efficiency of a Spark Ignition Engine Through the Application of an Automotive Thermoelectric Generator. *J. Electron. Mater.* **2016**, *45*, 4028–4037. [CrossRef]

4.   Mendera, K. Deliberations on thermodynamic cycle of the engine with extended expansion. In Proceedings of the KONMOT '76 Automotive Conference, Krakow, Poland, 20–23 April 1976; pp. 253–263. (In Polish)

5.   Noga, M.; Sendyka, B. Determination of the theorethical and total efficiency of the five-stroke SI engine. *Int. J. Automot. Technol.* **2014**, *15*, 1083–1089. [CrossRef]

6.   Cummins, C.L. *Internal Fire*, 3rd ed.; Carnot Press: Wilsonville, OR, USA, 2000; pp. 219–223.

7.   Miller, R.H. Supercharging and internal cooling cycle for high output. *ASME Trans.* **1947**, *69*, 453–464.

8.   Goto, T.; Hatamura, K.; Takizawa, S.; Hayama, N.; Abe, H.; Kanesaka, H. *Development of V6 Miller Cycle Gasoline Engine*; SAE Technical Paper 940198; SAE International: Warrendale, PA, USA, 1994.

9.   Malcev, V.; Bozhenov, A.; Schwab, R.; Müther, M. High Power Density High Speed Diesel. *MTZ Ind.* **2016**, *2*, 14–21. [CrossRef]

10.   Harada, S.; Shinagawa, T.; Kondo, T.; Togawa, K.; Kudo, M.; Matsubara, W. The New Toyota 1.2-Liter Inline 4-Cylinder ESTEC D-4T Engine. *Toyota Tech. Rev.* **2016**, *10*, 54–59.

11.   Grab-Rogaliński, K.; Szwaja, S. Influence of Intake Valve Closure Angle on IC Engine Indicated Parameters. *J. KONES* **2015**, *22*, 29–35.

12.   Pielecha, I.; Czajka, J.; Borowski, P.; Wisłocki, K. Thermodynamic indexes of Atkinson cycle combustion engine operation under transient conditions. *Combust. Engines* **2013**, *154*, 517–524.

13.   Schmitz, G. Five-Stroke Internal Combustion Engine. U.S. Patent 6553977, 29 April 2003. Available online: http://www.google.com/patents/US6553977 (accessed on 27 December 2016).

14.   Kéromnès, A.; Delaporte, B.; Schmitz, G.; Le Moyne, L. Development and validation of a 5 stroke engine for range extenders application. *Energy Convers. Manag.* **2014**, *82*, 259–267. [CrossRef]

15.   Krebs, R.; Böhme, J.; Dornhöfer, R.; Wurms, R.; Friedmann, K.; Helbig, J.; Hatz, W. The new Audi 2.0T FSI engine—The first direct injection turbo gasoline engine from Audi. In Proceedings of the 25th International Vienna Motor Symposium, Vienna, Austria, 29–30 April 2004; VDI-Verlag: Düsseldorf, Germany, 2004; pp. 224–246.

16.   Noga, M.; Sendyka, B. New design of the five stroke SI engine. *J. KONES* **2013**, *20*, 239–246. [CrossRef]

17.   Noga, M.; Sendyka, B. Increase of efficiency of SI engine through the implementation of thermodynamic cycle with additional expansion. *Bull. Pol. Acad. Sci. Tech. Sci.* **2014**, *62*, 349–355. [CrossRef]

18.   Li, T.; Zheng, B.; Yin, T. Fuel conversion efficiency improvements in a highly boosted spark-ignition engine with ultra-expansion cycle. *Energy Convers. Manag.* **2015**, *103*, 448–458. [CrossRef]

19.   Li, T.; Wang, B.; Zheng, B. A comparison between Miller and five-stroke cycles for enabling deeply downsized, highly boosted, spark-ignition engines with ultra expansion. *Energy Convers. Manag.* **2016**, *123*, 140–152. [CrossRef]

20.   Palanivendhan, M.; Modi, H.; Bansal, G. Five Stroke Internal Combustion Engine. *Int. J. Control Theory Appl.* **2016**, *13*, 5855–5862.

21.   Wlodarczyk, M.T. Fiber optic-based in-cylinder pressure sensor for advanced engine control and monitoring. *Combust. Engines* **2012**, *151*, 3–8.

Article

Computational Methodology for Knocking Combustion Analysis in Compression-Ignited Advanced Concepts

José Ramón Serrano, Ricardo Novella *, Josep Gomez-Soriano
and Pablo José Martinez-Hernandiz

CMT-Motores Térmicos, Universitat Politècnica de València, Camino de Vera s/n, 46022 Valencia, Spain;
jrserran@mot.upv.es (J.R.S.); jogoso1@mot.upv.es (J.G.-S.); pabmarh2@mot.upv.es (P.J.M.-H.)
* Correspondence: rinoro@mot.upv.es; Tel.: +34-96-387-9650

Received: 5 September 2018; Accepted: 17 September 2018; Published: 20 September 2018

Abstract: In the present work, a numerical methodology based on three-dimensional (3D) computational fluid dynamics (CFD) was developed to predict knock in a 2-Stroke engine operating with gasoline Partially Premixed Combustion (PPC) concept. Single-cycle Unsteady Reynolds-Averaged Navier Stokes (URANS) simulations using the renormalization group (RNG) $k - \varepsilon$ model were performed in parallel while the initial conditions are accordingly perturbed in order to imitate the variability in the in-cylinder conditions due to engine operation. Results showed a good agreement between experiment and CFD simulation with respect to cycle-averaged and deviation of the ignition timing, combustion phasing, peak pressure magnitude and location. Moreover, the numerical method was also demonstrated to be capable of predicting knock features, such as maximum pressure rise rate and knock intensity, with good accuracy. Finally, the CFD solution allowed to give more insight about in-cylinder processes that lead to the knocking combustion and its subsequent effects.

Keywords: gasoline PPC concept; 2-stroke engine; knocking combustion; CFD modelling; cycle-to-cycle variation

1. Introduction

The automotive industry is currently confronted with the hard challenge of achieving a compromise between performance and sustainability [1]. Most research efforts are focused on further developing spark-ignited (SI) engines and the exploration of new advanced combustion modes due to their advantages in terms of pollutant emissions [2]. In both concepts, knocking combustion is a major drawback to achieving higher thermal efficiency.

The overall tendency to knock is highly dependent on engine operating conditions as well as other aspects such as fuel anti-knock properties or combustion chamber design. It is, therefore, critical to gain a better understanding of knock generation mechanisms in order to develop robust knock mitigation strategies.

Owing to the clear propensity to generate extremely high burning rates, Low Temperature Combustion (LTC) concepts such HCCI (Homogeneous Charge Compression Ignition) [3,4], PCCI (Premixed Charge Compression Ignition) [5,6] or PPC (Partially Premixed Combustion) [7,8] could mean a significant improvement [9]. However, the extreme thermodynamic conditions achieved inside the combustion chamber due to the higher compression ratios and boosting pressures increase the knock propensity, thereby being an important constraint for their application to commercial vehicles.

In particular, PPC operated with low reacting fuels, such as gasoline, have shown encouraging results to achieve very low pollutant emissions while maintaining, or even improving, the thermal efficiency [10–12]. Indeed, this combustion concept operated in an innovative 2-Stroke high speed direct injection (HSDI) compression-ignited (CI) engine [13] offers a good flexibility to control the combustion timing and to extent the load range [14–16].

This concept operates between completely premixed and fully diffusive conditions, whereby low pollutant emissions may be attained. However, to achieve these conditions while retaining an accurate combustion timing control with the injection event remains as the main drawback of this particular system when operating under transient conditions.

Despite the attractive benefits of this engine system, its complexity due to the large number of parameters to be managed requires the use of optimization techniques which ensure greater flexibility, speed and lower costs than purely experimental procedures.

In this framework, the use of computational fluid dynamics (CFD) simulations is nowadays widely established in both the research community and the automotive industry. Here, aspects such as the simulation of turbulence and how it couples with the chemistry [17] are still the main limiting factors for reproducing the reacting flow field accurately. Since the requirements in both fields tend to differ, specially in terms of time available, the approaches used are also usually different. While in the industry sector, simulations are based on Unsteady Reynolds-averaged Navier–Stokes (URANS) turbulence modelling [18,19] and flamelet-based combustion models [20] owing its lower computational demands, the research community prefer to resort to high-fidelity combustion models and more complex turbulence schemes such as Large Eddy Simulations (LES) [21] or Direct Numerical Simulations (DNS) [22].

Although the industry standard tends to simplify the simulations as much as possible, reproducing only the most pertinent phenomena while dismissing the irrelevant ones, it is not always possible to meet all requirements in complex situations where there is no clearly dominant phenomenon. For instance, knocking combustion in SI engines appears as result of a particular situation in which local thermodynamic conditions are critical. In this context, the recreation of the in-cylinder temperature field (hot-spots, chamber inhomogeneities, etc.), turbulence field (flow velocity and at the spark plug) and species distribution (EGR, fuel injection, etc.) is determinant for the knock [9] and CCV prediction.

Misdariis et al. [23] has proven the suitability of multi-cycle LES and dual heat transfer to reproduce the cycle-to-cycle pressure variations (CCV) under knocking-like conditions [24]. However, requirements in terms of mesh resolution and runtime are prohibitive for practical applications. Furthermore, reproducing pressure effects of knock—local pressure oscillations—is extremely demanding [25–27], further compromising its application to the industry environment.

The main objective of this paper is therefore to develop and validate a robust knock simulation methodology based on 3D CFD modelling that allow to include knock systematically in the design process of future automotive engines. In order to achieve this target, several simplifications should be applied. For example, resorting to single-cycle simulations, using URANS turbulence schemes, perturbing the initial conditions and other strategies based on the experimental observation.

This paper is organised as follows. First, the engine and test cell specifications are briefly described. Then, the methodology is widely explained, along with the details of the CFD model. Subsequently, results from the validation are presented and discussed. A detailed analysis of the knock onset is also included hindsight. Finally, the paper concludes with a summary of the main findings.

2. Experimental Set-Up

The experimental facility already detailed in previous studies [28,29] was used to obtain the required data for the CFD model validation. Since the engine specifications and test cell features are widely described in the literature, only a brief description of them are included in this paper. Nonetheless, full details about the experimental facility can be found in the following documents [16,30].

A research version of a 2-stroke HSDI CI engine was used as baseline engine. The main specifications of the engine and the injector configuration are summarized in Table 1.

Table 1. Main engine specifications of the engine.

Engine	2-Stroke HSDI CI
Fuel (-)	RON95 gasoline
Number of cylinders (-)	1
Displacement (cm^3)	365
Bore–Stroke (mm)	76.0–80.5
Compression ratio (geometric)	17.8:1
Compression ratio (effective)	from 13.0:1 to 8.8:1
Number of valves (-)	2 intake and 2 exhaust

The control of the poppet valves relies on a hydraulic cam-driven Variable Valve Timing (VVT) system that allows to modify the opening/closing valve timings. Thanks to its flexibility to adjust the overlap among intake and exhaust periods, and the effective compression-expansion ratios, this system ensures an adequate in-cylinder flow pattern and optimizes the scavenging of burnt gases by reducing the short-circuit losses.

The equipped fuel injection system grants a maximum rail pressure of 110 MPa when operates with RON95 gasoline fuel. This is a prototype Delphi injection system which comes from a common-rail DFI 1.5 system.

The engine was installed in a completely instrumented test cell which supplies all required fluids for the engine operation through multiple auxiliary devices. The installation have independent water and oil cooling circuits and a low EGR system to supply arbitrary levels of cooled exhaust gas in any operating condition.

The in-cylinder pressure was measured with a Kistler 6061B pressure transducer flush-mounted between the intake and exhaust valves. A different piezo-resistive pressure sensor located at the cylinder liner close to Bottom Dead Center (BDC) was used to reference this pressure signal. Instantaneous signals, such as in-cylinder pressure, were sampled using a state-of-the-art acquisition system and recorded during 100 engine cycles for each operation condition.

3. Numerical Methodology

The main limiting aspect of traditional methodologies for the validation of CI engines operation, which are based on reproducing the in-cylinder pressure profile averaged from a given number of recorded cycles using simple turbulence schemes (URANS), is the incapacity to assess the cyclical deviation and its subsequent effects on combustion.

The use of more complex turbulence approaches (LES) and the simulation of several consecutive cycles is currently the standard approach to reproduce the CCV in spark-ignited (SI) engines [31,32] in which the combustion variability due to the larger turbulent scales is traditionally considered as the main cause of the cycle-to-cycle variation [33]. However, the huge increase of computational burden hardly compromises the application of LES modelling in most of the industry cases. Furthermore, it is not clear that pure stochastic fluctuations could explain the whole variation of flow conditions among consecutive cycles in gasoline PPC mode.

Available literature on CCV [34] claims that CCV observed in CI engines is mostly originated from the casuistic variability in fuel mass or trapped-gas conditions, rather than from arising in random variations in the turbulent flow and combustion process.

Therefore, other authors [35,36] employed URANS schemes to analyse CCV sources in several LTC strategies, where CCV is larger than in conventional Diesel engines but still far from SI engines. They investigated the cyclical dispersion from the point of view of the operation uncertainties. The basis of their approach is to perform many parallel URANS simulations in which one or more key operating conditions are perturbed about the reference conditions.

Here, a methodology based on this latter approach was used in an attempt of reproducing the experimental CCV and its effects on the combustion. Therefore, conditions at the intake valves closing (IVC) of a baseline cycle simulated were considered as the reference. These conditions, after being validated to ensure realistic in-cylinder conditions, were artificially modified in order to imitate the variation of the injected fuel, trapped mass, injection timing, etc.

The selection of those parameters that significantly affect the combustion was made by taking into account the conclusions obtained by Klos and Kokjohn [35] to reduce the number of parallel simulations as much as possible. Klos and Kokjohn found clear relationships among the dispersion of three operating parameters (EGR, mean gas temperature at IVC and fuel mass) and the combustion behaviour. Thus, these parameters and their recommended variations were utilized for modifying IVC baseline conditions.

Nonetheless, an additional parameter must be considered in this study since the 2-stroke engine has a particular in-cylinder flow motion. While in traditional CI engines the swirl flow motion is clearly dominant, the tumble motion—characteristic of SI engines—prevails in this particular engine design. As reported by Vermorel et al. [31], the tumble intensity significantly changes among consecutive cycles. They observed from LES simulations data that the magnitude of this rotational velocity could vary up to 25% at IVC. Therefore, this parameter was also included together with the other three determined by Klos. The parameters and their maximum variations used to perturb baseline conditions are displayed in Table 2. Note that EGR value was replaced by the percentage of combustion products within the cylinder at IVC, this is an equivalent parameter that define the amount of inert gases available during the closed-cycle.

Table 2. Baseline operation parameters and ranges of variation for the alteration of the IVC conditions.

Parameter	Baseline	$\pm$	[%]
Combustion products (%)	36.70	1.0	2.7
Temperature (K)	406.30	2.0	0.5
Tumble intensity (-)	4.31	1.1	25.0
Fuel mass (mg/cyc)	19.07	0.191	1.0

In addition to the baseline simulated case, ten parallel closed-cycle simulations were performed whereas the initial conditions at IVC were accordingly perturbed by randomly distributed variations of these four parameters to sample the four-dimensional domain.

This procedure was applied to the operation condition detailed in Table 3. This is defined by a medium speed (1500 rpm) and medium-high engine load (1.04 MPa of indicated mean effective pressure) that shows moderate levels of knock.

Table 3. Operating settings used for the model validation.

Engine speed (rpm)	1500
Torque (Nm)	49.9
IMEP (MPa)	1.04
Number Injections (-)	3 (pilot + main + post)
SoE_{main} (cad aTDC)	-42.0
Injection pressure (MPa)	85
Intake pressure (MPa)	0.275
EGR (%)	43.66

3.1. Numerical Model Set-Up

The numerical model of the engine was developed using the commercial CFD code CONVERGE [37]. The numerical domain, as shown in Figure 1, included the complete single cylinder geometry and the intake/exhaust ports, allowing to perform complete cycle simulations.

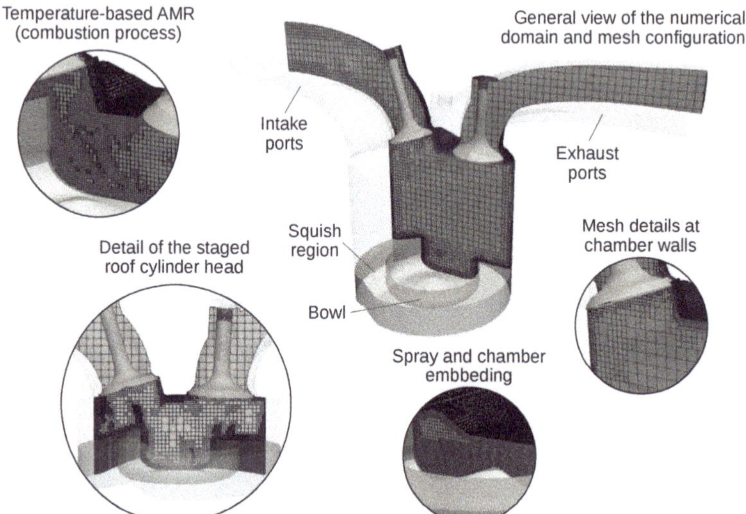

Figure 1. Numerical domain and mesh features of the engine.

The mesh discretization was done using the cut-cell Cartesian method available in the code. The base mesh size was 3 mm throughout the domain in the reference grid configuration. Three levels of fixed embedding (0.375 mm of cell size) were added to the walls of the combustion chamber, ports and region near the fuel injector, to improve boundary layer prediction and the accuracy of spray atomization, droplet breakup/coalescence, etc. Mesh size in the chamber was reduced by two levels of embedding (0.75 mm of cell size) after the start of combustion, for an improved recreation of the interaction and reflection of the pressure waves while avoiding undesired spatial aliasing effects. The code also employed adaptive mesh refinement (AMR) to increase grid resolution by three levels of additional refinement (up to 0.375 mm minimum cell size) based on the velocity and temperature sub-grid scales of 1 ms^{-1} and 2.5 K, respectively. As a result, the total number of cells varied between 1.5 million at BDC and 0.5 million at Top Dead Centre (TDC). This mesh configuration was achieved after a grid sensitivity analysis [27], offering a grid-independent solution when simulating combustion and its produced unsteady pressure field in internal combustion engines.

The Mach Courant-Friedrich-Lewy was set to 1.0 during the combustion in order to properly capture the local pressure oscillations caused by combustion. Thereby, the calculation time-step was gathered between 0.05 and 0.5 μs. This value was also obtained from the previously referred work performed by Torregrosa et al. [27], whose demonstrated that the energy of the unsteady pressure field is highly sensitive to CFL Mach number, but also that the energy does not change when this parameter is close to or below one.

The renormalization group (RNG) $k - \varepsilon$ URANS model [18] coupled with the heat transfer approach proposed by Angelberger et al. [38] was chosen for simulating the turbulent properties of the flow.

For combustion modelling, the SAGE detailed chemistry solver [39] was employed along with a multi-zone (MZ) approach, with bins of 5 K in temperature and 0.05 in equivalence ratio [40]. This combustion model, despite not using an explicit turbulent combustion closure [17], has demonstrated to perform well for simulating spray combustion under URANS schemes in previous works [41]. A reduced chemical kinetic mechanism for primary reference fuels (PRF) based on Brakora et al. [42] was used in this work to account for fuel chemistry. A blend of 5% n-heptane and 95% iso-octane was utilized for predicting the physical properties of the gasoline fuel, being a suitable surrogate for predicting the ignition features of the RON95 gasoline used in experiments. Activating iso-octane reactions, the chemical mechanism resulted in a 45 species and 152 reactions.

Coupled with appropriate turbulence models, Kodavasal et al. [43] demonstrated that a similar combustion approach allows an accurate reproduction of gasoline autoignition, while other researchers [29,44] established realistic rates of heat release under gasoline compression-ignited conditions.

The fuel injection was described by the standard Discrete Droplet Model (DDM) [45] and Kelvin Helmholtz (KH)–Rayleigh Taylor (RT) breakup model was employed to model spray atomization [46]. The injection rate was determined from the experimental injector characterization. This process measures the mass flow rate and spray momentum flux in a specific test rig [47,48] for the injection configurations defined a priori in similar real test conditions, in order to provide the most realistic injection features.

Cylinder wall temperatures were assumed to be constant and estimated using a lumped heat transfer model [49]. The instantaneous pressure signals registered at intake/exhaust manifolds were used to fix the inflow/outflow boundaries located at the end of the intake/exhaust ports. The temperature at these boundaries was considered constant and equal to the time-averaged value registered at the same manifolds.

This configuration has proven to accurately reproduce the in-cylinder pressure field oscillations over a wide range of operation conditions and combustion regimes [26,50,51].

4. Results and Discussion

In this section, results from the application of the proposed methodology are presented an discussed. First, the validity of the numerical solution is verified. Then, the knocking combustion is visualized and further analysed using several visualization techniques.

4.1. Validation

Following the guidelines depicted during the methodology description in Section 3, a unique engine cycle was calculated and validated prior to run the multiple parallel executions with the modified initial conditions. The target of this preliminary step is to obtain a numerical solution that reproduces the mean behaviour of the experiments.

In this way, experimental and simulated cylinder pressure profiles of the baseline simulation are compared in Figure 2. As it can be seen, the CFD model correctly predicts the in-cylinder pressure. Indeed, the predicted pressure shows a similar deviation as measurements dispersion. Although RoHR traces show that the combustion phasing is slightly delayed and its peak value is minimally overestimated, they do not affect the maximum peak pressure in a significant way. Given that the calculation of this parameter depends on material deformations and blow-by losses that the model does not take into account, this effect may be attributed to a minor underestimation of the effective compression ratio.

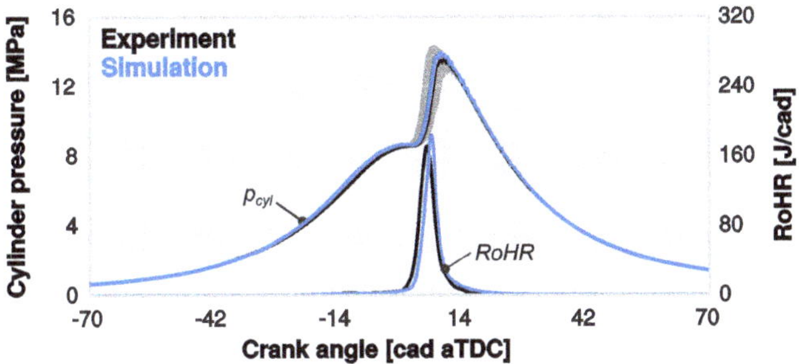

Figure 2. Comparison between measured and CFD calculated results of baseline test, measurements dispersion is represented by their standard deviation (SD).

Once that the baseline simulation can be considered representative of the real operating conditions, ten parallel executions were performed with randomly distributed variations of the IVC conditions. In Figure 3, the same comparison previously done in Figure 2 can be seen but including now the CCV spread of this latter simulations. As can be seen from the graph, the standard deviation of simulations is similar to that observed in the measurements proving the validity of the methodology in this particular case of study.

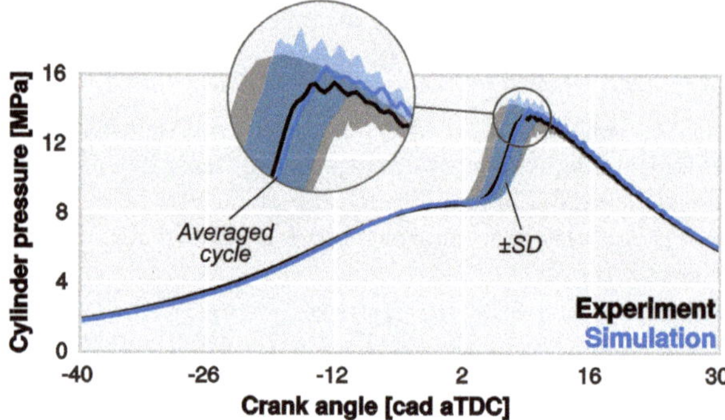

Figure 3. Comparison between measured and CFD calculated in-cylinder pressure traces of baseline test, dispersion due to CCV is represented by their standard deviation value.

However, the interest of the analysis is not only to check the performance of the methodology to reproduce a realistic CCV but also to examine the effects of this variability on the traditional combustion/knock parameters. In order to accomplish this target, the maximum pressure rise rate and the Maximum Amplitude Pressure Oscillation (MAPO) are plotted at each measured and simulated engine cycle in Figure 4. The standard deviation is also included to evaluate the capability of the methodology to reproduce CCV effects. Again, comparable results were obtained in terms of mean and dispersion levels for both parameters.

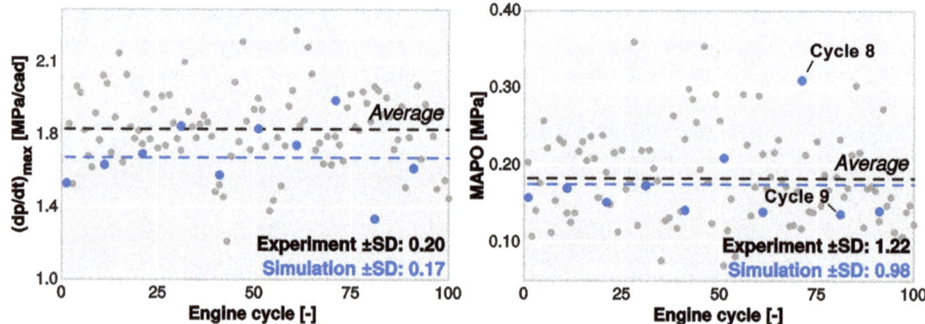

Figure 4. Comparison between measured and CFD calculated results of baseline test, dispersion due to CCV is represented by their standard deviation value. The maximum pressure rise rate and MAPO are plotted to compare experimental and simulation results.

4.2. Knocking Combustion Visualization

After the validation of the methodology, an analysis of the knock onset is performed in this section. The study is conducted for two extreme knocking cases. On the one hand, the simulated cycle number 8, highlighted in Figure 4, is selected as the upper limit since it exhibit the higher MAPO value. And, on the other hand, cycle number 9 is chosen as the lower limit, being the cycle with the slightest knock.

A series of snapshots were conscientiously chosen and plotted together in Figure 5 for studding the combustion process. They were specifically selected considering different stages of combustion to allow a proper visualization of the endgas auto-ignition.

In this figure, the combustion tracking is done by clipping the temperature field at 2000 K and colouring it by the fuel energy release, thus showing the location of the reaction zones. Besides, pressure profiles registered at the transducer position are included to distinguish at which cycle step is located each snapshot while allowing to relate the combustion with its corresponding pressure effects [51,52].

As can be seen in both sequences of snapshots, combustion starts spontaneously as a result of a first auto-ignition event located inside the piston bowl where local temperature and mixing conditions are more favourable. Then, the combustion rapidly progresses by consuming the charge located within the bowl. At this point, the pressure rise due to combustion compresses the unburned gases at the squish region, causing the appearance of additional hot spots as a result of a second auto-ignition event. Looking at the colour scale, the rate of energy released by these hot spots is very localized and noticeably greater than that released during the previous combustion phase.

Examination of the pressure profiles also reveals that the onset of these hot spots coincides with the instant at which pressure oscillations become apparent, showing again that this abnormal combustion event has many similarities with the traditional knocking combustion in SI engines. Furthermore, it is possible to see that the knock intensity decreases as combustion is shifted towards the expansion stroke.

Besides the visualization of the combustion progress, Figure 5 shows a comparison between two cycles with heavy and slight knocking conditions. It is clear from this figure that, while the ignition is produced 2 cad after TDC in the cycle with heavy knock, combustion is delayed almost 2 cad in the slight knock cycle. Moreover, the burning rate is significantly higher in the former case since temperature iso-volumes grow faster. Thereby, the endgas auto-ignition is produced closer to the TDC, contributing to increase the pressure rise and its associated local oscillations to a greater extent. However, despite these differences, it is important to note that spatial patterns are quite similar at the knock onset.

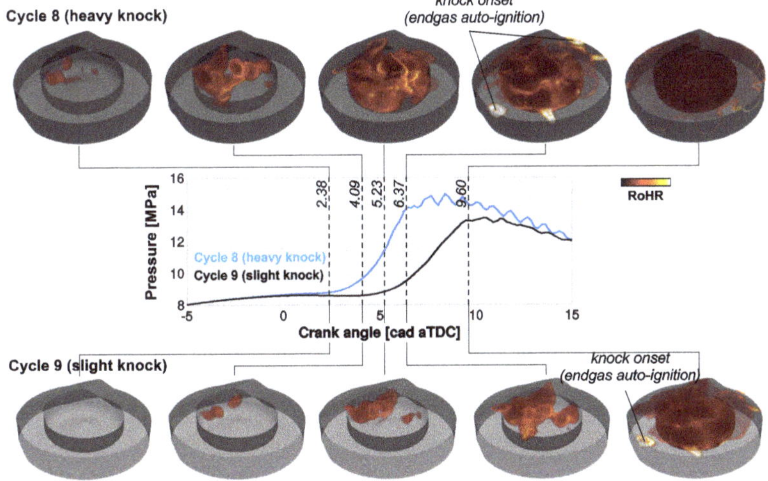

Figure 5. Visualization of the combustion process. A sequence of snapshots is depicted in order to identify the differences between two extreme cases with a remarkable knock level variation. The combustion process is visualized by clipping the temperature field at 2000 K and colouring it by the fuel energy release.

5. Conclusions

In view of the challenges to manage the gasoline partially premixed combustion in compression-ignited engines, this paper proposes a numerical approach based on multi-dimensional CFD in order to improve the knowledge and understanding of this particular combustion concept. In particular, the presented methodology was specifically developed to capture knocking combustion, allowing an comprehensive analysis of all involved phenomena and their undesired effects.

The proposed methodology allows a realistic estimation of both the cycle-averaged and dispersion values of the main combustion/knock metrics while keeping the computational burden under reasonable values. Thereby it offers the chance to include these parameters in the design process, optimizing them altogether with the rest of relevant emissions and performance metrics.

Combining distinct visualization methods, such as iso-surfaces of temperature and energy release contours, allowed to identify the differences in the combustion process among two extreme cycles, thereby enhancing the understanding the knock phenomenon.

Results revealed that the propensity of the knock onset is reduced as the combustion is delayed towards the expansion stroke. In this sense, cycles with lower burning rates lead to a slight knocking combustion whereas high burning speed cycles tends to significantly increase the knock.

Nonetheless, further efforts must be taken to confirm, through LES simulations, that URANS is capturing all relevant phenomena in the knocking combustion. Moreover, further analysis should be done for providing more insight about the combustion process itself and its related knocking generation mechanisms.

Author Contributions: All authors discussed and agreed on the contents of the manuscript. R.N. managed the work defining the objectives of the proposed methodology, guiding the technical discussion of the results and contributing to the critical review of the manuscript. J.G.-S. conducted the research tasks and leaded the investigation process; developed the methodology, performed the simulations, analysed the results and wrote the initial draft. P.J.M.-H. contributed to the manuscript preparation and presentation. J.R.S. discussed the results, contributed to the manuscript review and provided technical guidance in the design of methodology.

Funding: The work has been partially supported by the Spanish Ministerio de Economía y Competitividad through grant number TRA2016-79185-R. The equipment used in this work has been partially supported by FEDER project funds "Dotación de infraestructuras científico técnicas para el Centro Integral de Mejora Energética y Medioambiental de Sistemas de Transporte (CiMeT), (FEDER-ICTS-2012-06)" from the operational program of unique scientific and technical infrastructure of the Spanish Ministerio de Economía y Competitividad. In addition, J. Gomez-Soriano is partially supported by an FPI contract (FPI-S2-2016-1353) of the "Programa de Apoyo para la Investigación y Desarrollo (PAID-01-16)" of the Universitat Politècnica de València.

Acknowledgments: The authors want to express their gratitude to CONVERGENT SCIENCE Inc. and Convergent Science GmbH for their kind support for the CFD calculations with the CONVERGE software.

Conflicts of Interest: The authors declare no conflict of interest. The founding sponsors had no role in the design of the study; in the collection, analyses or interpretation of data; in the writing of the manuscript; nor in the decision to publish the results.

Abbreviations

The following abbreviations are used in this manuscript:

AMR	Adaptive Mesh Refinement
aTDC	after Top Dead Centre
BDC	Bottom Dead Centre
cad	Crank Angle Degree
CCV	Cycle-to-Cycle Variation
CFD	Computational Fluid Dynamics
CI	Compression-Ignited
DDM	Discrete Droplet Model
DI	Direct Injection
DNS	Direct Numerical Simulation
EGR	Exhaust Gas Recirculation

HCCI	Homogeneous Charge Compression Ignition
HSDI	High Speed Direct Injection
ICE	Internal Combustion Engine
IMEP	Indicated Mean Effective Pressure
IVC	Intake Valves Closing
KH	Kelvin-Helmholtz
LES	Large Eddy Simulation
LTC	Low Temperature Combustion
MAPO	Maximum Amplitude Pressure Oscillation
MZ	Multi-Zone
NO_x	Nitrous Oxides (NO and NO_2)
PCCI	Premixed Charge Compression Ignition
PPC	Partially Premixed Combustion
PRF	Primary Reference Fuel
RANS	Reynolds-averaged Navier–Stokes
RNG	Re-Normalized Group
RoHR	Rate of Heat Release
RON	Research Octane Number
RT	Rayleigh-Taylor
SAGE	Detailed Chemistry Solver
SD	Standard Deviation
SI	Spark-Ignited
SoE_m	Start of Energizing of the Injector (main injection)
TDC	Top Dead Centre
URANS	Unsteady Reynolds-averaged Navier–Stokes
VVT	Variable Valve Timing

References

1. Serrano, J.R. Imagining the Future of the Internal Combustion Engine for Ground Transport in the Current Context. *Appl. Sci.* **2017**, *7*, 1011, doi:10.3390/app7101001. [CrossRef]
2. Bermúdez, V.; Serrano, J.R.; Piqueras, P.; Sanchis, E.J. On the impact of particulate matter distribution on pressure drop of wall-flow particulate filters. *Appl. Sci.* **2017**, *7*, 234, doi:10.3390/app7030234. [CrossRef]
3. Takeda, Y.; Keiichi, N.; Keiichi, N. *Emission Characteristics of Premixed Lean Diesel Combustion with Extremely Early Staged Fuel Injection*; SAE Technical Paper; SAE: Warrendale, PA, USA, 1996; doi:10.4271/961163.
4. Ryo, H.; Hiromichi, Y. *HCCI Combustion in a DI Diesel Engine*; SAE Technical Paper 2003-01-0745; SAE: Warrendale, PA, USA, 2003; doi:10.4271/2003-01-0745.
5. Torregrosa, A.J.; Broatch, A.; García, A.; Mónico, L.F. Sensitivity of combustion noise and NOx and soot emissions to pilot injection in PCCI Diesel engines. *Appl. Energy* **2013**, *104*, 149–157, doi:10.1016/j.apenergy.2012.11.040. [CrossRef]
6. Torregrosa, A.J.; Broatch, A.; Novella, R.; Gomez-Soriano, J.; Mónico, L.F. Impact of gasoline and Diesel blends on combustion noise and pollutant emissions in Premixed Charge Compression Ignition engines. *Energy* **2017**, *137*, 58–68, doi:10.1016/j.energy.2017.07.010. [CrossRef]
7. Boyarski, N.J.; Reitz, R.D. Premixed Compression Ignition (PCI) Combustion with Modeling-Generated Piston Bowl Geometry in a Diesel Engine. In Proceedings of the SAE 2006 World Congress & Exhibition, Detroit, MI, USA, 3 April 2006; doi:10.4271/2006-01-0198.
8. Okude, K.; Mori, K.; Shiino, S.; Moriya, T. *Premixed Compression Ignition (PCI) Combustion for Simultaneous Reduction of NOx and Soot in Diesel Engine*; SAE Technical Paper 2004-01-1907; SAE: Warrendale, PA, USA, 2004; doi:10.4271/2004-01-1907.
9. Wang, Z.; Liu, H.; Reitz, R.D. Knocking combustion in spark-ignition engines. *Prog. Energy Combust. Sci.* **2017**, *61*, 78–112, doi:10.1016/j.pecs.2017.03.004. [CrossRef]
10. Hanson, R.; Splitter, D.; Reitz, R. *Operating a Heavy-Duty Direct-Injection Compression-Ignition Engine with Gasoline for Low Emissions*; SAE Technical Paper 2009-01-1442; SAE: Warrendale, PA, USA, 2009; doi:10.4271/2009-01-1442.

11. Manente, V.; Johansson, B.; Tunestal, P.; Cannella, W. Effects of Different Type of Gasoline Fuels on Heavy Duty Partially Premixed Combustion. *SAE Int. J. Engines* **2009**, *2*, 71–88, doi:10.4271/2009-01-2668. [CrossRef]

12. Lewander, M.; Johansson, B.; Tunestål, P. *Investigation and Comparison of Multi Cylinder Partially Premixed Combustion Characteristics for Diesel and Gasoline Fuels*; SAE Technical Paper 2011-01-1811; SAE: Warrendale, PA, USA, 2011; doi:10.4271/2011-01-1811.

13. Tribotte, P.; Ravet, F.; Dugue, V.; Obernesser, P.; Quechon, N.; Benajes, J.; Novella, R.; De Lima, D. Two Strokes Diesel Engine—Promising Solution to Reduce CO_2 Emissions. *Procedia* **2012**, *48*, 2295–2314, doi:10.1016/j.sbspro.2012.06.1202. [CrossRef]

14. Laget, O.; Ternel, C.; Thiriot, J.; Charmasson, S.; Tribotté, P.; Vidal, F. *Preliminary Design of a Two-Stroke Uniflow Diesel Engine for Passenger Car*; SAE Technical Paper 2013-01-1719; SAE: Warrendale, PA, USA, 2013; doi:10.4271/2013-01-1719.

15. Benajes, J.; Novella, R.; De Lima, D.; Tribotté, P.; Quechon, N.; Obernesser, P.; Dugue, V. Analysis of the combustion process, pollutant emissions and efficiency of an innovative 2-stroke HSDI engine designed for automotive applications. *Appl. Therm. Eng.* **2013**, *58*, 181–193, doi:10.1016/j.applthermaleng.2013.03.050. [CrossRef]

16. Benajes, J.; Molina, S.; Novella, R.; De Lima, D. Implementation of the Partially Premixed Combustion concept in a 2-stroke HSDI diesel engine fueled with gasoline. *Appl. Energy* **2014**, *122*, 94–111, doi:10.1016/j.apenergy.2014.02.013. [CrossRef]

17. Pal, P.; Keum, S.; Im, H.G. Assessment of flamelet versus multi-zone combustion modeling approaches for stratified-charge compression ignition engines. *Int. J. Engine Res.* **2016**, *17*, 280–290, doi:10.1177/1468087415571006. [CrossRef]

18. Yakhot, V.; Orszag, S. Renormalization group analysis of turbulence. *J. Sci. Comput.* **1986**, *1*, 3–51, doi:10.1007/BF01061452. [CrossRef]

19. Wilcox, D.C. Formulation of the k-w turbulence model revisited. *AIAA J.* **2008**, *46*, 2823–2838, doi:10.2514/1.36541. [CrossRef]

20. Chen, J.H.; Hawkes, E.R.; Sankaran, R.; Mason, S.D.; Im, H.G. Direct numerical simulation of ignition front propagation in a constant volume with temperature inhomogeneities: I. Fundamental analysis and diagnostics. *Combust. Flame* **2006**, *145*, 128–144, doi:10.1016/j.combustflame.2005.09.017. [CrossRef]

21. Pope, S.B. Ten questions concerning the large-eddy simulation of turbulent flows. *New J. Phys.* **2004**, *6*, 35, doi:10.1088/1367-2630/6/1/035. [CrossRef]

22. Pillai, A.L.; Kurose, R. Numerical investigation of combustion noise in an open turbulent spray flame. *Appl. Acoust.* **2018**, *133*, 16–27, doi:10.1016/j.apacoust.2017.11.025. [CrossRef]

23. Misdariis, A.; Vermorel, O.; Poinsot, T. LES of knocking in engines using dual heat transfer and two-step reduced schemes. *Combust. Flame* **2015**, *162*, 4304–4312, doi:10.1016/j.combustflame.2015.07.023. [CrossRef]

24. Broatch, A.; López, J.J.; García-Tíscar, J.; Gomez-Soriano, J. Experimental Analysis of Cyclical Dispersion in Compression-Ignited Versus Spark-Ignited Engines and Its Significance for Combustion Noise Numerical Modeling. *J. Eng. Gas Turbines Power* **2018**, *140*, 102808, doi:10.1115/1.4040287. [CrossRef]

25. Pal, P.; Kolodziej, C.; Choi, S.; Broatch, A.; Gomez-Soriano, J.; Wu, Y.; Lu, T.; See, Y.C.; Som, S. *Development of a Virtual CFR Engine Model for Knocking Combustion Analysis*; SAE Technical Paper 2018-01-0187; SAE: Warrendale, PA, USA, 2018; doi:10.1016/2018-01-0187.

26. Broatch, A.; Novella, R.; Gomez-Soriano, J.; Pal, P.; Som, S. *Numerical Methodology for Optimization of Compression-Ignited Engines Considering Combustion Noise Control*; SAE Technical Paper 2018-01-0193; SAE: Warrendale, PA, USA, 2018; doi:10.1016/2018-01-0193.

27. Torregrosa, A.J.; Broatch, A.; Gil, A.; Gomez-Soriano, J. Numerical approach for assessing combustion noise in compression-ignitied Diesel engines. *Appl. Acoust.* **2018**, *135*, 91–100, doi:10.1016/j.apacoust.2018.02.006. [CrossRef]

28. Benajes, J.; Broatch, A.; Garcia, A.; Muñoz, L.M. *An Experimental Investigation of Diesel-Gasoline Blends Effects in a Direct-Injection Compression-Ignition Engine Operating in PCCI Conditions*; SAE Technical Paper; SAE: Warrendale, PA, USA, 2013, doi:10.4271/2013-01-1676.

29. Benajes, J.; Novella, R.; De Lima, D.; Tribotte, P. Investigation on multiple injection strategies for gasoline PPC operation in a newly designed 2-stroke HSDI compression ignition engine. *SAE Int. J. Engines* **2015**, *8*, 758–774, doi:10.4271/2015-01-0830. [CrossRef]

30. Benajes, J.; García, A.; Domenech, V.; Durrett, R. An investigation of partially premixed compression ignition combustion using gasoline and spark assistance. *Appl. Therm. Eng.* **2013**, *52*, 468–477, doi:10.1016/j.applthermaleng.2012.12.025. [CrossRef]

31. Vermorel, O.; Richard, S.; Colin, O.; Angelberger, C.; Benkenida, A.; Veynante, D. Towards the understanding of cyclic variability in a spark ignited engine using multi-cycle LES. *Combust. Flame* **2009**, *156*, 1525–1541, doi:10.1016/j.combustflame.2009.04.007. [CrossRef]

32. Vermorel, O.; Richard, S.; Colin, O.; Angelberger, C.; Benkenida, A.; Veynante, D. *Multi-Cycle LES Simulations of Flow and Combustion in a PFI SI 4-Valve Production Engine*; SAE Technical Paper 2007-01-0151; SAE: Warrendale, PA, USA, 2007; doi:10.4271/2007-01-0151.

33. Granet, V.; Vermorel, O.; Lacour, C.; Enaux, B.; Dugué, V.; Poinsot, T. Large-Eddy Simulation and experimental study of cycle-to-cycle variations of stable and unstable operating points in a spark ignition engine. *Combust. Flame* **2012**, *159*, 1562–1575, doi:10.1016/j.combustflame.2011.11.018. [CrossRef]

34. Fansler, T.D.; Wagner, R.M. Cyclic dispersion in engine combustion-Introduction by the special issue editors. *Int. J. Engine Res.* **2015**, *16*, 255–259, doi:10.1177/1468087415572740. [CrossRef]

35. Klos, D.; Kokjohn, S.L. Investigation of the sources of combustion instability in low-temperature combustion engines using response surface models. *Int. J. Engine Res.* **2015**, *16*, 419–440, doi:10.1177/1468087414556135. [CrossRef]

36. Jia, M.; Dempsey, A.B.; Wang, H.; Li, Y.; Reitz, R.D. Numerical simulation of cyclic variability in reactivity-controlled compression ignition combustion with a focus on the initial temperature at intake valve closing. *Int. J. Engine Res.* **2015**, *16*, 441–460, doi:10.1177/1468087414552088. [CrossRef]

37. CONVERGENT SCIENCE Inc. *CONVERGE 2.2 Theory Manual*; CONVERGENT SCIENCE Inc.: Madison, WI, USA, 2015.

38. Angelberger, C.; Poinsot, T.; Delhay, B. *Improving Near-Wall Combustion and Wall Heat Transfer Modeling in SI Engine Computations*; SAE Technical Paper 2003-01-0542; SAE: Warrendale, PA, USA, 1997; doi:10.4271/972881.

39. Senecal, P.K.; Pomraning, E.; Richards, K.J.; Briggs, T.E.; Choi, C.Y.; McDavid, R.M.; Patterson, M.A. *Multi-Dimensional Modeling of Direct-Injection Diesel Spray Liquid Length and Flame Lift-off Length Using CFD and Parallel Detailed Chemistry*; SAE Technical Paper 2003-01-1043; SAE: Warrendale, PA, USA, 2003; doi:10.4271/2003-01-1043.

40. Babajimopoulos, A.; Assanis, D.N.; Flowers, D.L.; Aceves, S.M.; Hessel, R.P. A fully coupled computational fluid dynamics and multi-zone model with detailed chemical kinetics for the simulation of premixed charge compression ignition engines. *Int. J. Engine Res.* **2005**, *6*, 497–512, doi:10.1243/146808705X30503. [CrossRef]

41. Pal, P.; Probst, D.; Pei, Y.; Zhang, Y.; Traver, M.; Cleary, D.; Som, S. Numerical Investigation of a Gasoline-Like Fuel in a Heavy-Duty Compression Ignition Engine Using Global Sensitivity Analysis. *SAE Int. J. Fuels Lubr.* **2017**, *10*, 56–68, doi:10.4271/2017-01-0578. [CrossRef]

42. Brakora, J.; Reitz, R.D. *A Comprehensive Combustion Model for Biodiesel-Fueled Engine Simulations*; SAE Technical Paper 2013-01-1099; SAE: Warrendale, PA, USA, 2013; doi:10.4271/2013-01-1099.

43. Kodavasal, J.; Kolodziej, C.P.; Ciatti, S.A.; Som, S. Computational fluid dynamics simulation of gasoline compression ignition. *J. Energy Resour. Technol.* **2015**, *137*, 032212, doi:10.1115/1.4029963. [CrossRef]

44. Benajes, J.; Novella, R.; Lima, D.D.; Thein, K. Impact of injection settings operating with the gasoline Partially Premixed Combustion concept in a 2-stroke HSDI compression ignition engine. *Appl. Energy* **2017**, *193*, 515–530, doi:10.1016/j.apenergy.2017.02.044. [CrossRef]

45. Dukowicz, J.K. A particle-fluid numerical model for liquid sprays. *J. Comput. Phys.* **1980**, *35*, 229–253, doi:10.1016/0021-9991(80)90087-X. [CrossRef]

46. Reitz, R.D.; Beale, J.C. Modeling spray atomization with the Kelvin-Helmholtz/Rayleigh-Taylor hybrid model. *Atomization Sprays* **1999**, *9*, 623–650, doi:10.1615/AtomizSpr.v9.i6.40. [CrossRef]

47. Payri, R.; García, J.M.; Salvador, F.; Gimeno, J. Using spray momentum flux measurements to understand the influence of diesel nozzle geometry on spray characteristics. *Fuel* **2005**, *84*, 551–561, doi:10.1016/j.fuel.2004.10.009. [CrossRef]

48. Payri, R.; Salvador, F.J.; Gimeno, J.; Bracho, G. A new methodology for correcting the signal cumulative phenomenon on injection rate measurements. *Exp. Tech.* **2008**, *32*, 46–49, doi:10.1111/j.1747-1567.2007.00188.x. [CrossRef]

49. Torregrosa, A.J.; Olmeda, P.; Degraeuwe, B.; Reyes, M. A concise wall temperature model for DI Diesel engines. *Appl. Therm. Eng.* **2006**, *26*, doi:10.1016/j.applthermaleng.2005.10.021. [CrossRef]

50. Torregrosa, A.J.; Broatch, A.; Margot, X.; Gomez-Soriano, J. Towards a Predictive CFD Approach for Assessing Noise in Diesel Compression Ignition Engines. Impact of the Combustion Strategies. In Proceedings of the 9th COMODIA International Conference on Modeling and Diagnostics for Advanced Engine systems, Okayama, Japan, 25 July 2017.

51. Torregrosa, A.J.; Broatch, A.; García-Tíscar, J.; Gomez-Soriano, J. Modal decomposition of the unsteady flow field in compression-ignited combustion chambers. *Combust. Flame* **2018**, *188*, 469–482, doi:10.1016/j.combustflame.2017.10.007. [CrossRef]

52. Broatch, A.; Margot, X.; Novella, R.; Gomez-Soriano, J. Impact of the injector design on the combustion noise of gasoline partially premixed combustion in a 2-stroke engine. *Appl. Therm. Eng.* **2017**, *119*, 530–540, doi:10.1016/j.applthermaleng.2017.03.081. [CrossRef]

 *applied sciences*

Article

On the Impact of Particulate Matter Distribution on Pressure Drop of Wall-Flow Particulate Filters

Vicente Bermúdez, José Ramón Serrano, Pedro Piqueras * and Enrique José Sanchis

CMT-Motores Térmicos, Universitat Politècnica de València, Camino de Vera s/n, Valencia 46022, Spain; bermudez@mot.upv.es (V.B.); jrserran@mot.upv.es (J.R.S.); ensanpac@mot.upv.es (E.J.S.)
* Correspondence: pedpicab@mot.upv.es; Tel.: +34-96-387-7650

Academic Editor: Antonio Ficarella
Received: 26 January 2017; Accepted: 24 February 2017; Published: 2 March 2017

Abstract: Wall-flow particulate filters are a required exhaust aftertreatment system to abate particulate matter emissions and meet current and incoming regulations applying worldwide to new generations of diesel and gasoline internal combustion engines. Despite the high filtration efficiency covering the whole range of emitted particle sizes, the porous substrate constitutes a flow restriction especially relevant as particulate matter, both soot and ash, is collected. The dependence of the resulting pressure drop, and hence the fuel consumption penalty, on the particulate matter distribution along the inlet channels is discussed in this paper taking as reference experimental data obtained in water injection tests before the particulate filter. This technique is demonstrated to reduce the particulate filter pressure drop without negative effects on filtration performance. In order to justify these experimental data, the characteristics of the particulate layer are diagnosed applying modeling techniques. Different soot mass distributions along the inlet channels are analyzed combined with porosity change to assess the new properties after water injection. Their influence on the subsequent soot loading process and regeneration is assessed. The results evidence the main mechanisms of the water injection at the filter inlet to reduce pressure drop and boost the interest for control strategies able to force the re-entrainment of most of the particulate matter towards the inlet channels' end.

Keywords: internal combustion engines; emissions; particulate matter; wall-flow particulate filter; pressure drop; soot distribution; particulate layer

---

1. Introduction

Pollutant regulations applying to compression ignition engines have focused on particulate matter emissions as one of the main pollutant emissions with which to deal. In Europe, particulate matter emissions have been restricted from 140 mg/km in Euro 1 to 4.5 mg/km in current Euro 6b. Besides stringent mass constraints, particulate number is also regulated by Euro 5b applying both mass and number regulations also to direct injection spark ignition engines. Similar trends are found worldwide becoming wall-flow particulate filters in a required exhaust aftertreatment device already present in diesel engines and being progressively adopted in gasoline-powered vehicles.

The reason for the implantation of the wall-flow particulate filters is related to its high filtration efficiency in the whole range of particle size [1]. However, soot collection into and on the porous substrate produces a noticeable pressure drop, increasing as soot loading does. In standardized post-turbine placement, this pressure drop is multiplied by the turbine pressure ratio to define the increase in engine back-pressure directly damaging pumping work. The result is a non-negligible fuel consumption and carbon dioxide (CO_2) emission penalty [2]. In addition, the typically low temperature downstream of the turbine makes the use of active regeneration strategies necessary involving additional periodic fuel consumption.

Although most of these drawbacks may be overcome with a pre-turbine Diesel Particulate Filter (DPF) placement [3], which enables conditions for passive regeneration and highly reduces DPF pressure drop impact on the fuel consumption penalty [4], the need for advanced boosting architectures to guarantee fast dynamic response [5] is postponing its further development. In the meantime, efforts are being driven to reduce aftertreatment volume by combining several abatement functions [6]. Thus, new devices, such as the Selective Catalytic Reduction Filter (SCRF) system, are gaining in interest. SCRF consists of the combination into a single wall-flow monolith of soot filtration and nitrogen oxides (NO_x) abatement capability by coating the porous substrate with a NO_x Selective Catalytic Reduction (SCR) catalyst [7]. Despite the potential improvement of this solution in terms of DPF passive regeneration, SCR light-off and conversion efficiency at low temperature, the combined chemical behavior of soot and NO_x is still suggesting doubts concerning the overlapping of temperature ranges, which can lead to a final slight loss of NO_x abatement and passive regeneration capability [8]. Regardless of the need for further understanding on new particulate filters with extended chemical functions, this kind of solution does not involve improvements concerning the fluid-dynamic behavior, i.e., pressure drop. The current context relies on cell geometry or porous wall optimization concepts, such as asymmetrical cell designs with different inlet channels geometry [9] or a two-layer substrate combining different micro-geometry properties [10]. These solutions contribute to partially mitigate the DPF impact on engine performance by changing the pressure drop to soot loading dependence. However, these parameters are still closely related, and the effects concerning fuel consumption are in the best case only slightly displaced to higher soot loading [11].

Under this context, pre-DPF water injection emerges as a technique able to reduce pressure drop with respect to a baseline DPF, making it independent of the particulate matter loading [12]. It leads to clear advantages in terms of fuel economy and CO_2 emission. Secondarily, the particulate matter accumulation capacity is also increased. This feature is beneficial both in terms of ash through the reduction of maintenance requirements and soot, whose active regeneration can become exclusively controlled by soot mass loading instead of pressure drop criteria, thus avoiding an excessive regeneration temperature. Previous works [13] have hypothesized the water drag of the particulate matter towards the inlet channels end as the main cause of the pressure drop reduction without soot mass removal. This kind of restructuring is in agreement with findings about the influence of the ash deposition pattern along the inlet channels. While ash deposition mixed with soot on the particulate layer produces a great pressure drop [14], ash deposition in the rear end region leads to lower pressure drop. Rear end deposition of ash is explained by exhaust gas drag during long-term engine operation [15]. This kind of mechanism, i.e., soot restructuring during engine operation, would also contribute to explain why different pressure drops are usually found for the same engine operating conditions and soot loading under real driving operation. In the case of pre-DPF water injection, the drag process is forced by controlled periodic injection events affecting both soot and ash.

In this work, the impact of the particulate layer characteristics along the inlet channel is discussed. The analysis is guided by a computational study carried out with a one-dimensional gas dynamic code for wall-flow DPFs [16]. Parametric studies focus on the effect of the particulate layer thickness profile (soot mass distribution) combined with porosity to identify the solution domain that would provide the pressure drop obtained experimentally and its increased rate after pre-DPF water injection. The conclusions of the theoretical analysis are also supported by visualization of monolith substrates analyzed after pre-DPF water injection and in baseline conditions. As a final step, the influence of the particulate layer restructuring on the regeneration process is also explored by means of the modeling of experimental data obtained during active regeneration.

2. Wall-Flow DPF Model

The computational study performed in this work is based on the use of a wall-flow DPF model developed in previous works, which is next briefly described. The model is integrated into OpenWAM™(Version 2.2, CMT-Motores Térmicos, Valencia, Spain), which is an open-source gas dynamics software developed at CMT-Motores Térmicos [17,18]. The model solves the conservation equations in a single pair of inlet and outlet channels assuming non-homentropic one-dimensional unsteady compressible flow:

- Mass conservation:

$$\frac{\partial \left(\rho_j F_j\right)}{\partial t} + \frac{\partial \left(\rho_j u_j F_j\right)}{\partial x} = (-1)^j 4 \left(\alpha - 2w_{pl}j\right) \rho_j u_{w_j} \tag{1}$$

- Momentum conservation:

$$\frac{\partial \left(\rho_j u_j F_j\right)}{\partial t} + \frac{\partial \left(\rho_j u_j^2 F_j + p_j F_j\right)}{\partial x} - p_j \frac{dF_j}{dx} = -F_w \mu_j u_j \tag{2}$$

- Energy conservation:

$$\frac{\partial \left(e_{0j}\rho_j F_j\right)}{\partial t} + \frac{\partial \left(h_{0j}\rho_j u_j F_j\right)}{\partial x} = q_j \rho_j F_j + (-1)^j 4 \left(\alpha - 2w_{pl}j\right) h_{0w}\rho_j u_{w_j} \tag{3}$$

- Chemical species conservation:

$$\frac{\partial \left(\rho_j Y_j F_j\right)}{\partial t} + \frac{\partial \left(\rho_j Y_j u_j F_j\right)}{\partial x} = (-1)^j 4 \left(\alpha - 2w_{pl}\right) \rho_j u_{wj} Y_j \tag{4}$$

In Equations (1)–(4), j identifies the type of monolith channel (0 = outlet, 1 = inlet) and takes into account the existence of the particulate layer. Figure 1 shows schematically the way in which the DPF channels are discretized in the axial direction, as well as the cross-section of an inlet monolith channel identifying the main geometrical characteristics of the cell.

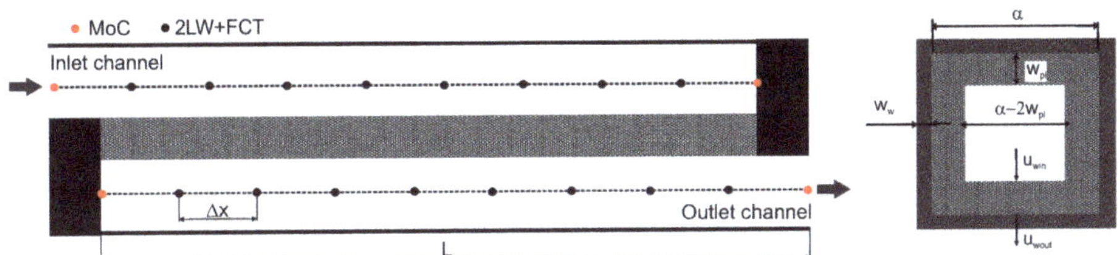

Figure 1. Scheme of the axial and cross-sections of Diesel Particulate Filter (DPF) channels.

The flow field of inlet and outlet channels is conditioned by the source terms related to flow across the porous media, which are governed by Darcy's equation applied along the particulate layer and the porous wall [16]. Thus, the filtration velocity at every axial node of the inlet channel can be calculated as a function of the pressure difference between the inlet and outlet channel, the cellular geometry and the permeability of every porous medium:

$$u_{win} = \frac{p_{in} - p_{out}}{\frac{\mu_{in}w_w}{k_w}\frac{p_{in}\left(\alpha - 2w_{pl}\right)}{p_{out}\alpha} + \frac{\mu_{in}\left(\alpha - 2w_{pl}\right)}{2k_{pl}}\ln\left(\frac{\alpha}{\alpha - 2w_{pl}}\right)} \tag{5}$$

Accordingly, the filtration velocity corresponding to the outlet channel is then obtained considering the continuity equation between the inlet and outlet interface of the porous media:

$$u_{wout} = \frac{u_{win}p_{in}\left(\alpha - 2w_{pl}\right)}{p_{out}\alpha} \tag{6}$$

The solution of the governing equations is obtained applying finite difference methods. In particular, the two-step Lax and Wendroff method (2LW) [19] adapted to porous medium channels is coupled with the Flux-Corrected Transport (FCT) technique [20]. The monolith channels are coupled to inlet and outlet volumes, which are included to account for the inertial pressure drop contribution because of flow contraction and expansion. The volumes are solved by a filling and emptying method and its connection to the monolith applying the Method of Characteristics (MoC) [21] adapted to solve the boundary conditions of inlet and outlet porous channels [22], as indicated in Figure 1.

According to Equation (5), the filtration velocity is dependent on the gas and porous media properties and the monolith meso-geometry. Both the porous wall and particulate layer permeabilities are obtained as a function of the porosity of the medium (ε), the collector unit diameter (d_c) and the slip-flow correction given by the Stokes–Cunningham Factor (SCF) as [23]:

$$k = f\left(\varepsilon\right) d_c^2 SCF \tag{7}$$

In particular, the permeability of the porous wall is calculated considering that the soot penetration is only partial [24,25], so that Equation (7) is applied to a soot-loaded porous wall thickness and to the complementary one, which is considered to be kept fully clean:

$$k_{w,e} = \frac{k_w k_{w_0}}{f_w k_{w_0} + (1 - f_w) k_w} \tag{8}$$

In Equation (8), f_w represents the fraction of porous wall thickness where soot is collected; k_{w_0} is the permeability of the clean porous wall; and k_w is the permeability of the loaded porous wall. The properties of the soot loaded porous wall are obtained from clean conditions applying the packed spherical particles theory [26]. The diameter of the collector unit as soot is collected is obtained applying Equation (9)

$$d_{c,w} = 2 \left(\frac{d_{c,w_0}^3}{8} + \frac{3 m_{s_{cell}}}{4\pi \chi \rho_{s,w}} \right)^{\frac{1}{3}}, \tag{9}$$

where the apparent density of the collected soot is defined as the product of the density of soot aggregates of mean fractal dimension ($\rho_{s,w}$) [27] and a shape factor (χ) representing the irregular deposition of the soot around the collector unit [23]. The variation of the collector unit diameter as the porous wall is loaded involves the change of the porosity. Knowing the cell unit diameter ($d_{cell,w}$) from clean conditions,

$$d_{cell,w} = \frac{d_{c,w_0}}{(1 - \varepsilon_{w_0})^{\frac{1}{3}}} \tag{10}$$

the porosity under soot loading conditions is obtained as:

$$\varepsilon_w = 1 - \frac{d_{c,w}^3}{d_{cell,w}^3}. \tag{11}$$

Besides permeability, the change in porous wall properties also determines the fluid-dynamic field in the inlet channels as the soot loading takes places and governs the variation in filtration efficiency [28]. Brownian diffusion, interception and inertial deposition mechanisms are considered to compute the filtration efficiency of an isolated collector unit. Integrating within the packed bed control volume using the pore velocity as characteristic velocity for the particles due to the proximity among collectors [29], the filtration efficiency of the porous wall yields:

$$E_{f,w} = 1 - e^{-\frac{3\eta_{DRI}(1-\varepsilon_w)w_w f_w S_c}{2\varepsilon_w d_{c,w}}} \tag{12}$$

where η_{DRI} is the filtration efficiency of an isolated collector unit due to the combination of the related collection mechanisms.

Once the transition from deep bed to cake filtration regimes is completed, the porous wall properties remain constant, and the particulate layer acts as a barrier filter. Thus, all collected soot is assumed to be deposited on the particulate layer, varying its thickness.

The amount of soot loading is determined every time-step accounting for the balance between filtrated and regenerated soot mass. Incomplete soot oxidation due to oxygen (O_2) and nitrogen dioxide (NO_2) is considered [30]. The variation of these reagents is solved separately across the particulate layer and the porous wall thickness for every reagent [31] as:

$$\frac{\partial X_n}{\partial z} = -\frac{S_p k_n \alpha_n}{u_w},$$
(13)

where subscript n identifies O_2 or NO_2, X is the molar fraction, S_p is the soot specific surface, α_n the stoichiometric coefficient and k_n the kinetic constant, which is temperature dependent according to an Arrhenius-type equation. Knowing the depletion rate of every gaseous reagent across the particulate layer and the porous wall, the amount of regenerated soot per time-step and control volume can be obtained as:

$$\Delta n_n = \Delta X_n u_w A_f C_{gas} \Delta t$$
(14)

$$m_{s,reg} = M_C \left(-\frac{\Delta n_{NO_2}}{\alpha_{NO_2}} - \frac{\Delta n_{O_2}}{\alpha_{O_2}} \right)$$
(15)

where n represents the reagent moles, A_f is the filtration area, C_{gas} is the gas concentration and M_C is the soot molecular weight, which is assumed to correspond to carbon.

The rate of heat generated by the soot oxidation is included in the thermal balance solved in the porous substrate of every control volume in which the channels are axially discretized. The heat transfer model [32] is based on a bi-dimensional discretisation of the porous media between a pair of inlet and outlet channels. Besides the regeneration heat source, the model accounts for thermal inertia, convection gas to solid heat transfer, heat conduction across the substrate both in the axial and the tangential directions, as well as radial conduction towards the external canister, whose wall temperature is also computed taking into account heat transfer to the ambient environment.

3. Experimental Setup

The main details of the experimental setup and tests shown in this work are next briefly described. Further description can be found in [12]. In this work, the results of all tests to which a wall-flow DPF was subjected aimed to characterize its response against the use of pre-DPF water injection are presented. The main characteristics of the wall-flow DPF are summarized in Table 1, where it is identified as DPF #A.

Table 1. Characteristics of wall-flow DPFs.

		#A [33]	#B [34]
Diameter (D)	(mm)	132	140
Channel length (L)	(mm)	200	230
Honeycomb cell size (α)	(mm)	1.48	1.42
Porous wall thickness (w_w)	(mm)	0.31	0.46
Cell density (σ)	(cpsi)	200	180
Porosity (ε_{w_0})	(-)	0.41	0.41
Mean pore diameter ($d_{p_{w_0}}$)	(μm)	12.1	18.55
Permeability (k_{w_0})	($\times 10^{-13}$ m^2)	2.49	5.85

The test campaign was conducted in a Euro 4 turbocharged diesel engine for passenger car use. The main characteristics of the engine are shown in Table 2. The engine was installed in a completely instrumented test cell equipped with all of the required auxiliary facilities for its operation and control. Engine speed and torque were controlled by connecting the engine to an asynchronous dynamometer both under steady-state and transient operating conditions. The engine was also instrumented with sensors to measure the main magnitudes of operation, such as temperature and mean pressure along the intake and exhaust lines, air mass flow, fuel consumption and turbocharger speed. All of these data were completed with the electronic control unit and test cell parameters.

Table 2. Basic engine characteristics.

Type	HSDI Diesel Passenger Car
Emission standards	Euro 4
Displacement	1997 cm^3
Bore	85 mm
Stroke	88 mm
Number of cylinders	4 in line
Number of valves	4 per cylinder
Compression ratio	18:1
Maximum power @ speed	100 kW @4000 rpm
Maximum torque @ speed	320 Nm @ 1750 rpm
Aftertreatment	Close-coupled DOC + Underfloor DOC-DPF

In the particular case of the DPF, its pressure drop was measured by placing two piezoresistive transducers in the inlet and outlet cones of the DPF canning. The temperature was also measured in these locations with K-type thermocouples. The water was injected at the DPF inlet by means of a simple calibrated nozzle [12]. The assessment of the filtration efficiency was performed based on the particle concentration measurements upstream and downstream of the DPF with a TSI™EEPS (Engine Exhaust Particle Sizer) spectrometer.

3.1. Tests

Several types of tests were performed in order to evaluate the pre-DPF water injection technique impact on the DPF performance. Every kind of test was repeated twice, thus comparing the test under baseline engine and DPF operation (without pre-DPF water injection) against the test in which pre-DPF water injection was applied.

Several DPF soot loading tests were performed to analyze the effects on the DPF pressure drop of the pre-DPF water injection technique under controlled conditions. In particular, a soot loading tests up to 30 g (11 g/L) in soot mass has been selected as a basis for the pressure drop and filtration efficiency study presented in this work. During all soot loading tests, the engine was run at 2500 rpm and 28% in engine load being the exhaust gas recirculation rate 16%.

After every soot loading test, the engine was subjected to different operating conditions in order to verify the pressure drop decrease and fuel economy benefit in a wider range of operation. The study covered driving cycles, motoring conditions at several engine speeds and steady-state operating conditions of low engine load. Finally, the DPF was regenerated applying an active regeneration based on an in-cylinder post-injection strategy. A regeneration process has been also selected for its modeling in the present work. In the selected test, the DPF was firstly loaded up to 60 g followed by two consecutive New European Driving Cycles (NEDC) before regeneration. This DPF loading process was repeated twice covering baseline operation and use of pre-DPF water injection, thus allowing the comparison of the corresponding regeneration processes.

4. Discussion of the Results

4.1. Pressure Drop

Figure 2 shows the DPF pressure drop during the soot loading tests up to 30 g in soot mass. According to the experimental analysis presented in [12], the restructuring of the soot deposits in the particulate layer was hypothesized as the main cause to explain the decrease in pressure drop after every injection event and the capability to limit maximum pressure drop regardless of the amount of collected soot. This hypothesis has been analyzed in this work diagnosing the main particulate layer properties. A variety of soot mass distributions in the particulate layer with different effective porosity have been computed applying the wall-flow DPF model. The porous wall has been kept saturated with a soot penetration thickness of 2%. These characteristics are based on the results obtained from the modeling of the soot loading process up to the first water injection event, which are described in [28]. Therefore, the objective has been to identify the main trends in particulate layer properties' variation that provide the pressure drop after a pre-DPF injection event.

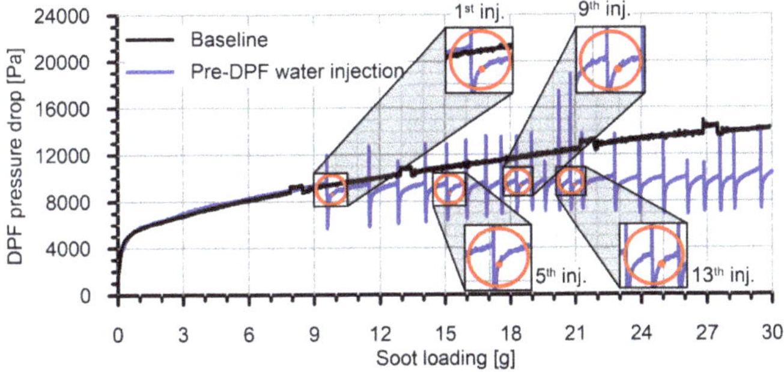

Figure 2. Soot loading test in DPF #A defining the conditions of the parametric studies.

In particular, the pressure drop at the end of the injections marked with a red circle in Figure 2 has been analyzed. The selected pressure drop value is the one after the end of the thermal transient that follows a pre-DPF water injection. This value determines the benefit in pressure drop reduction [12]. For the sake of simplicity, the properties of all inlet channels have been assumed to be the same, which implies the assumption of a homogeneous water distribution within the monolith cross-section. Therefore, the results of the parametric study must be understood as lumped representative properties of the collected soot.

According to the results presented in [28] and as the boundary condition for the modeling work, the particulate layer porosity before the first water injection event is known to be 0.65. This data were obtained assuming that the representative collector unit diameter in the particulate layer is that of the mode of the particle size distribution (69 nm). Nevertheless a range of particulate layer porosity from 0.4–0.97 has been considered in this work. It provides a wide range to analyze the possible effect of particulate layer compaction at the same time that the maximum values of porosity reported in the literature [35] have been covered.

Concerning the soot mass distribution, an increase of the particulate layer thickness has been assumed along the inlet channels till the plug end. In every axial location, the thickness of the particulate layer is assumed to be homogenous on all of the channel walls. Linear and parabolic laws to define the rate of thickness increase have been explored. In addition, the particulate layer thickness is assumed to be very thin and constant from the inlet cross-section up the a given distance from which the increasing thickness profile is imposed. This distance, which will be referred to as the onset of the soot mass distribution (δ_{pl}) from now on, has been also varied in order to determine its impact on the pressure drop. This study is based on the conclusions obtained from Scanning Electron Microscope (SEM) analysis [34]. Figure 3 shows SEM pictures corresponding to two different samples of DPF #B, which was loaded with 44.6 g. The pictures show the cross-section of an inlet channel at two different distances from the monolith entrance. One of the samples was subjected to a pre-DPF water injection. Both cases show that the penetration is very small, as concluded from experimental [25] and modeling [23] studies, just affecting the porous wall rugosity. On the one hand, the baseline sample shows a similar thickness of the soot cake both at the inlet and rear end. On the other hand, the case of the sample subjected to water injection shows an irregular thin particulate layer at the inlet region of the channel. The soot tends to be accumulated in the end region of the channel close to the plug. This is confirmed by Figure 4, which shows three pictures of the rear end (21.5 cm from monolith inlet) in different inlet channels. A clear random accumulation of soot layer fragments can be observed. Its effect can be assumed equivalent to a fast increase of the particulate layer thickness in this region, thus decreasing the inlet channel effective cross-section area.

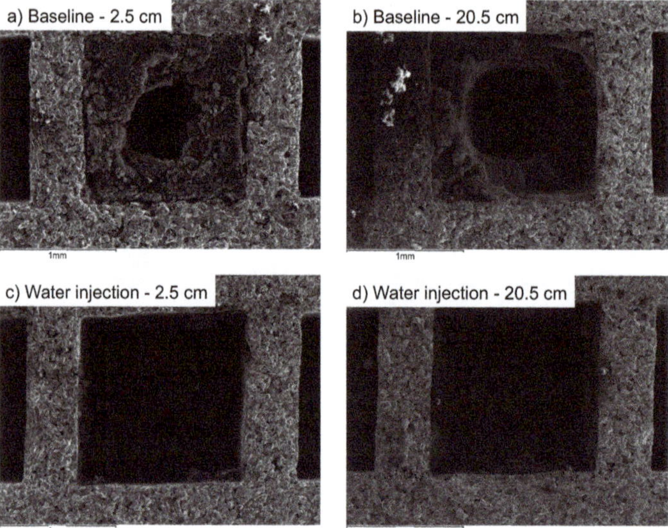

Figure 3. Scanning Electron Microscope (SEM) pictures of the cross-section of DPF #B inlet channels with the same soot loading at different locations in baseline and after water injection conditions: (**a**) baseline at 2.5 cm; (**b**) baseline at 20.5 cm; (**c**) pre-DPF water injection at 2.5 cm; (**d**) pre-DPF water injection at 20.5 cm.

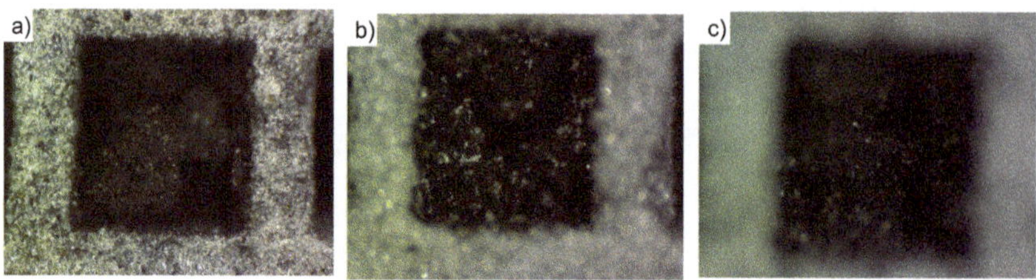

Figure 4. Pictures of soot agglomerates accumulation in the rear end of three inlet channels (21.5 cm from monolith inlet) after water injection in DPF #B.

The swept-in particulate layer porosity and onset of the soot mass distribution provide an extensive family of particulate layer structures. Figure 5a shows how the particulate layer porosity changes the soot thickness along the inlet channels. The grey dashed series represents the particulate layer thickness just before the first injection event, the particulate layer porosity being 0.65. In all remaining cases, the soot mass distribution is exactly the same, i.e., at a given point, the amount of soot is the same at any particulate layer porosity. In these examples, the onset of the soot mass distribution is at $\delta_{pl} = 0.1$ m from the inlet monolith cross-section imposing a parabolic profile. Therefore, the change in porosity is the responsible of the different cake thickness according to Equation (16):

$$w_{pl,i} = \frac{\alpha_{in} - \sqrt{\alpha_{in}^2 - \frac{m_{s,pl,i}}{\Delta x \rho_{pl}}}}{2} \tag{16}$$

where subscript *i* identifies the node of computation along the channel. The density of the particulate layer (ρ_{pl}) is a function of the carbon density and the porosity of the particulate layer:

$$\rho_{pl} = \rho_C \left(1 - \varepsilon_{pl}\right) \tag{17}$$

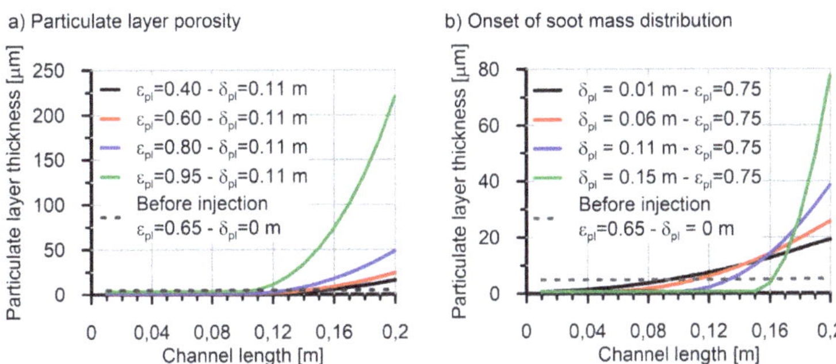

Figure 5. Effect of particulate layer porosity and soot mass distribution onset on the particulate layer thickness profile in DPF #A.

Below a porosity of 0.6, the particulate layer thickness is very thin along the whole channel. However, as the porosity increases, the thickness undergoes a faster growth, as clearly observed for 0.95 in porosity. The differences in thickness become evident from the very beginning of the soot mass distribution onset and increase towards the inlet channel rear end. In contrast to porosity, the onset of the soot mass distribution in the particulate layer, whose effect is represented in Figure 5b, gives rise to quite homogeneous cake thickness, most of the differences being concentrated in the rear end region. This is especially evident as the soot mass distribution is moved back, which produces a sharp rate of thickness increase from 0.15 m in the onset length.

The DPF pressure drop resulting from the parametric study imposing experimental inlet flow conditions and soot loading after the water injection is shown in Figure 6 for Injections 1, 5, 9 and 13. A parabolic soot mass distribution is considered in these computations. In all plots, the white line represents the solution domain corresponding to the experimental pressure drop value for every injection.

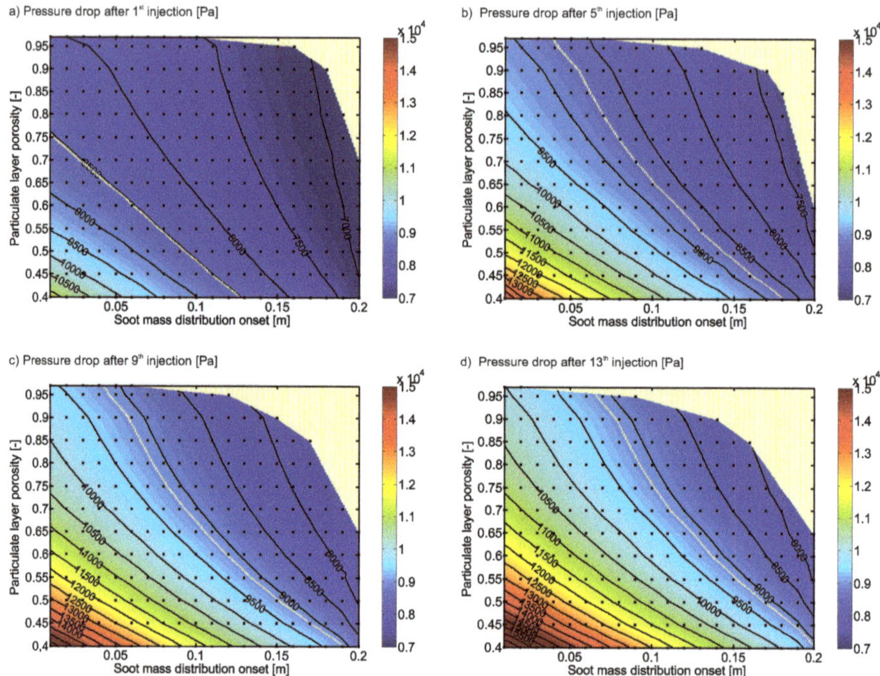

Figure 6. DPF pressure drop as a function of the particulate layer porosity and the soot mass distribution onset after different injection events in DPF #A.

As observed in all cases, moving back the soot mass distribution onset provides an almost linear decrease of the pressure drop regardless of the particulate layer porosity. This is shown in Figure 7, which presents the pressure drop dependence on the onset of the soot mass distribution for the parabolic profile and different water injections. The maximum pressure drop is always found when the onset of the particulate layer is placed close to the monolith inlet. This means that the worst loading conditions of the DPF are determined by an homogeneous particulate layer along the whole channel, which is the natural profile towards soot loading processes' convergence [28]. These results evidence the interest for soot and ash accumulation in the rear end of the inlet channels. On the other hand, given any onset for the growth of the particulate layer thickness, the pressure drop decreases as the porosity increases. These kinds of conditions would be caused by an engine operating at high mass flow, thus at medium-low temperature, thus resulting in a high Peclet number [35]. In addition, the impact is greater in homogenous soot mass distribution and as soot loading increases. Higher soot loading correlates with the injection number according to Figure 2, i.e., the 13th water injection takes place at higher soot loading conditions.

Keeping as a reference a particulate layer porosity of 0.65, the analysis of the plots in Figure 6 reveals that the experimental pressure drop obtained in DPF #A after the first pre-DPF water injection can be only attained provided that the particulate layer begins its growth several centimeters after the channel inlet (∼4.5 cm). This trend is more clear as the number of injections increases, even taking into account that the pressure drop after the thermal transient related to the injection event grows up from 8500 Pa–9200 Pa. In the 13th water injection, the growth of the particulate layer should begin at 12.5 cm from the monolith inlet, assuming that the particulate layer porosity is kept in 0.65. Therefore, the onset of the soot mass distribution is progressively moved towards the channel end as the amount of soot increases. This fashion in soot mass distribution in the particulate layer explains the need to increase the target pressure drop after the 13th water injection that was necessary to impose during the soot loading test shown in Figure 2. In order to ensure the effectiveness of the water injection technique [12], it is necessary to allow the particulate layer thickness to grow again along the inlet channels before to perform the next water injection. In addition, it is worth noting how the rear soot accumulation should be more intensive if the porosity of the particulate layer decreases as a result of a water compaction process. According to Darcy's law, this effect indicates that the permeability decrease caused by the porosity reduction has much more negative impact than the benefits brought by a thinner particulate layer. Figure 8 represents the particulate layer permeability as a function of the porosity, according to Equation (7). In contrast, the thickness is a function of the porosity, the soot distribution profile and, consequently, the axial location along the inlet channel, in agreement with Figure 5.

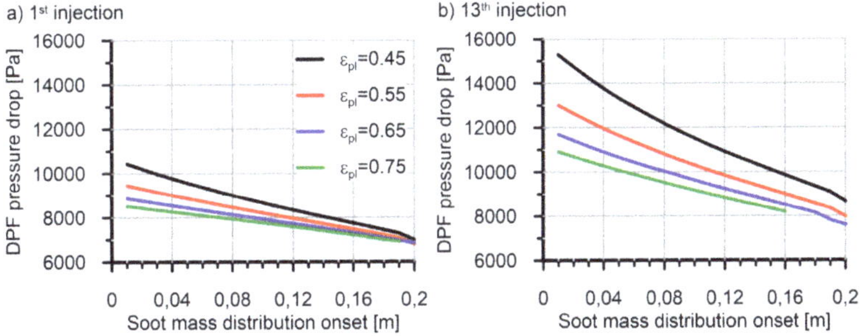

Figure 7. Impact of the onset of the soot mass distribution on the DPF pressure drop as a function of the particulate layer porosity and the soot mass loading (number of injections) in DPF #A.

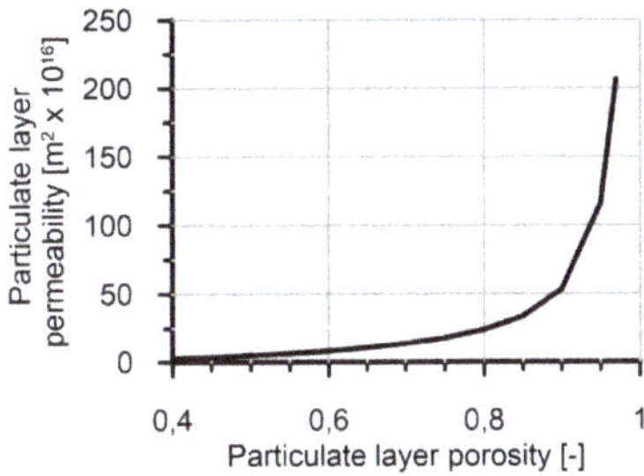

Figure 8. Particulate layer permeability as a function of the porosity.

Based on these magnitudes, the resulting pressure drop is finally defined by the filtration velocity. Figure 9a shows the filtration velocity profile along the inlet channel for a particular onset of the soot mass distribution ($\delta_{pl} = 0.11$ m) as a function of the particulate layer porosity after the first water injection. As the porosity decreases, the filtration velocity gets reduced in the initial inlet channel region. However, the filtration velocity is higher for low porosities in the rear end region. The flow tends to accumulate in the rear end of the inlet channel increasing the gas pressure because of the lower permeability related to low porosity. As a consequence, more mass flow passes across the thicker particulate layer region at higher velocity when the porosity decreases, thus leading to higher pressure drop. Complementarily, the filtration velocity profile for a particular porosity ($\varepsilon_{pl} = 0.75$) is represented in Figure 9b as a function of the soot mass distribution onset. In this case, the filtration velocity gets reduced in the thin particulate layer region as the onset is moved back. Despite an intermediate region with the highest filtration velocity, it gets the minimum value also in the rear end, where the particulate layer has the maximum thickness. This kind of profile leads progressively to the almost linear pressure drop decrease shown in Figure 7 as the soot mass distribution onset is moved back.

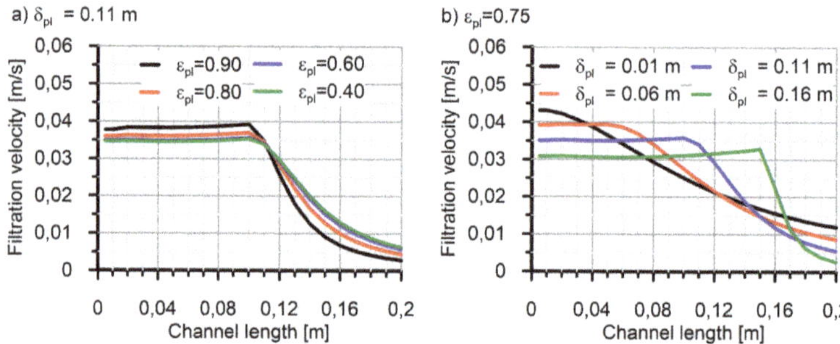

Figure 9. Filtration velocity profile as a function of the soot mass distribution onset and the particulate layer porosity after the first water injection event in DPF #A.

Several combinations of values of particulate layer porosity and onset of the soot mass distribution provide the experimental pressure drop, i.e., the white line represented in the plots of Figure 6. This is due to the influence of these variables on filtration velocity and particulate layer permeability and thickness. Figure 10 represents a set of filtration velocity profiles determined by combinations of particulate layer porosity and the onset of soot mass distribution that reproduce the experimental pressure drop after every water injection event. In all cases, the compaction of the particulate layer must be accompanied by moving back the soot mass distribution since it leads to lower filtration velocity both in the thin layer region and in the rear end region, reducing the length of the transition from fast to slow velocities. This trend is more apparent as the amount of soot and the number of injections increase, since the soot is progressively dragged towards the rear end of the inlet channels.

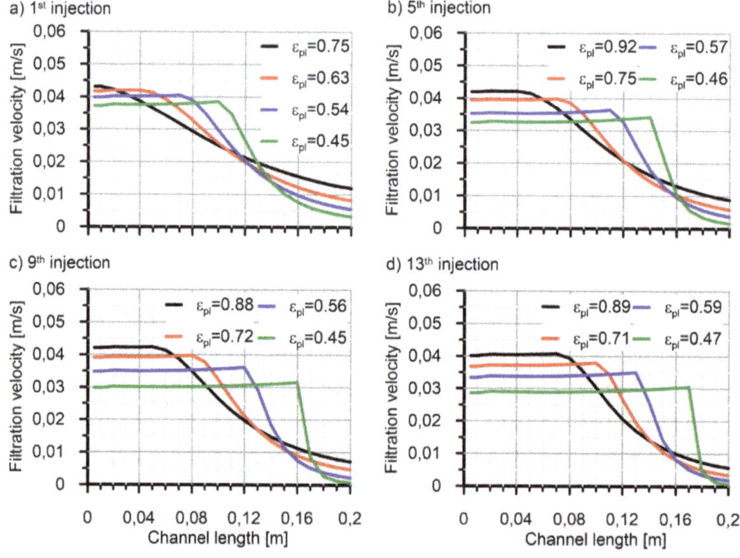

Figure 10. Filtration velocity profiles for different soot mass distribution onsets and particulate layer porosities providing the experimental pressure drop after every modeled injection event in DPF #A.

The analysis of the pressure drop after the water injection events provides general trends on the change of the particulate layer properties. In order to describe with higher detail its characteristics after every injection, several pairs of particulate layer porosity and soot mass distribution onset were selected to model the soot loading process following the injection event. Figure 11 represents the cases for the first and the 13th water injections. As shown in (a), the slope of the pressure drop after the first water injection is very sensitive to the properties of the particulate layer. In fact, the best fitting for the soot loading process is obtained for the case of no effects on the particulate layer porosity, whose reference value before injection is 0.65, just a minimum drag of the particulate layer with an onset of the soot mass distribution in 0.045 m being required. Shorter drag would have minor effects on the pressure drop increasing rate, but requiring higher porosity of the particulate layer than baseline. By contrast, noticeable drag of the particulate layer after the first injection would require great compaction of the particulate layer, leading to small porosity values and to an excessive slope of the pressure drop increase as a function of the soot loading. As the number of injections increases, the slope of the pressure drop loss sensitivity to the particulate layer properties since the onset of the soot mass distribution increases for all possible porosities. As shown in the 13th water injection, in which the effectiveness of the injection process is limited, the onset of the soot mass distribution is within a small range. Nevertheless, it is interesting to note that there is still a valid solution for the baseline porosity, which confirms that the compaction effect of the water can be considered negligible.

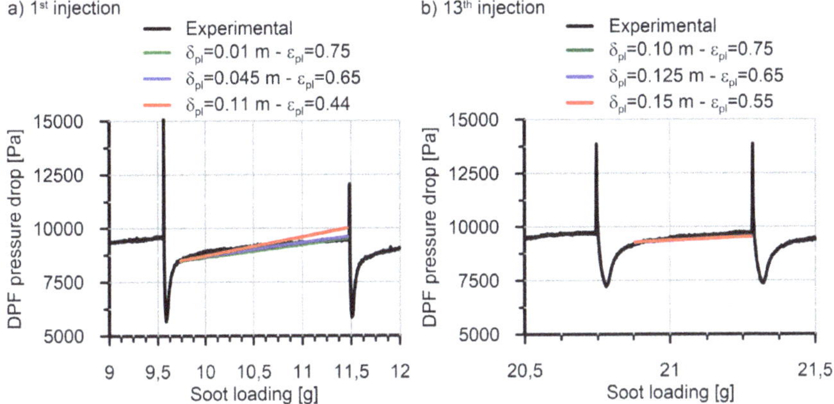

Figure 11. Influence of the soot mass distribution onset and the particulate layer porosity on the rate of increase of the pressure drop after water injection events in DPF #A.

To finish the analysis of the pressure drop reduction causes, Figure 12 represents the obtained results for the first and 13th water injections imposing a linear soot mass distribution instead of a parabolic profile. As observed, all of the general trends previously described can be considered independent of the kind of soot mass distribution law defining the particulate layer. The main difference is related to the onset of the soot mass distribution, which should be delayed in the case of a linear distribution. This is why this kind of soot distribution imposes, for a particular onset, less soot moved towards the rear end in comparison to a parabolic distribution. Nevertheless, the order of magnitude of the solutions is very similar and proves that a lumped representation of the inlet channels cross-section provides good accuracy to understand the mechanism leading to pressure drop reduction after pre-DPF water injection.

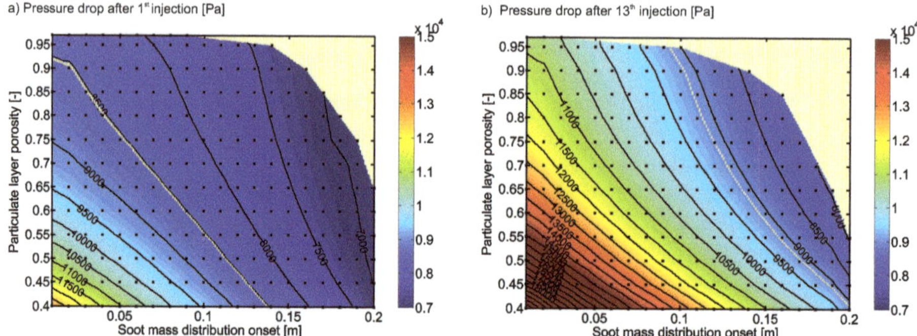

Figure 12. DPF pressure drop as a function of the particulate layer porosity and the soot mass distribution onset in DPF #A imposing a linear soot mass distribution.

4.2. Filtration Efficiency and Regeneration

Experimental data confirmed that the use of pre-DPF water injection does not affect the filtration efficiency of the DPF [36], which was between 99.35% and 99.85% (number-based filtration efficiency) during the soot loading process shown in Figure 2. According to the theoretical and visualization results shown in Section 4.1, this behavior can be justified based on the lack of variation of the soot penetration into the porous wall, keeping a saturated portion that acts as a barrier filter of high collection efficiency. It ensures high filtration efficiency in the entrance region before the onset of the particulate layer. In addition, the sparse thin particulate layer along the inlet channels and its concentration towards the end region also contribute to ensure the filtration performance of the DPF. Thus, the theoretical mass-based filtration efficiency is over 99.95% for all considered particulate layer characteristics, as shown in Figure 13 for operating conditions after the first and 13th water injections.

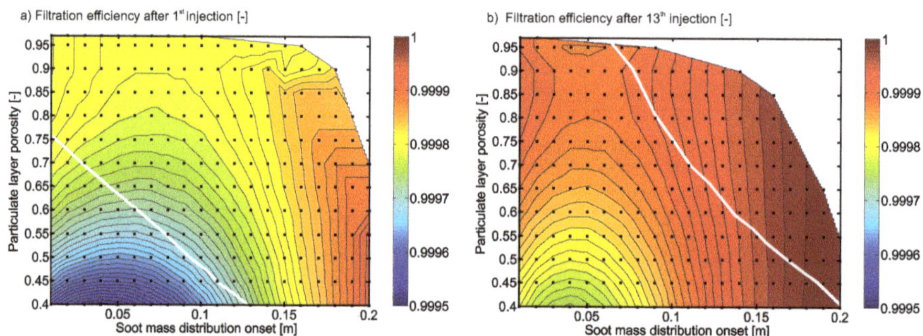

Figure 13. Filtration efficiency as a function of the particulate layer porosity and the soot mass distribution onset after different injection events.

With respect to the regeneration behavior, previous works showed the lack of relevant influence of pre-DPF water injection on passive and active regeneration processes based on pressure drop and outlet gas temperature evolution [12]. In this work, an active regeneration process have been modeled. Figure 14 shows in (a) the experimental and modeled pressure drop variation along the active regeneration, while (b) is devoted to the temperature fashion. In both tests (baseline and applying consecutive water injections), the DPF was previously loaded up to 60 g. After the soot loading, two consecutive NEDCs were performed before active regeneration. Details on the soot loading can be consulted in [12]. At the beginning of the regeneration, lower pressure drop in the case of the soot loading applying pre-DPF water injections can be clearly observed. This is even obtained with higher temperature along the monolith (higher outlet gas temperature), whose effect is the pressure drop increase (lower gas density). The interest for these tests is in fact such a difference in outlet temperature at the beginning of the regeneration. The outlet gas temperature is higher in the case of the DPF sample subjected to water injections. This is due to the fact that a longer thermal stabilization period took place in this case before the regeneration started. Being that the inlet gas temperature is equal during the two regenerations and the outlet one higher in the pre-DPF water injection sample, a faster regeneration process is expected for this last case. This result is also deduced from the pressure drop dynamics, which is properly predicted by the model for both tests. The temperature peak at the DPF outlet is almost equal in both tests, obtaining also consistent modeled results. The gas temperature within the monolith, which is shown in Figure 15, is also very similar in both regenerations. The temperature increase is faster in the pre-DPF water injection case mainly due to higher initial temperature. Nevertheless, this situation does not promote hot spots' appearance keeping the maximum temperature in the same order of magnitude as the baseline regeneration.

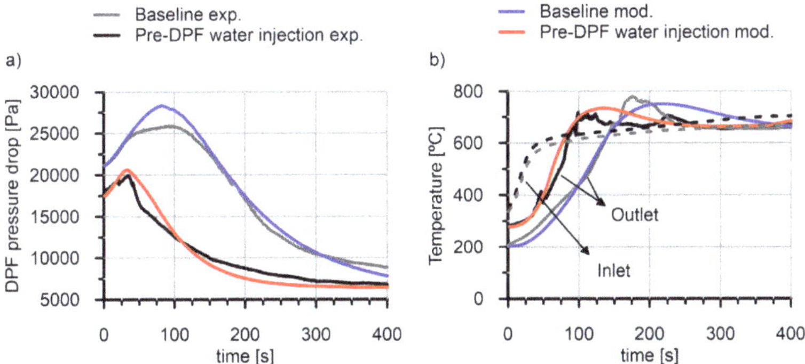

Figure 14. Caption can be rewritten as: Comparison between experimental and modeled (**a**) DPF pressure drop and (**b**) gas temperature during active regeneration: baseline vs. pre-DPF water injection.

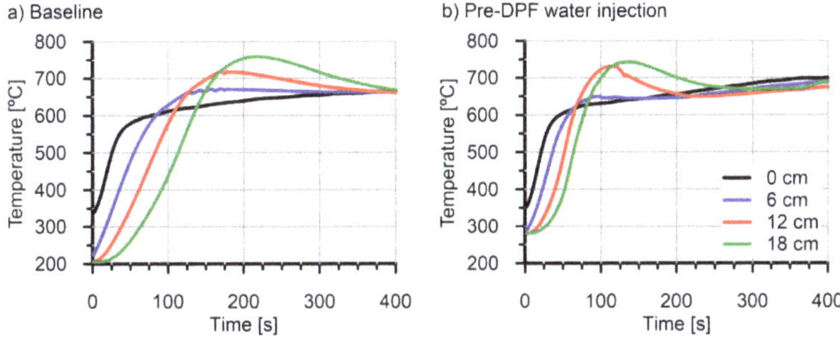

Figure 15. Gas temperature evolution during the regeneration process in different locations of the monolith.

Nevertheless, a more detailed analysis of the modeling results provides interesting insights. Figure 16a shows the evolution of the DPF soot mass during the regeneration process. The soot mass depletion rate is higher in the case of the pre-DPF water injection during the first seconds of the regeneration, but it gets progressively slower. In fact, the amount of soot still accumulated into the DPF in the case of the pre-DPF water injection is clearly higher from 200 s on. Therefore, the fast reduction in pressure drop during this test can be only explained by non-uniformity in the soot depletion rate along the monolith. Figure 16b shows the variation in particulate layer thickness during the regeneration process in different channels' locations. In the baseline regeneration, the change in particulate layer thickness follows the same dynamics at all distances observing just some delay towards the monolith rear end governed by the thermal transient. This is confirmed by the soot depletion rate and the O_2 outlet mass fraction shown in (c) and (d) of Figure 16, respectively. This behavior is governed by the filtration velocity, which is shown in Figure 16e. It is very homogenous along the inlet channels, thus leading to similar gas mass flow and dwell time.

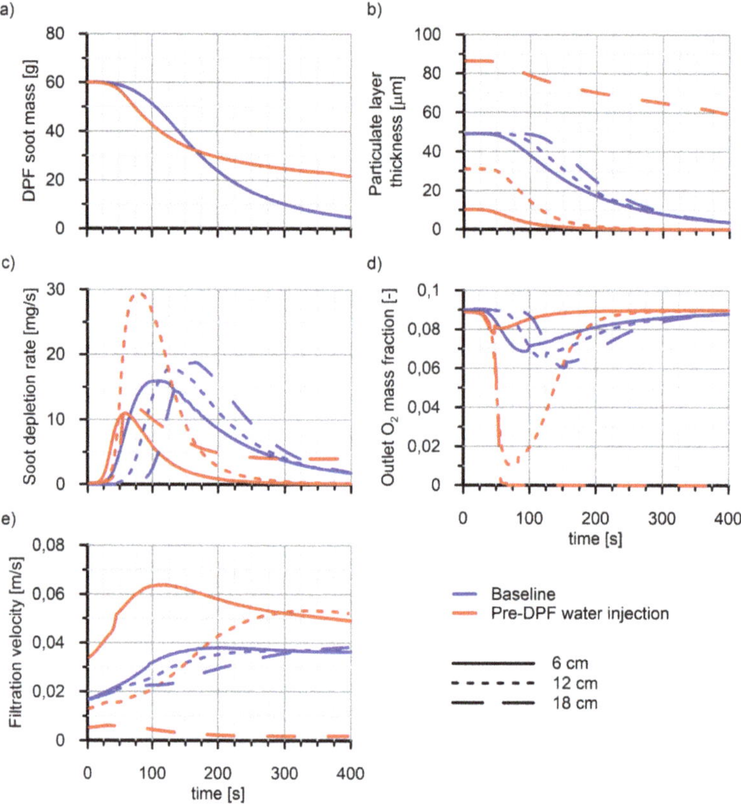

Figure 16. Comparison of the evolution of (**a**) soot mass; (**b**) particulater layer thickness; (**c**) soot depletion rate; (**d**) outlet O2 mass fraction and (**e**) filtration velocity during active regeneration: baseline vs. pre-DPF water injection.

Nevertheless, the different initial particulate layer thickness in pre-DPF water injection case conditions its subsequent reduction dynamics along the regeneration process. Figure 16b confirms how the particulate layer reaches a thin thickness and disappears very fast up to the intermediate inlet channel region. It explains the fast pressure drop reduction inducing to conclude that the regeneration is close to the end. However, the oxidation of soot in the rear end region, where most of the soot has been dragged, is very slow. It is interesting to note that according to Figure 16c, the depletion rate is very similar in the inlet and rear end region, being maximum in the middle region. This is explained by the filtration velocity. It is very high at 6 cm from the very beginning because of the thin particulate layer. Consequently, the soot depletion rate is determined by a reduced dwell time, but high gas temperature and O_2 concentration (outlet mass fraction shown in Figure 16d). As a result, the soot depletion rate is high enough to quickly remove the particulate layer. As the particulate layer is progressively removed along the inlet channels, the filtration velocities tend to coincide, as deduced from the analysis of the filtration velocity at 6 cm and 12 cm. Compared to these distances, the soot depletion rate in the rear end (18 cm), where the particulate layer is thicker, is as high as at 6 cm. However, the filtration velocity is very small. It provides high dwell time, favoring soot oxidation, but the small total amount of O_2 mass (despite high inlet concentration) produced is all consumed, thus limiting the soot depletion rate. This behavior is more penalized as soot is oxidized in the inlet and middle channel regions because the flow tends to go across the porous wall in these sections. Consequently, the filtration velocity is further reduced in the channel rear end, slowing down the final regeneration phase.

5. Conclusions

A modeling work conducted on the understanding of the causes of pre-DPF water injection effects on wall-flow DPF pressure drop, filtration and regeneration response has been presented. A one-dimensional wall-flow DPF model has been applied to explore how the soot mass distribution in the particulate layer formed on the inlet channels walls and the porosity of the particulate layer influence on pressure drop. Good agreement has been found between experimental data, optical visualization of the monolith channels and modeling of pressure drop. In fact, the modeling results of the pressure drop after pre-DPF water injections events have confirmed that the soot mass on the particulate layer must be moved back as a required condition to reduce the DPF pressure drop.

Although the onset of the soot mass distribution is dependent on the particulate layer porosity, the modeling of the soot loading after every injection event has demonstrated that the particulate layer compaction has a negligible impact. Likewise, parabolic and linear soot mass distribution profiles have been computed obtaining just minor changes to the onset of the soot mass distribution. The main outcomes on the mechanism governing the pressure drop reduction are not sensitive to this parameter.

In agreement with experimental data obtained in previous works, modeling results indicate that the filtration efficiency is not modified by pre-DPF water injection. The reason lies in the fact that the porous wall is kept saturated acting as a barrier filter. However, the conclusions on the regeneration dynamics during active processes obtained from experimental data have been contradicted by the modeled results. The results obtained in this work evidence that the soot depletion rate when the DPF has been subjected to water injection is non-uniform along the inlet channels, in contrast to the baseline case. In fact, the entering and middle regions are quickly regenerated. This makes the pressure drop rapidly decrease. However, the rear end region of the inlet channels behaves as a plug end, the filtration velocity (mass flow) being very small along this porous wall region. Consequently, the soot oxidation gets mass transfer limited, leading to a slower regeneration end phase than the baseline operating conditions. This behavior points out that optimization of active regeneration strategies should be required in order to take maximum advantage from pre-DPF water injection benefits.

Beyond particular concerns on pre-DPF water injection, the results obtained in this study also contribute to highlight the importance of the soot loading process and how engine operation history can determine the soot structure (recurrent engine stops with water condensation, highly dynamic engine operation, etc.). It would explain pressure drop variability under the same operating conditions, which can cover up the real DPF state and lead to additional engine fuel consumption penalty and substrate durability issues, especially when pressure drop-based control is considered.

Acknowledgments: This work has been partially supported by the Spanish Ministry of Economy and Competitiveness through Grant No. TRA2016-79185-R. Additionally, the Ph.D. student Enrique José Sanchis has been funded by a grant from Universitat Politècnica de València with the reference FPI-2016-S2-1355. This support is gratefully acknowledged by the authors.

Author Contributions: All authors discussed and agreed on the contents of the manuscript. Pedro Piqueras coordinated the work defining the objectives of the experimental and modeling tasks, guiding the technical discussion of the results and manuscript writing. Enrique José Sanchis contributed to the design of the theoretical study, performed the calculations, analyzed the results and collaborated in the manuscript preparation. José Ramón Serrano and Vicente Bermúdez discussed the results, contributed to manuscript writing and provided technical guidance in experimental phases from the first stages of the project development.

Conflicts of Interest: The authors declare no conflict of interest. The founding sponsors had no role in the design of the study; in the collection, analyses or interpretation of data; in the writing of the manuscript; nor in the decision to publish the results.

Abbreviations

The following abbreviations are used in this manuscript:

A_f	filtration area
C_{gas}	gas concentration
d_c	collector unit diameter
d_{cell}	cell unit diameter
d_p	mean pore diameter
D	diameter
e_0	specific stagnation internal energy
$E_{f,w}$	porous wall filtration efficiency
f_w	saturated fraction of porous wall thickness
F	area
F_w	momentum transfer coefficient for square channels
h_0	specific stagnation enthalpy
k	permeability
k_n	kinetic constant of reagent n
L	channel length
m_s	soot mass
M_C	carbon molecular weight
p	pressure
q	heat per unit of time and mass
S_p	specific surface
t	time
u	velocity
u_w	filtration velocity
w_{pl}	particulate layer thickness
w_w	porous wall thickness
x	axial dimension
X	molar fraction
Y	mass fraction
z	tangential dimension

Greek letters

α	honeycomb cell size
α_n	stoichiometric coefficient of reagent n
δ_{pl}	soot mass distribution onset
Δ	variation
ε	porosity
η_{DRI}	single sphere filtration efficiency
μ	dynamic viscosity
ρ	gas density
ρ_C	carbon density
$\rho_{s,w}$	soot packing density inside the porous wall
σ	cell density
χ	shape factor

Subscripts

cell	referred to the cell unit
e	effective
in	referred to the inlet channel
j	channel type
n	reagent (O_2, NO_2)
out	referred to the outlet channel
pl	referred to the particulate layer
reg	referred to the regeneration
w	referred to the porous wall
w_0	referred to the clean porous wall

Abbreviations

2LW	Two-step Lax and Wendroff method
DOC	Diesel Oxidation Catalyst
DPF	Diesel Particulate Filter
EEPS	Engine Exhaust Particle Sizer
FCT	Flux-Corrected Transport
HSDI	High-Speed Direct Injection
MoC	Method of Characteristics
NEDC	New European Driving Cycle
SCF	Stokes–Cunningham Factor
SCR	Selective Catalytic Reduction
SCRF	Selective Catalytic Reduction Filter
SEM	Scanning Electron Microscope

1. Johnson, T.V. Review of vehicular emissions trends. *SAE Int. J. Engines* **2015**, *8*, 1152–1167.
2. Bermúdez, V.; Serrano, J.R.; Piqueras, P.; García-Afonso, O. Assessment by means of gas dynamic modeling of a pre-turbo diesel particulate filter configuration in a turbocharged HSDI diesel engine under full-load transient operation. *Proc. Inst. Mech. Eng. D J. Automob. Eng.* **2011**, *225*, 1134–1155.
3. Subramaniam, M.N.; Joergl, V.; Keller, P.; Weber, O.; Toyoshima, T.; Vogt, C.D. Feasibility Assessment of a pre-turbo after-treatment system with a 1D modeling approach. *SAE Tech. Pap.* **2009**, doi:10.4271/2009-01-1276.
4. Luján, J.M.; Bermúdez, V.; Piqueras, P.; García-Afonso, O. Experimental assessment of pre-turbo aftertreatment configurations in a single stage turbocharged Diesel engine. Part 1: Steady-state operation. *Energy* **2015**, *80*, 599–613.
5. Luján, J.M.; Serrano, J.R.; Piqueras, P.; García-Afonso, O. Experimental assessment of a pre-turbo aftertreatment configuration in a single stage turbocharged Diesel engine. Part 2: Transient operation. *Energy* **2015**, *80*, 614–627.
6. Lee, J.H.; Michael, J.; Brown, D.B. Evaluation of Cu-based SCR/DPF technology for diesel exhaust emission control. *SAE Int. J. Fuels Lubr.* **2009**, *1*, 96–101.
7. Watling, T.C.; Ravenscroft, M.R.; Avery, G. Development, validation and application of a model for an SCR catalyst coated diesel particulate filter. *Catal. Today* **2012**, *188*, 32–41.
8. Marchitti, F.; Nova, I.; Tronconi, E. Experimental study of the interaction between soot combustion and NH3–SCR exhaust reactivity over a Cu–Zeolite SDPF catalyst. *Catal. Today* **2016**, *217*, 110–118.
9. Konstandopoulos, A.G.; Kostoglou, M. Analysis of asymmetric and variable cell geometry wall-flow particulate filters. *SAE Int. J. Fuels Lubr.* **2014**, *7*, 489–495.
10. Bollerhoff, T.; Markomanolakis, I.; Koltsakis, G. Filtration and regeneration modeling for particulate filters with inhomogeneous wall structure. *Catal. Today* **2012**, *188*, 24–31.
11. Iwata, H.; Konstandopoulos, A.G.; Nakamura, K.; Ogiso, A.; Ogyu, K.; Shibata, T.; Ohno, K. Further experimental study of asymmetric plugging layout on DPFs: Effect of Wall thickness on pressure drop and soot oxidation. *SAE Tech. Pap.* **2015**, doi:10.4271/2015-01-1016.

12. Serrano, J.R.; Bermúdez, V.; Piqueras, P.; García-Afonso, O. Pre-DPF water injection technique for loaded DPF pressure drop reduction and control. *Appl. Energy* **2015**, *140*, 234–245.

13. Serrano, J.R.; Bermúdez, V.; Piqueras, P.; Angiolini, E. Application of pre-DPF water injection technique for pressure drop limitation. *SAE Tech. Pap.* **2015**, doi:10.4271/2015-01-0985.

14. Wang, Y.; Wong, V.; Sappok, A.; Munnis, S. The sensitivity of DPF performance to the spatial distribution of ash inside DPF inlet channels. *SAE Tech. Pap.* **2013**, doi:10.4271/2013-01-1584.

15. Sappok, A.; Govani, I.; Kamp, C.; Wang, Y.; Wong, V. In-situ optical analysis of ash formation and transport in diesel particulate filters during active and passive regeneration processes. *SAE Int. J. Fuels Lubr.* **2013**, *6*, 336–349.

16. Torregrosa, A.J.; Serrano, J.R.; Arnau, F.J.; Piqueras, P. A fluid dynamic model for unsteady compressible flow in wall-flow Diesel particulate filters. *Energy* **2011**, *36*, 671–684.

17. OpenWAM Website. CMT-Motores Tèrmicos (Universitat Politècnica de València). 2017. Available online: www.openwam.org (accessed on 15 January 2017).

18. Galindo, J.; Serrano, J.R.; Arnau, F.J.; Piqueras, P. Description and analysis of a one-dimensional gas-dynamic model with Independent Time Discretization. In Proceedings of the ASME Internal Combustion Engine Division 2008 Spring Technical Conference ICES 2008, Chicago, IL, USA, 27–30 April 2008.

19. Lax, P.; Wendroff, B. Systems of conservation laws. *Commun. Pure Appl. Math.* **1964**, *17*, 381–398.

20. Serrano, J.R.; Arnau, F.J.; Piqueras, P.; García-Afonso, O. Application of the two-step Lax&Wendroff-FCT and the CE-SE method to flow transport in wall-flow monoliths. *Int. J. Comput. Math.* **2014**, *91*, 71–84.

21. Benson, R.S. *The Thermodynamics and Gas Dynamics of Internal-Combustion Engines*; Clarendon Press: New York, NY, USA, 1982; Volume 1.

22. Desantes, J.M.; Serrano, J.R.; Arnau, F.J.; Piqueras, P. Derivation of the method of characteristics for the fluid dynamic solution of flow advection along porous wall channels. *Appl. Math. Model.* **2012**, *36*, 3144–3152.

23. Serrano, J.R.; Arnau, F.J.; Piqueras, P.; Garcia-Afonso, O. Packed bed of spherical particles approach for pressure drop prediction in wall-flow DPFs (diesel particulate filters) under soot loading conditions. *Energy* **2013**, *58*, 644–654.

24. Murtagh, M.; Sherwood, D.; Socha, L. Development of a diesel particulate filter composition and its effect on thermal durability and filtration performance. *SAE Tech. Pap.* **1994**, doi:10.4271/940235.

25. Fino, D.; Russo, N.; Millo, F.; Vezza, D.; Ferrero, F.; Chianale, A. New tool for experimental analysis of diesel particulate filter loading. *Top Catal.* **2009**, *52*, 13–20.

26. Konstandopoulos, A.G.; Johnson, J.H. Wall-flow diesel particulate filters—Their pressure drop and collection efficiency. *SAE Tech. Pap.* **1989**, doi:10.4271/890405.

27. Lapuerta, M.; Ballesteros, R.; Martos, F. A method to determine the fractal dimension of diesel soot agglomerates. *J. Coll. Interface Sci.* **2006**, *303*, 149–158.

28. Serrano, J.R.; Climent, H.; Piqueras, P.; Angiolini, E. Filtration modeling in wall-flow particulate filters of low soot penetration thickness. *Energy* **2016**, *112*, 883–888.

29. Logan, B.E.; Jewett, D.G.; Arnold, R.G.; Bouwer, E.J.; O'Melia, C.R. Clarification of clean-bed filtration models. *J. Environ. Eng.* **1995**, *121*, 869–873.

30. Koltsakis, G.C.; Stamatelos, A.M. Modes of catalytic regeneration in Diesel particulate filters. *Ind. Eng. Chem.* **1997**, *36*, 4155–5165.

31. Bisset, E.J. Mathematical model of the thermal regeneration of a wall-flow monolith diesel particulate filter. *Chem. Eng. Sci.* **1984**, *39*, 1233–1244.

32. Galindo, J.; Serrano, J.R.; Piqueras, P.; García-Afonso, O. Heat transfer modeling in honeycomb wall-flow diesel particulate filters. *Energy* **2012**, *43*, 201–213.

33. Payri, F.; Broatch, A.; Serrano, J.R.; Piqueras, P. Experimental–theoretical methodology for determination of inertial pressure drop distribution and pore structure properties in wall-flow diesel particulate filters (DPFs). *Energy* **2011**, *36*, 6731–6744.

34. Angiolini, E. Contribution to the Understanding of Filtration and Pressure Drop Phenomena in Wall-Flow DPFs. Ph.D. Thesis, Universitat Politècnica de València, València, Spain, 2017.

35. Konstandopoulos, A.G.; Skaperdas, E.; Masoudi, M. Microstuctural properties of soot deposits in diesel particulate traps. *SAE Tech. Pap.* **2002**, doi:10.4271/2002-01-1015.
36. Bermúdez, V.; Serrano, J.R.; Piqueras, P.; Campos, D. Analysis of the influence of pre-DPF water injection technique on pollutants emission. *Energy* **2015**, *89*, 778–792.

 *applied sciences*

Article

The Impact of Diesel/LPG Dual Fuel on Performance and Emissions in a Single Cylinder Diesel Generator

Mustafa Aydin [1,*], Ahmet Irgin [2] and M. Bahattin Çelik [3]

[1]   Department of Mechatronics Engineering, Faculty of Technology, Karabuk University,
     78050 Karabuk, Turkey
[2]   Department of Motor Vehicles and Transportation Technologies, Kure Vocational School, Kastamonu
     University, 37900 Kastamonu, Turkey; airgin@kastamonu.edu.tr
[3]   Department of Mechanical Engineering, Faculty of Engineering, Karabuk University, 78050 Karabuk, Turkey;
     mbcelik@karabuk.edu.tr
*   Correspondence: m.aydin@karabuk.edu.tr; Tel.: +90-505-696-0616

Received: 18 April 2018; Accepted: 18 May 2018; Published: 20 May 2018

Abstract: Compared to other engines of the same size, diesel engines are more economical in addition to their ability to generate high power. For this reason, they are widely used in many fields such as industry, agriculture, transportation, electricity generation. The increasing environmental concerns and diminishing oil resources led researchers to improve fuel consumption and emissions. In this context, the usage of Liquefied Petroleum Gas (LPG) fuel in diesel engines is one of the important research subjects that has been keeping up to date. This paper investigates the effects of LPG direct injection towards the end of air inlet period on engine emissions and performance characteristics. A four-stroke, air cooled, single cylinder diesel engine was modified to direct injection of LPG for diesel/LPG dual fuel operation. An Electronic Control Unit (ECU) was designed and used to adjust LPG injection timing and duration. LPG injection rates were selected as 30%, 50% and 70% on a mass base. The test engine was operated at 3000 rpm constant engine speed under varying load conditions. Throughout the experiments, it was observed that smoke density significantly reduced on the dual-fuel operation, compared to the pure diesel operation. Carbon Monoxide (CO) and Hydrocarbon (HC) emissions decreased by 30% and 20%, respectively. Brake Specific Fuel Consumption (BSFC) decreased by 8%. Nitrogen Oxide (NO_x) emissions increased by 6% while effective efficiency increased up to 1.25%.

Keywords: diesel engine; LPG direct injection; diesel/LPG dual fuel; performance; emissions

---

1. Introduction

Because of the rapidly increasing human population around the world, mechanization and energy requirements have increased in many fields such as transportation, agriculture, electric generation and heavy industry. The diesel engine has a very high utilization rate in those fields due to producing high power at low cost when compared with other engine types in the same size range. Depending on the widespread use of diesel engines, the essential research titles related to the diesel engines are improving the performance and reducing the harmful emissions while the fuel consumption is decreasing. Because of exhausting lifetimes of fossil fuels and tightening emissions standards around the world, developing eco-friendly fuels and fuel systems for diesel engines has been keeping up its importance. In diesel engines, there are many studies to improve engine performance and reduce harmful emissions by using alternative fuels [1–4].

The primary fuel used in diesel engines is diesel as well as many liquid or gaseous fuels are used as alternative fuels. Biodiesel produced from various sources such as vegetable oil, animal fat, waste plastics and waste cooking oils, Tire Derived Fuel (TDF) obtained from waste tires, and various alcohol mixtures are preferred alternative liquid fuels [5–13]. In addition to the liquid fuels, the gas fuels such as hydrogen, Compressed Natural Gas (CNG), Diesel Methyl Ester (DME), biogas and LPG can be used in diesel engines [14–19]. Liquefied Petroleum Gas (LPG) and CNG have currently easiest accessibility and usability among the gas fuels.

LPG fuel can be used as gas or liquid phase in diesel engines. In the gas phase, it is fumigated in the air intake and the LPG-air mixture is formed in the intake manifold [19–22]. When LPG is the liquid phase, it mixes with diesel fuel under higher pressure than 0.5 MPa. Liquefied LPG is mixed with diesel fuel and pressurized by the high-pressure pump. The high-pressure pump delivers diesel/LPG blends to the injector [23–26]. The liquid phase LPG is injected either as LPG-diesel mixture by a single injector or separately by a second injector [27].

In diesel engines operating with LPG in the gas phase, the vapored LPG is taken into the cylinder with the intake air and LPG-air mixture is compressed like in a conventional diesel engine. The LPG-air mixture does not auto-ignite because of its high self-ignition temperature. A small amount of diesel fuel called pilot is injected for ignition of LPG-air mixture. The pilot diesel fuel, which is injected by the conventional diesel injection equipment, normally contributes only a small fraction of the engine output power [28]. LPG usage in the gas phase has been extensively studied. It leads to better engine performance, low particulate and smoke emissions [20,28,29].

Ciniviz [19] investigated the effect of diesel/LPG dual fuel in diesel engine on performance and emissions. They designed gas adjustment valve system in order to deliver the LPG with 30% rate to the intake manifold. Experimental results showed that the engine power, engine torque, and specific fuel consumption were improved with dual fuel run. As a result, the dual fuel operation when compared with the single operation, engine moment and power were increased 5.8%, and NO_x emission and k factor were decreased 5.9% and 1/9 respectively. Additionally, they showed that CO_2 emissions were lower than single fuel mode because CO emissions could not be converted to CO_2 in dual fuel mode.

Alam et al. [20] studied the performance and emissions of a direct injection diesel engine operated on 100% butane LPG. They added di-tertiary-butyl peroxide (DTBP) and aliphatic hydrocarbon (AHC) to the LPG fuel in order to enhance the cetane number. A stable diesel engine operation in wide engine load range was possible with the cetane improved LPG. A few different LPG blended fuels were obtained by changing the concentration of DTBP and AHC. According to experimental results, LPG and only AHC blended fuels increased the NO_x emission compared to diesel fuel operation. Experimental result showed that the thermal efficiency of LPG powered diesel engine was comparable to pure diesel fuel operation. In terms of exhaust emissions, the NO_x and smoke could be considerably reduced with using the various blend of LPG, DTBP, and AHC.

Saleh [21] focused on the effect of propane ratio changes in LPG content on emissions and performance in dual-fuel diesel engines. In the study, LPG with various propane contents was delivered to a diesel engine with EGR capability. The best engine efficiency was achieved with a 40% propane ratio. Depending on the LPG content, high butane ratio led to the decreasing of NO_x emissions and high propane content caused the reduction of CO emission as well. In a mixture of 30% butane and 70% propane content, the engine performance remained at the same level as pure diesel fuel. NO_x emissions were reduced about 27% at full load in 70% propane and 30% butane mixture.

Rao et al. [30] conducted a performance evaluation of a diesel/LPG dual fuel engine. 10%, 20%, 30%, 40% and 50% of LPG were sent to intake manifold of the single-cylinder test engine. Experiments were carried out at constant 1500 rpm engine speed at different loads. The 50% LPG fuel ratio could only be used up to 40% of the engine load. In all LPG fuel mixture ratios, the effective efficiency had increased when it was compared with pure diesel fuel. They proved that smoke emission and specific fuel consumption were reduced gradually while the LPG ratio of fuel mixture was increasing.

Ergenç and Koca [31], studied the usage of LPG in the diesel engines experimentally. They used an LPG injector mounted in the intake manifold. The measurements were performed in 10%, 20% and 25% LPG ratios. The maximum improvements in engine power, engine torque, and specific fuel consumption were achieved with 25% LPG ratio. In terms of exhaust emissions, NO_x and HC emissions decreased with all LPG ratios while CO and CO_2 emissions increased.

Lata et al. [32], investigated the influence of hydrogen and LPG addition on the efficiency and emissions of a dual-fuel diesel engine. They showed that the efficiency was increased with the usage of LPG at high loads while HC, NO_x, and smoke emissions were reduced. They observed a serious knocking in the test engine at the 70% LPG ratio. The best engine performance was obtained at 40% LPG ratio.

Mirgal et al. [33] studied on the diesel/LPG dual fuel engine. The gas phase LPG fuel was delivered to the intake manifold of a single-cylinder diesel engine. The experiments were carried out at 50% engine load and constant 1500 rpm engine speed. Experimental results were recorded at 35%, 67%, 73% and 90% LPG fuel ratio approximately. As the LPG fuel ratio increased, NO_x emissions decreased, and HC emissions increased regularly. Additionally, CO emissions increased at first and then decreased slightly. It was seen that there was a slight decrease in the cylinder pressure due to the increase in the LPG fuel ratio.

LPG in the liquid phase is mixed with diesel fuel and delivered to the high-pressure pump when the liquid phase LPG fuel is used in diesel engines. The mixture of liquid LPG and diesel fuel is injected into the cylinder with the diesel injector at high pressure. The liquid phase LPG can change to the gas phase easily when it is injected into the cylinder because of the low boiling point of LPG. The quick evaporation of the LPG in diesel/LPG blend can improve the atomization of the fuel spray. The increase of LPG content in the fuel blend will decrease the cetane number of diesel/LPG blend, and this will lead to the increase of ignition delay. In addition, the latent heat of evaporation and the Lower Heating Value (LHV) of diesel/LPG blend give a slight increase in ignition delay. Addition of the LPG in diesel fuel can accomplish a good spray atomization and contributes the fuel–air mixing process, however, the high proportion of LPG in the blends may induce engine knock or combustion noise [23–26].

Cao et al. [23] studied comparison of the LPG and diesel fuels in diesel engines. The LPG in the liquid phase and diesel fuel were transferred to the high-pressure pump as a mixture. LPG-diesel fuel mixture injected into cylinder between 180 and 260 bar pressure by a common injector. They performed experiments with %100 diesel, %10 and %30 LPG ratios. They observed that engine power and torque remained the same level in used fuel ratios under constant 1800 rpm speed. The best CO, NO_x, and smoke emissions were achieved using %30 LPG ratio. On the other hand, the best HC emission was obtained with %100 diesel fuel.

Qi et al. [24] investigated combustion and emission characteristics of diesel/LPG dual fuel in a compression ignition engine. They mixed diesel fuel and liquid phase LPG with 10%, 20%, 30%, and 40% ratios and injected with a common injector. The tests were carried out at engine speeds of 1500 rpm and 2000 rpm under between 15% and 90% engine loads. In all load and cycling conditions, it was observed that the cylinder pressure decreased while the LPG ratio increased. In terms of emissions, NO_x decreased, and HC increased at both engine speeds and at all loads when the LPG ratio was increased. The main reason is that the cylinder gas temperature is lower for blended fuel operation at the low engine load with the increase of LPG mass flow rate, and the more aromatic hydrocarbons in the LPG content, which are too stable to burn out entirely. On the other hand, a good spray can reduce blended fuel close to the cylinder chamber wall, thus HC emissions greatly reduces. There was not a great change in CO emission with pure diesel fuel operation. Smoke emissions decreased gradually, and best smoke emission was achieved by using 40% LPG ratio.

Ma et al. [26] studied the effect of diesel and diesel-propane blends on fuel injection timing in a single cylinder compression ignition engine. Propane rate, maximum heat release rate, premixed heat release, maximum cylinder gas temperature and NO_x emissions increased for the same engine speed, engine load, and injection advance while total combustion time, CO, HC, and smoke emissions reduced.

The objective of this study is to observe the effect of diesel/LPG dual fuel, on engine performance and exhaust emissions of a DI small diesel engine at constant engine speed and different loads. For this purpose, a conventional small diesel engine was converted to direct injection diesel/LPG dual fuel engine. The test engine was operated at constant 3000 rpm speed and different engine loads changing from 500 to 1500 W. For each fuel blend and engine load, the impact of LPG direct injection on a single cylinder diesel engine was investigated on engine performance (effective efficiency, fuel consumption, BSFC, EGT) and emissions (NO_x, HC, CO, smoke). Although many studies of LPG/diesel dual fuel have been found in the literature, these studies usually show that LPG is injected into the intake manifold and sent into the cylinder together with the intake air. It is believed that this study could help fill the gap in the literature about the direct injection of LPG on diesel engines.

2. Materials and Methods

2.1. Experimental Fuels

In the experiments, diesel and LPG fuels were used as test fuels. LPG content was 70% propane and 30% butane. The properties of the test fuels are given in Table 1. Diesel fuel meets requirements of EN590 for cetane index, density, and viscosity. The test engine was operated with four different fuel types as D-100 (pure diesel), LPG-30 (%70 diesel + %30 LPG) LPG-50 (%50 diesel + %50 LPG), and LPG-70 (%30 diesel + %70 LPG). The pilot diesel quantity was changed by varying the flow rate of LPG for each load condition at 3000 rpm constant speed. LPG flow rate was adjusted by changing Gasoline Direct Injection (GDI) injector trigger pulse duration via the designed ECU. The starting of GDI injector trigger pulse was set to 250 Crank Angle (CA) towards the end of the intake stroke. Thus, it was prevented that the mixing of evaporated LPG with intake air and reducing the amount of intake air. The width of the trigger pulse of the GDI injector for D-100 fuel was set to zero. In this case, it was ensured that the fuel sent into the cylinder was only diesel fuel. For LPG-30 fuel the duration of the GDI injector trigger pulse was gradually increased starting from 1 ms. The mass flows of the diesel and LPG fuels within a certain time interval were measured for each GDI injector trigger pulse duration. Thus, the fuel sent into the cylinder was set 70% diesel + 30% LPG. The same procedure was repeated for the LPG-50 and LPG-70 fuel ratios respectively. The LPG ratio in total fuel is calculated by using the following equation.

$$LPG_{ratio} = \frac{m_{LPG}}{m_{diesel} + m_{LPG}} * \%100 \tag{1}$$

Table 1. Diesel and Liquefied Petroleum Gas (LPG) fuel properties [34,35].

Fuel	Diesel	LPG
Chemical Structure	$C_{13}H_{28}$	$\%30\ C_3H_8 + \%70\ C_4H_{10}$
Lower heating value (kJ/kg)	42,500	45,908
Autoignition temperature (°C)	240	454
Boiling point (°C)	160–370	−13
Cetane index	52	8
Viscosity-kinematic @ 40 °C	3	0.32
Density (15 °C) (kg/L)	0.83	0.560
Latent heat of evaporation (MJ/kg)	0.260	0.383
Carbon/Hydrogen Ratio (C/H)	~0.47	0.39

2.2. Test Engine

In this study, a single-cylinder diesel generator was used which technical specifications were given in Table 2. The engine used in the generator is a naturally aspirated, air-cooled, single cylinder, direct injection diesel engine.

Table 2. Generator specifications.

Engine Specifications		Alternator Specifications	
Manufacturer	Katana	Manufacturer	Katana
Engine Type	Km 178 Fe	Model	KD 4500 E
Diameter × Stroke	78 × 62	Maximum Power	2.4 kVA
Cylinder Volume	296 cm^3	Power	6.3 kVA
Maximum Output Power	7.6 hp	Phase	1
Continuous Output Power	0.6 hp	Voltage	230 VAC
Engine Speed	3000 rpm	Frequency	50 Hz

2.3. Installation Modifications

The test engine was modified to run diesel/LPG dual fuel. A second injector housing was placed on the diesel engine cylinder head and the Magneti Marelli brand GDI injector was installed. LPG and diesel injector locations are shown in Figure 1. In order to control GDI injector, the angular piston position data which was obtained from an encoder coupled to the crankshaft was used. The ECU controlled the LPG injection timing and duration by evaluating the angular position data. To measure the instantaneous mass consumption of diesel and LPG fuels, electronic scales with precise measuring capability were used. A nitrogen cylinder with 200 bar operating pressure and pressure regulator were used to adjust LPG injection pressure. The liquid phase LPG that accumulates at the bottom of the LPG tank is transmitted to the GDI injector through a high-pressure fuel line. Thus, LPG injection in the liquid phase is provided via GDI injector.

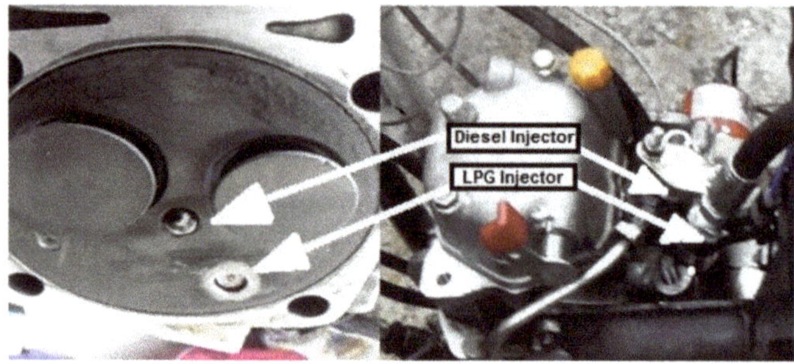

Figure 1. LPG and diesel injector locations.

An Opkon brand PRI 50 model incremental optic rotary encoder coupled to the crankshaft was used for determining the angular piston position. As shown in Figure 2, the encoder has three output channels, A, B, and Z. Channel B leads Channel A by 90 degrees phase shift. Channel Z produces a pulse in every full turn of the encoder. A and B channels are used for rotation direction and position information while Z channels are used for rpm measurement. Direction sensing is determined by generating separate pulse trains for CW and CCW direction. A pulse train is generated by checking for falling edges on A or B pulses when other pulses are high. For position measurement, one pulse train adds to a count register and the other subtracts from a count register. Thus, a precise position detection was ensured, and the LPG injector was controlled with an appropriate timing. The GDI injector was triggered either by the injection of LPG into the cylinder or just after that ignition of the diesel fuel injected into the cylinder.

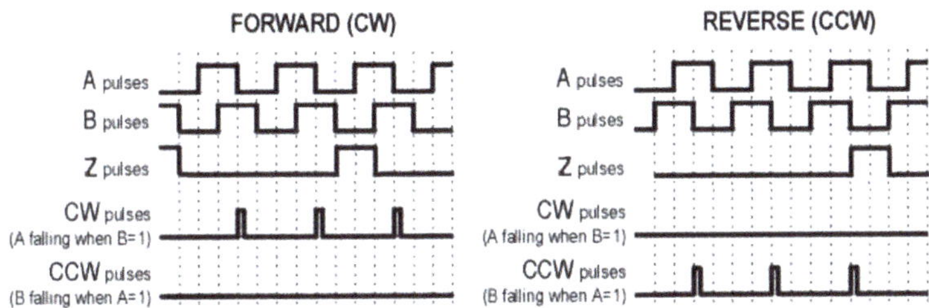

Figure 2. Encoder output signals.

2.4. Test Procedure

The schematic view of the experimental setup is shown in Figure 3. A lamp load unit was built that had 500, 750, 1000, 1250 and 1500 W lamps in order to determine engine performance and emissions under different load states at constant 3000 rpm.

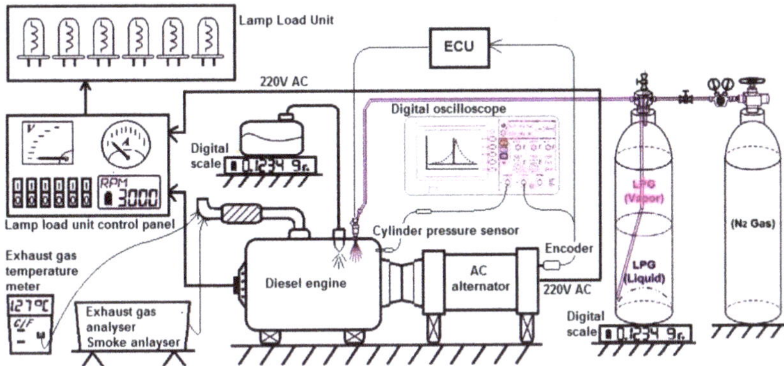

Figure 3. Schematic view of the experimental setup.

The technical specifications of SPIN ITALO PLUS exhaust emission device are given in Table 3. The exhaust emission analyzer can measure CO (% vol) with 0.001 sensibility, HC (ppm), NO_x (ppm) and opacity (%) values. The exhaust gas analyzer device measurement method is based on laser absorption spectroscopy technology, which determines the gas concentration and temperature from the optical absorption at a specific wavelength. The instantaneous exhaust gas concentration and temperature are obtained by the analysis of laser light falling on the measurement sensor after passing through the exhaust gas. A K-type thermocouple with a temperature indicator was mounted to the exhaust pipe for measurement of the Exhaust Gas Temperature (EGT).

Table 3. Exhaust gas analyzer and opacimeter technical specifications.

Measured Parameter	Measuring Method	Measuring Range	Accuracy
CO (%, v/v)	NDIR	0~9.99	0.01
HC (ppm)	NDIR	0~2500	1
NO_x (ppm)	CLD	0~2000	1
Opacity (%)	NDIR	0~99	±2
Operating temperature (°C)		5–40	
Operating Voltage (Vdc)		12	

The digital scales having 1 mg sensitive and a digital chronometer were used to measure both the diesel and LPG fuels flow by weight difference in a constant period. The specific fuel consumption was calculated. The calculated mass fuel consumption is compared to the engine power and the brake specific fuel consumption was found. The Brake Specific Fuel Consumption (BSFC) is estimated in g/(kW h) by using Equation (2).

$$\text{BSFC} = \frac{mf_{\text{diesel}} + mf_{\text{LPG}}}{P_e} \tag{2}$$

The effective efficiency was calculated with Equation (3) by considering the lower heating value and mass flow rate of both fuels diesel and LPG.

$$\eta = \frac{P_e}{(mf * \text{LHV})_{\text{diesel}} + (mf * \text{LHV})_{\text{LPG}}} \tag{3}$$

Before the data collection, the test engine was run until it reached engine operating temperature of 90 °C with experimental fuels. Experiments were conducted on stable operation modes by loading with lamp load unit. For each fuel blends (D-100, LPG-30, LPG-50, and LPG-70) used in experiments, engine performance and emissions were measured and recorded according to entire load conditions. Experimental measurements were repeated for three times for each operation point and obtained results were averaged. In the experiments, effective efficiency, EGT, fuel consumption, BSFC, and exhaust emissions (NO_x, Smoke, HC, CO) were measured.

3. Results and Discussion

The effect of experiment fuels on the effective efficiency depends on the engine load is shown in Figure 4. The highest effective efficient was achieved about 28.34% using LPG-70 fuel at 1250 W engine load. The ignition delay, uncontrolled combustion, and post-combustion phases were occurred in a short time because of better atomization of the LPG fuel in the cylinder. The lower heating value and C/H ratio of LPG lead to higher flame temperature and effective combustion. In addition, LPG direct injection does not affect intake air and maximum air charging results in the air intake stroke. This phenomenon led to decreasing the fuel consumption and increasing the effective efficiency overall LPG fuel ratios. Effective efficiency results are in concordance with other studies [21,22,36,37].

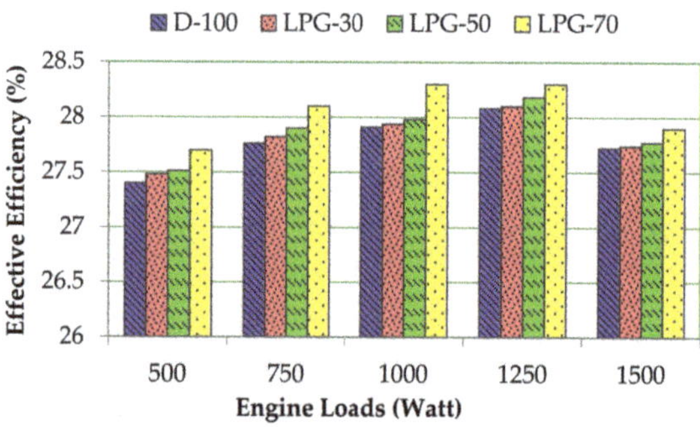

Figure 4. Effect of engine loads change on effective efficiency.

The changes in EGT depends on the engine load are shown in Figure 5. In the experiments, the amount of consumed fuel and EGT increased continuously through the increase of engine load. In addition, EGT presented incremental behavior for all fuel types. However, as the engine load increased, the increase in the LPG ratio reflected more on the EGT. At the lower engine loads ignition delay period of diesel fuel increases and also, fine ignition and combustion of LPG do not occur due to low temperature and pressure inside the combustion chamber. However, at the higher engine loads ignition delay period of diesel fuel decreases and also pressure and the temperature inside the cylinder become higher, an increase in EGT occurs due to this fine ignition and combustion of LPG. In addition, the low C/H ratio of LPG, high combustion rate, and better fuel atomization than diesel fuel have improved combustion process. These similar results were also observed by other researchers [21,34–38].

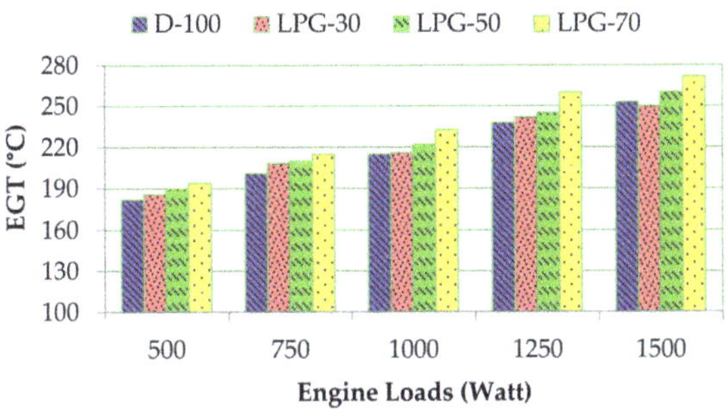

Figure 5. Effect of engine loads change on Exhaust Gas Temperature (EGT).

The effect of engine loads change on the amount of the fuel consumption is given in Figure 6. The fuel consumption was reduced by means of occurring better combustion because the lower heat value of LPG is higher than diesel fuel about 8%. This situation may be explained by increasing of the LPG flow rate increases the heat release because of the overall equivalence ratio and the combustion is inclined to be more complete, leading to high in-cylinder pressures and increased power output. The fuel consumption results are in concordance with those of other researchers [21,30,32,37,39].

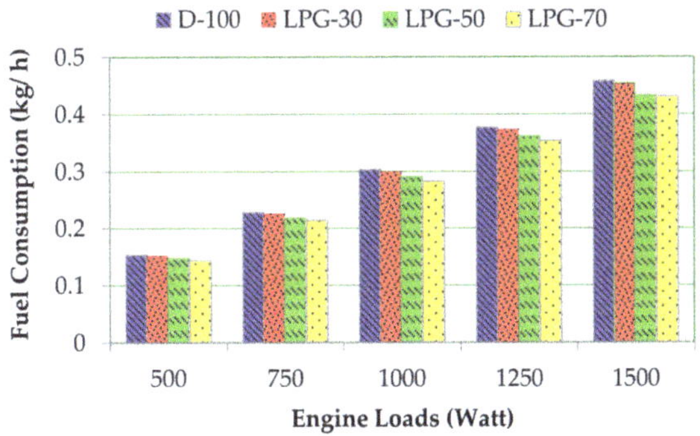

Figure 6. Effect of engine loads change on fuel consumption.

The BSFC for experiment fuels is given as a function of engine loads in Figure 7. The lowest BSFC was achieved using LPG-70 fuel at 1000-Watt engine load. When the BSFC was compared in terms of D-100 and LPG-70, it was seen that BSFC demonstrated reducing behavior by about 6%. Thus, the BSFC decreased because of the lower heat value of LPG was higher than pure diesel. Similar results were reported in other studies [21,30,40–42].

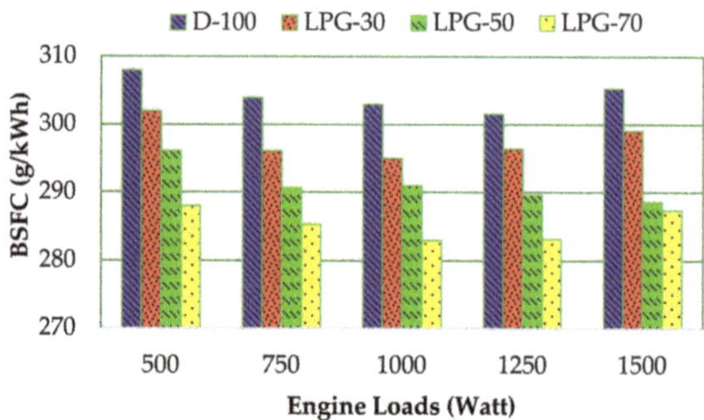

Figure 7. Variation of Brake Specific Fuel Consumption (BSFC) depending on engine loads.

The NO_x emission is given as a function of engine load in Figure 8. Although better combustion in the cylinder and increasing effective efficiency is desired in the automotive industry, they cause to rise NO_x emission directly. It is well known that NO_x emissions are the result of nitrogen reacts with oxygen at the high temperature in the cylinder. The cylinder peak pressure, the maximum heat release rate, the maximum cylinder mean gas temperature, the proportion of the premixed heat release, and NOx emission increase while increasing the propane proportion in the fuel blends [42,43]. As the load is increased, the richer mixture results in higher temperatures which in turn results in higher NO_x emissions. Due to locally rich combustion, NO_x emissions of the diesel engine are less sensitive to temperature increases resulting from increasing load [31–33]. For the diesel/LPG dual fuel engine, the maximum pressure is always higher than pure diesel fuel operation, due to the combustion and extra heat released from gaseous fuel [1]. The higher LPG ratio in dual fuel operation leads to two effects. First, the premixed combustion and the speed of flame propagation increases but the mixing-controlled combustion for the liquid fuel reduces. Second, the reduced amount of pilot injection causes the smaller size of the ignition sources, therefore increases the path that the flame needs to propagate to consume all the premixed mixture in the chamber [44]. This may be postulated to higher ignition delay of the liquid fuel and/or the lower self-ignition temperature of the gaseous fuel. The LPG fuel has lower cetane number and this can increase the ignition delay period of the fuel compared to pure diesel. In addition, the LPG has a high self-ignition temperature compared to diesel fuel. Therefore, it is expected that diesel/LPG blend would exhibit a longer delay for pure diesel fuel to ignite and the lower self-ignition temperature of LPG can increase the rate of pressure rise during the combustion. Thus, NO_x will increase due to the excessive change in pressure per unit CA and the increase of maximum temperature of the cycle.

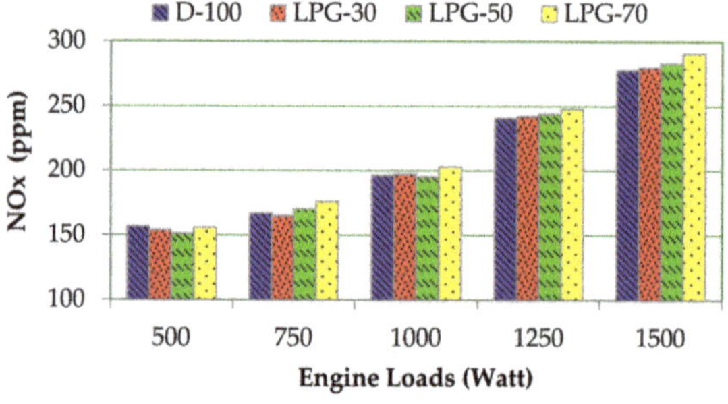

Figure 8. Variation of NO_x emission depending on engine loads.

In the case of diesel/LPG dual fuel operation, the injected more diesel fuel ratio generates bigger initial flame and this produces smoother combustion of the gaseous fuel when the load is increased. The increasing of gaseous fuel ratio which is admitted into the cylinder causes the fuel to burn at higher rates. The oxygen in the excess air taken into the cylinder at the air intake stroke combines with nitrogen due to the high burnt gas temperature and occurs increased NO_x emissions.

In the experiments, NO_x emission increased as both the engine load and the proportion of LPG were increased. This situation has occurred because LPG exhibited a better combustion reaction and higher temperature at the end of combustion than D-100. These results are in concordance with other studies [21,26,38,41,42].

Smoke emission values of the test engine are shown in Figure 9. LPG is a cleaner fuel than diesel fuel because LPG has lower carbon content and can be mixed with air homogeneously. In addition, increasing the LPG fuel ratio and the combustion temperature in the cylinder led to a decrease in smoke emission effectively. Additionally, because of LPG fuel has a lower C/H ratio than diesel fuel, it exhibits lower smoke emissions. For this reason, the diesel/LPG dual fuel operation reduces the smoke emission at all engine load conditions as compared to pure diesel operation. Diesel/LPG dual fuel keep the engine clean and smoke-free. Smoke emissions are in concordance with other studies [24,30,34,41–43,45,46].

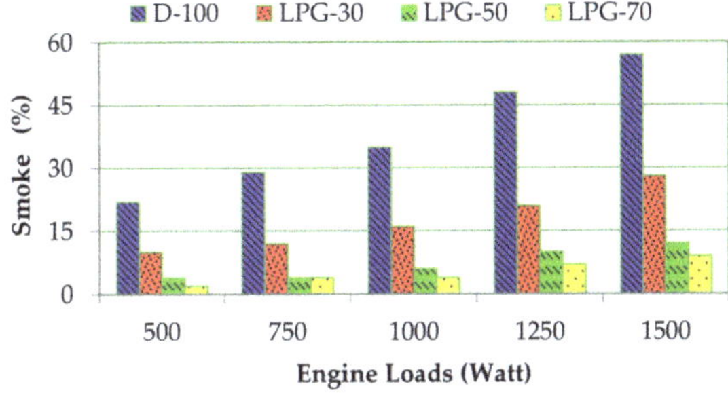

Figure 9. Variation of smoke emission depending on engine loads.

The measured HC emission values depend on engine load ratios are given in Figure 10. HC emissions using different LPG ratio decreased by means of getting better combustion reaction with diesel pilot fuel and performing more effective combustion reaction with the help of LPG. The essential factors of forming HC emissions are lower combustion temperature in the extremely poor mixture, insufficient oxygen and limited reaction time in the excessively rich mixture. The direct injection of LPG has better air charging performance and this phenomenon led to decrease unburned fuel as well. In the experimental setup, using of extremely poor and excessively rich mixtures were prevented by ECU while LPG in the liquid phase was injecting into the cylinder. Thus, HC emissions of diesel/LPG operation were reduced about 20% when it was compared with D-100 fuel. Similar results were obtained by other studies [24,26,36,46].

The measured CO emission values as a function of engine load are shown in Figure 11. In general, CO_2 emissions were produced because of full combustion of a large amount of fuel in the cylinder. On the other hand, CO emissions occurred when the remained fuel from full combustion was burned inadequately. HC emissions were formed by the non-combustible portion of the fuel. The reasons of forming CO and HC emissions are very similar. The direct injection of LPG fuel into the cylinder under high pressure by an injector caused to reach better atomization level of the LPG compared to the diesel fuel. Thus, improvement of the combustion reaction using LPG fuel provided reducing CO emissions. The better combustion and higher calorific value of the LPG improve the flame propagation and oxidation reactions that reduce the HC and CO emissions slightly. Moreover, the lower C/H ratio of LPG deceases HC and CO emissions. CO emission decreased using LPG-70 fuel about 30% than D-100 fuel. These results are in concordance with other studies [21,23,26,40].

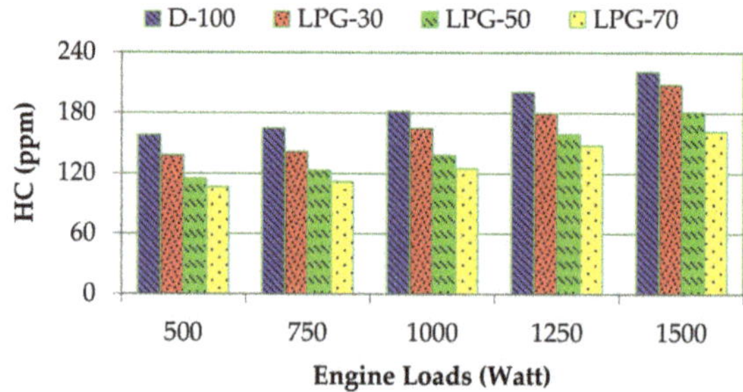

Figure 10. Variation of HC emission depending on engine loads.

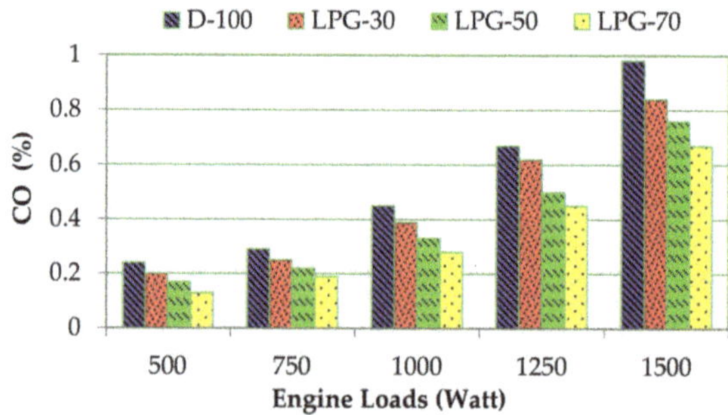

Figure 11. Variation of CO emission depending on engine loads.

4. Conclusions

In this study, the effect of direct injection of LPG into the cylinder on performance and emissions has been investigated experimentally. The tests were performed using different fuel compositions and engine loads at a constant 3000 rpm engine speed. D-100, LPG-30, LPG-50, and LPG-70 fuels were used in the experiments. The test engine was loaded with 500, 750, 1000, 1250 and 1500 W loads through the loading unit. Depending on those parameters, fuel consumption and exhaust emissions were measured. As a result;

The best effective efficiency was reached using LPG-70 fuel. It was increased about 1% than D-100 fuel. The BSFC was reduced linearly depends on the lower heat value which increased with the LPG fuel ratios. The BSFC was decreased as 6% at 1000 W engine load when the LPG-70 fuel was compared with D-100 fuel.

In general, the EGT values for all LPG fuel ratios were higher than D-100 fuel. LPG fuels produced more temperature about 10% than D-100 fuel. NO_x emissions increased about max. 4% when the engine operated with LPG fuels. CO and HC emissions were improved because of the low carbon content of LPG and increasing the in-cylinder temperature. The most emission improvement was gathered for smoke emission. The smoke emission reduced gradually by means of adding LPG content in the fuel mixture. In addition, the increasing pilot diesel fuel ratio caused the increasing smoke emission significantly. The smoke emission decreased as 70% for LPG-50 and 80% for LPG-70 when it was compared with D-100 results.

Appl. Sci. **2018**, *8*, 825

The dual fuel engine improves fuel economy and exhaust emissions, but it has some disadvantages like a second injector, separate fuel line, electronic and software modifications on fuel injection systems. Additionally, it can be reversed to a pure diesel engine easily when it is necessary.

Author Contributions: All authors designed the experimental setup, analyzed the data, discussed the results and implications and commented on the manuscript at all stages. M.A. and A.I. performed the experiments and wrote the paper. M.B.C. led the development of the paper.

Conflicts of Interest: The authors declare no conflict of interest.

Nomenclature

BSFC	Brake specific fuel consumption (g/kWh)
CA	Crank Angle
CLD	Chemiluminescence detector
CO	Carbon monoxide
CW	Clock wise
CCW	Counter clock wise
CNG	Compressed natural gas
DME	Diesel Methyl Ester
D-100	Diesel fuel
ECU	Electronic control unit
EGT	Exhaust gas temperature
HC	Hydrocarbon
GDI	Gasoline direct injection
LPG	Liquefied petroleum gas
LPG-30	%30 LPG + %70 Diesel fuel
LPG-50	%50 LPG + % 50 Diesel fuel
LPG-70	%70 LPG + %30 Diesel fuel
NDIR	Non-dispersive infrared
NOx	Nitrogen oxide
PM	Particulate matter
TDF	Tire Derived Fuel
η	Effective efficiency (%)
mf	Fuel consumption per hour (kg/h)
LVH	Lower heating value (kJ/kg)
Pe	Effective engine power (kW)

References

1. Selim, M.Y.; Radwan, M.S.; Saleh, H.E. Improving the performance of dual fuel engines running on natural gas/LPG by using pilot fuel derived from jojoba seeds. *Renew. Energy* **2008**, *33*, 1173–1185. [CrossRef]
2. Ramadhas, A.S.; Muraleedharan, C.; Jayaraj, S. Performance and emission evaluation of a diesel engine fueled with methyl esters of rubber seed oil. *Renew. Energy* **2005**, *30*, 1789–1800. [CrossRef]
3. Rakopoulos, C.D.; Antonopoulos, K.A.; Rakopoulos, D.C.; Hountalas, D.T.; Giakoumis, E.G. Comparative performance and emissions study of a direct injection diesel engine using blends of diesel fuel with vegetable oils or bio-diesels of various origins. *Energy Convers. Manag.* **2006**, *47*, 3272–3287. [CrossRef]
4. Qasim, M.; Ansari, T.M.; Hussain, M. Combustion, Performance, and Emission Evaluation of a Diesel Engine with Biodiesel Like Fuel Blends Derived from a Mixture of Pakistani Waste Canola and Waste Transformer Oils. *Energies* **2017**, *10*, 1023. [CrossRef]
5. Qi, D.H.; Chen, H.; Geng, L.M.; Bian, Y.Z. Experimental studies on the combustion characteristics and performance of a direct injection engine fueled with biodiesel/diesel blends. *Energy Convers. Manag.* **2010**, *51*, 2985–2992. [CrossRef]
6. Usta, N.; Öztürk, E.; Can, Ö.; Conkur, E.S.; Nas, S.; Con, A.H.; Topcu, M. Combustion of biodiesel fuel produced from hazelnut soapstock/waste sunflower oil mixture in a diesel engine. *Energy Convers. Manag.* **2005**, *46*, 741–755. [CrossRef]

7. Dorado, M.P.; Ballesteros, E.; Arnal, J.M.; Gomez, J.; Lopez, F.J. Exhaust emissions from a Diesel engine fueled with transesterified waste olive oil. *Fuel* **2003**, *82*, 1311–1315. [CrossRef]

8. Ghobadian, B.; Rahimi, H.; Nikbakht, A.M.; Najafi, G.; Yusaf, T.F. Diesel engine performance and exhaust emission analysis using waste cooking biodiesel fuel with an artificial neural network. *Renew. Energy* **2009**, *34*, 976–982. [CrossRef]

9. Ge, J.C.; Yoon, S.K.; Kim, M.S.; Choi, N.J. Application of canola oil biodiesel/diesel blends in a common rail diesel engine. *Appl. Sci.* **2016**, *7*, 34. [CrossRef]

10. Doğan, O.; Çelik, M.B.; Özdalyan, B. The effect of tire derived fuel/diesel fuel blends utilization on diesel engine performance and emissions. *Fuel* **2012**, *95*, 340–346. [CrossRef]

11. Rakopoulos, D.C.; Rakopoulos, C.D.; Giakoumis, E.G.; Dimaratos, A.M.; Kyritsis, D.C. Effects of butanol–diesel fuel blends on the performance and emissions of a high-speed DI diesel engine. *Energy Convers. Manag.* **2010**, *51*, 1989–1997. [CrossRef]

12. Ramos, Á.; García-Contreras, R.; Armas, O. Performance, combustion timing and emissions from a light duty vehicle at different altitudes fueled with animal fat biodiesel, GTL and diesel fuels. *Appl. Energy* **2016**, *182*, 507–517. [CrossRef]

13. Bezergianni, S.; Dimitriadis, A.; Faussone, G.C.; Karonis, D. Alternative Diesel from Waste Plastics. *Energies* **2017**, *10*, 1750. [CrossRef]

14. Saravanan, N.; Nagarajan, G.; Sanjay, G.; Dhanasekaran, C.; Kalaiselvan, K.M. Combustion analysis on a DI diesel engine with hydrogen in dual fuel mode. *Fuel* **2008**, *87*, 3591–3599. [CrossRef]

15. Liu, J.; Yang, F.; Wang, H.; Ouyang, M.; Hao, S. Effects of pilot fuel quantity on the emissions characteristics of a CNG/diesel dual fuel engine with optimized pilot injection timing. *Appl. Energy* **2013**, *110*, 201–206. [CrossRef]

16. Papagiannakis, R.G.; Hountalas, D.T. Experimental investigation concerning the effect of natural gas percentage on performance and emissions of a DI dual fuel diesel engine. *Appl. Therm. Eng.* **2003**, *23*, 353–365. [CrossRef]

17. Arcoumanis, C.; Bae, C.; Crookes, R.; Kinoshita, E. The potential of di-methyl ether (DME) as an alternative fuel for compression-ignition engines: A review. *Fuel* **2008**, *87*, 1014–1030. [CrossRef]

18. Bari, S. Effect of carbon dioxide on the performance of biogas/diesel duel-fuel engine. *Renew. Energy* **1996**, *9*, 1007–1010. [CrossRef]

19. Ciniviz, M. Dizel Motorlarında Dizel Yakıtı ve LPG Kullanımının Performans ve Emisyona Etkisi. Ph.D. Thesis, Selçuk Üniversitesi Fen Bilimleri Enstitüsü, Konya, Turkey, 2001.

20. Alam, M.; Goto, S.; Sugiyama, K.; Kajiwara, M.; Mori, M.; Konno, M.; Oyama, K. *Performance and Emissions of a DI Diesel Engine Operated with LPG and Ignition Improving Additives (No. 2001-01-3680)*; SAE Technical Paper; SAE International: Warrendale, PA, USA, 2001.

21. Saleh, H.E. Effect of variation in LPG composition on emissions and performance in a dual fuel diesel engine. *Fuel* **2008**, *87*, 3031–3039. [CrossRef]

22. Abd Alla, G.H.; Soliman, H.A.; Badr, O.A; Abd Rabbo, M.F. Effect of pilot fuel quantity on the performance of a dual fuel engine. *Energy Convers. Manag.* **2000**, *41*, 559–572. [CrossRef]

23. Cao, J.; Bian, Y.; Qi, D.; Cheng, Q.; Wu, T. Comparative investigation of diesel and mixed liquefied petroleum gas/diesel injection engines. *J. Automob. Eng.* **2004**, *218*, 557–565. [CrossRef]

24. Qi, D.H.; Bian, Y.Z.H.; Ma, Y.Z.H.; Zhang, C.H.H; Liu, S.H.Q. Combustion and exhaust emission characteristics of a compression ignition engine using liquefied petroleum gas-diesel blended fuel. *Energy Convers. Manag.* **2007**, *48*, 500–509. [CrossRef]

25. Mancaruso, E.; Marialto, R.; Sequino, L.; Vaglieco, B.M.; Cardone, M. *Investigation of the Injection Process in a Research CR Diesel Engine Using Different Blends of Propane-Diesel Fuel (No. 2015-24-2477)*; SAE Technical Paper; SAE International: Warrendale, PA, USA, 2015.

26. Ma, Z.; Huang, Z.; Li, C.; Wang, X.; Miao, H. Effects of fuel injection timing on combustion and emission characteristics of a diesel engine fueled with diesel−propane blends. *Energy Fuels* **2007**, *21*, 1504–1510. [CrossRef]

27. Boretti, A. Numerical study of the substitutional diesel fuel energy in a dual fuel diesel-LPG engine with two direct injectors per cylinder. *Fuel Process. Technol.* **2017**, *161*, 41–51. [CrossRef]

28. Ashok, B.; Ashok, S.D.; Kumar, C.R. LPG diesel dual fuel engine–A critical review. *Alex. Eng. J.* **2015**, *54*, 105–126. [CrossRef]

29. Wagemakers, A.M.L.M.; Leermakers, C.A.J. *Review on the Effects of Dual-Fuel Operation, Using Diesel and Gaseous Fuels, on Emissions and Performance (No. 2012-01-0869)*; SAE Technical Paper; SAE International: Warrendale, PA, USA, 2012.
30. Rao, G.A.; Raju, A.V.S.; Govinda Rajulu, K.; Mohan Rao, C.V. Performance evaluation of a dual fuel engine (Diesel + LPG). *Indian J. Sci. Technol.* **2010**, *3*, 235–237.
31. Ergenç, A.T.; Koca, D.Ö. PLC controlled single cylinder diesel-LPG engine. *Fuel* **2014**, *130*, 273–278. [CrossRef]
32. Lata, D.B.; Misra, A.; Medhekar, S. Effect of hydrogen and LPG addition on the efficiency and emissions of a dual fuel diesel engine. *Int. J. Hydrogen Energy* **2012**, *37*, 3084–6096. [CrossRef]
33. Mirgal, N.; Kumbhar, S.; Ibrahim, M.M.; Chellapachetty, B. Experimental investigations on LPG—Diesel dual fuel engine. *J. Chem. Pharm. Sci.* **2017**, *10*, 211–214.
34. Jothi, N.M.; Nagarajan, G.; Renganarayanan, S. Experimental studies on homogeneous charge CI engine fueled with LPG using DEE as an ignition enhancer. *Renew. Energy* **2007**, *32*, 1581–1593. [CrossRef]
35. İpragaz, A.Ş. LPG'nin Kimyasal ve Fiziksel Özellikleri. Available online: www.ipragaz.com.tr/lpg-nedir.asp (accessed on 19 October 2017).
36. Ngang, E.A.; Abbe, C.V.N. Experimental and numerical analysis of the performance of a diesel engine retrofitted to use LPG as secondary fuel. *Appl. Therm. Eng.* **2018**, *136*, 462–474. [CrossRef]
37. Vinoth, T.; Vasanthakumar, P.; Krishnaraj, J.; ArunSankar, S.K.; Hariharan, J.; Palanisamy, M. Experimental Investigation on LPG+ Diesel Fuelled Engine with DEE Ignition Improver. *Mater. Today Proc.* **2017**, *4*, 9126–9132. [CrossRef]
38. Sigar, C.P.; Soni, S.L.; Sharma, D.; Mathur, J. Effect of LPG Induction on Performance and Emission Characteristics of Bio-diesel in a CI Engine. *Energy Sources Part A* **2008**, *30*, 1451–1459. [CrossRef]
39. Ayhan, V.; Parlak, A.; Cesur, I.; Boru, B.; Kolip, A. Performance and exhaust emission characteristics of a diesel engine running with LPG. *Int. J. Phys. Sci.* **2011**, *6*, 1905–1914.
40. Tiwari, D.R.; Sinha, G.P. Performance and emission study of LPG diesel dual fuel engine. *Int. J. Eng. Adv. Technol.* **2014**, *3*, 198–203.
41. Tira, H.S.; Herreros, J.M.; Tsolakis, A.; Wyszynski, M.L. Characteristics of LPG-diesel dual fuelled engine operated with rapeseed methyl ester and gas-to-liquid diesel fuels. *Energy* **2012**, *47*, 620–629. [CrossRef]
42. Oester, U.; Wallace, J.S. *Liquid Propane Injection for Diesel Engines (No. 872095)*; SAE Technical Paper; SAE International: Warrendale, PA, USA, 1987.
43. Kajitani, S.; Chen, C.L.; Oguma, M.; Alam, M.; Rhee, K.T. *Direct Injection Diesel Engine Operated with Propane-DME Blended Fuel (No. 982536)*; SAE Technical Paper; SAE International: Warrendale, PA, USA, 1998.
44. Wattanavichien, K. Visualization of LPG-PME Dual Fuel Combustion in an IDI CI Engine. In Proceedings of the Second TSME International Conference on Mechanical Engineering, Krabi, Thailand, 19–21 October 2011.
45. Vijayabalan, P.; Nagarajan, G. Performance, emission and combustion of LPG diesel dual fuel engine using glow plug. *IJMIE* **2009**, *2*, 105–110.
46. Goto, S.; Furutani, H.; Komori, M.; Yagi, M. LPG–Diesel engine. *Int. J. Veh. Des.* **1994**, *15*, 279–290.

 *applied sciences*

Article

Dual-Fuel Combustion for Future Clean and Efficient Compression Ignition Engines

Jesús Benajes, Antonio García *, Javier Monsalve-Serrano and Vicente Boronat

CMT-Motores Térmicos, Universitat Politècnica de València, Camino de Vera s/n, Valencia 46022, Spain; jbenajes@mot.upv.es (J.B.); jamonse1@mot.upv.es (J.M.-S.); viboco1@mot.upv.es (V.B.)
* Correspondence: angarma8@mot.upv.es; Tel.: +34-963-877-659

Academic Editor: Jose Ramon Serrano
Received: 24 November 2016; Accepted: 23 December 2016; Published: 29 December 2016

Abstract: Stringent emissions limits introduced for internal combustion engines impose a major challenge for the research community. The technological solution adopted by the manufactures of diesel engines to meet the NOx and particle matter values imposed in the EURO VI regulation relies on using selective catalytic reduction and particulate filter systems, which increases the complexity and cost of the engine. Alternatively, several new combustion modes aimed at avoiding the formation of these two pollutants by promoting low temperature combustion reactions, are the focus of study nowadays. Among these new concepts, the dual-fuel combustion mode known as reactivity controlled compression ignition (RCCI) seems more promising because it allows better control of the combustion process by means of modulating the fuel reactivity depending on the engine operating conditions. The present experimental work explores the potential of different strategies for reducing the energy losses with RCCI in a single-cylinder research engine, with the final goal of providing the guidelines to define an efficient dual-fuel combustion system. The results demonstrate that the engine settings combination, piston geometry modification, and fuel properties variation are good methods to increase the RCCI efficiency while maintaining ultra-low NOx and soot emissions for a wide range of operating conditions.

Keywords: reactivity controlled compression ignition; efficiency; EURO VI emissions; dual-fuel

1. Introduction

Pollutant emissions regulations have become more and more stringent in recent years, especially for conventional diesel combustion (CDC) engines. Due to the existing trade-off between NOx and soot emissions in compression ignition (CI) engines, manufacturers had to incorporate diesel particulate filter (DPF) and selective catalyst reduction (SCR) aftertreatment systems to remove these two pollutants from the exhaust gas down to the limits imposed by the emissions regulations, such as EURO VI. The addition of these systems to the engine unit directly impacts to the production costs and also implies a higher level of complexity [1]. In addition, the efficient operation of these systems requires the consumption of some extra fluids in the exhaust line, such as urea upstream of the SCR to reduce NOx emissions and diesel fuel downstream the turbine to rise the exhaust gas temperature during the heating-up period of the engine. Moreover, the aftertreatment systems generate an extra back pressure, increasing also the in-cylinder fuel consumption.

Several combustion strategies have been investigated with the aim of reducing the costs caused by using aftertreatment systems [2]. These strategies are intended to keep the benefits of CDC operation in terms of performance and improve the engine efficiency [3]. To achieve this, many researchers have focused on the low temperature combustion (LTC) strategies [4]. As the literature demonstrates, the LTC strategies mitigate the NOx and soot formation by promoting a highly diluted in-cylinder fuel-air mixture [5] and extended premixing time between fuel and air prior to combustion [6,7], which breaks the NOx-soot trade-off with CDC strategies. Moreover, the efficiency of the engine is improved due to the heat transfer reduction, among other factors [8].

Within the LTC strategies, homogeneous charge compression ignition (HCCI) was widely investigated at institutions worldwide [9]. HCCI relies on achieving a homogenous fuel-air mixture prior to combustion, which contributes to reducing the NOx and soot formation [10]. This promotes a fast heat release at the autoignition time, which results in higher thermal efficiency than CDC if the combustion event is well-phased on the engine cycle. However, despite that HCCI seems thermodynamically attractive, the combustion onset is entirely governed by chemical kinetics, i.e., in-cylinder pressure, temperature, equivalence ratio, and fuel properties. In addition, the fast heat release during HCCI combustion leads to high pressure gradients, which provoke mechanical stress on the engine as well as excessive combustion noise as load increases. For these reasons, HCCI was found to be limited to partial loads [11].

Another LTC strategy deeply studied is the partially premixed combustion (PPC) [12]. This strategy relies on using low reactivity fuels and more delayed injection timings to overcome the HCCI weaknesses in terms of combustion control and knocking at high loads [13]. Researchers have demonstrated that using gasoline instead of diesel allows improvement of control over the heat release rate, providing low levels of NOx and soot [14]. On the other hand, several PPC studies with different octane number fuels showed that the higher the research octane number (RON), the higher the unburning problems and dispersion cycle-to-cycle, being critical for gasolines with RON higher than 91 [15]. In this sense, several authors explored the possibility of using a spark plug to improve the combustion control and reduce the unburned products at low load [16], but the advantages of PPC in terms of NOx and soot emissions were lost [17]. Considering that diesel fuel ignites easier than gasoline, Park et al. [18] decided to explore the effects of fuel blends formed by diesel and gasolines. In that work it is stated that the addition of gasoline to the blend provides a reduction in the fuel density, kinetic viscosity, and surface tension, improving the atomization process. In addition, it provides also high ignition delays enhancing a more homogeneous blend formation. Thereby, the trade-off between NOx and soot is reduced. On the other hand, it was also found that carbon monoxide (CO) and unburned hydrocarbons (HC) emissions increased substantially.

In this sense, the study performed by Bessonette et al. [19] suggested that the optimum fuel for LTC strategies depends on the engine operating conditions. In particular, at low loads a highly reactive fuel in necessary, but at high loads a low reactivity fuel is needed. Based on this statement, Inagaki et al. [20] proposed a premixed dual-fuel strategy that used two fuels of different reactivity. In particular, gasoline-like fuels were injected by a port fuel injector (PFI) and diesel was direct injected (DI) into the combustion chamber as ignition trigger. This concept allows the in-cylinder mixing of both fuels with different ratios, which allows obtaining the desired reactivity. The combustion onset was managed by varying the reactivity of the fuel blend. As suggested by Bessonete et al., a clean and efficient operation was achieved by promoting a low cetane number in-cylinder ambient at high load and a high cetane number at low load.

More recently, Kokjohn et al. [21] continued developing this technique and proposed the reactivity controlled compression ignition (RCCI) concept [22], which follows the same injection method as the dual-fuel PCI concept proposed by Inagaki. In particular, port fuel Injection (PFI) is used to inject gasoline or other low reactivity fuel and direct injection (DI) is used for diesel fuel [23], so that it is possible to adjust the fuel reactivity for the requirements at the different engine loads [24]. The low reactivity fuel is injected generating a premixed blend of fuel, fresh air, and EGR. Then, diesel is injected in one or more injection pulses. Later, when the autoignition conditions at the combustion chamber are reached, the combustion starts and drives into the burning of the premixed charge at the regions with highest reactivity fuel and propagates down the reactivity gradient, as described by Hanson et al. [25].

Several authors have demonstrated that RCCI provides ultra-low NOx and soot emissions and the same time that improves the fuel consumption versus CDC [26]. This is achieved through the combination of high gasoline fractions and optimized injection timings for diesel fuel, which was proven to be crucial to attain a clean and efficient combustion [27]. A great part of the efficiency gain as compared to CDC was demonstrated to be caused by the lower heat transfer losses, which are explained due to lower in-cylinder temperature because of the highly diluted ambient [21]. However, despite the advantages of this concept, the research community still has to face several challenges such as the low combustion efficiency at low loads [28], which reduces the potential of this concept and also can compromise the efficiency of the diesel oxidation catalyst (DOC). Continuing the investigation in this line, the present study focuses on actions that could be taken to improve the RCCI concept. For this purpose, several studies have been carried to extract some guidelines for minimizing the energy losses with RCCI. In particular, three different paths are examined in this work: the engine settings combination as a method to improve the combustion efficiency at low load, a piston bowl geometry modification to diminish the heat transfer losses, and a low reactivity fuel type variation to improve combustion.

2. Materials and Methods

2.1. Test Cell and Engine Description

A single-cylinder diesel engine, representative of commercial truck engines, has been used in this study. The major difference to the standard unit production is the hydraulic variable valve actuation (VVA) system, which allows controlling of the timing, duration, and lift of each valve independently. Detailed specifications of the engine are given in   tab:applsci-07-00036-t001.

Table 1. Single cylinder engine specifications.

Engine Data	
Engine type	Single cylinder, 4 Stroke, DI, turbocharged
Bore × Stroke [mm]	123 × 152
Connecting rod length [mm]	225
Displacement [l]	1.806
Geometric compression ratio [-]	14.4:1
Bowl Type	Open crater
Number of Valves	4
IVO	375 CAD ATDC
IVC	535 CAD ATDC
EVO	147 CAD ATDC
EVC	347 CAD ATDC

To enable RCCI operation, the engine was equipped with a double injection system, one for each fuel used. This injection hardware enabled to vary the in-cylinder fuel blending ratio and fuel mixture properties according to the engine operating conditions. To inject the diesel fuel, the engine was equipped with a common-rail flexible injection hardware which is able to perform up to five injections per cycle. The main characteristic of this hardware is its capability to amplify common-rail fuel pressure for one of the injection events by means of a hydraulic piston directly installed inside the injector. Concerning the gasoline injection, an additional fuel circuit was in-house built including a reservoir, fuel filter, fuel meter, electrically driven pump, heat exchanger, and commercially available PFI. The mentioned injector was located at the intake manifold and was specified to be able to deliver all the gasoline mass into the cylinder during the intake stroke. Consequently, the gasoline injection timing was fixed at 10 CAD after the intake valve opening (IVO) to allow the fuel to flow along 160 mm length (distance from PFI location to the intake valves seats). Accordingly, this set-up avoided fuel pooling over the intake valve and the undesirable variability introduced by this phenomenon. The main characteristics of the diesel and gasoline injectors are depicted in tab:applsci-07-00036-t002.

Table 2. Diesel and gasoline fuel injector characteristics.

Diesel Injector		Gasoline Injector	
Actuation Type	Solenoid	Actuation Type	Solenoid
Steady flow rate @ 100 bar [cm^3/s]	28.56	Steady flow rate @ 3 bar [cm^3/s]	980
Number of Holes	7	Included Spray Angle [°]	30
Hole diameter [um]	194	Fuel Pressure [bar]	5.5
Included Spray Angle [°]	142	Start of Injection [CAD aTDC]	385

To carry out the experimental tests shown in Sections 3.1 and 3.2, commercially available diesel and 98 octane number (ON) gasoline were selected as high and low reactivity fuels (HRF and LRF), respectively. Their main properties are listed in   tab:applsci-07-00036-t003.   For convenience, the properties of the fuels used in Section 3.3 will be presented there.

Table 3. Physical and chemical properties of the fuels used along the study.

Properties	Gasoline	Diesel
Density [kg/m^3] (T = 15 °C)	772	824
Viscosity [mm^2/s] (T = 40 °C)	0.37	2.8
Octane number [-]	98	-
Cetane number [-]	-	52
Lower heating value [MJ/kg]	44.54	42.65

The engine was installed in a fully instrumented test cell, with all the auxiliary facilities required for its operation and control, as it is illustrated in Figure 1.

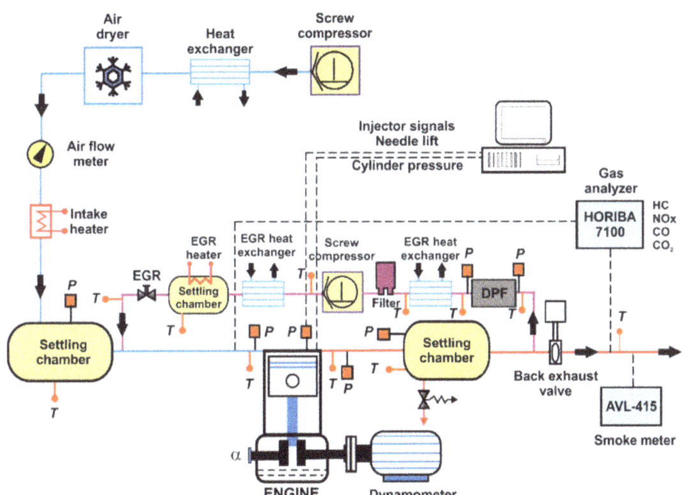

Figure 1. Test cell setup.

To achieve stable intake air conditions, a screw compressor supplied the required boost pressure before passing through an air dryer. The air pressure was adjusted within the intake settling chamber, while the intake temperature was controlled in the intake manifold after mixing with the exhaust gas recirculation (EGR) flow. The exhaust backpressure produced by the turbine in the real engine was replicated by means of a valve placed in the exhaust system, controlling the pressure in the exhaust settling chamber. Low pressure EGR was produced taking exhaust gases from the exhaust settling chamber. The EGR rate was calculated using the experimental measurement of intake and exhaust carbon dioxide (CO_2) concentration.

The concentrations of NOx, CO, unburned HC, intake, and exhaust CO_2, and oxygen (O_2) were analyzed with a five gas Horiba MEXA-7100 DEGR analyzer bench by averaging 40 s after attaining steady state operation. Smoke emissions were measured with an AVL 415S Smoke Meter and averaged between three samples of a 1 liter volume each with paper-saving mode off, providing results directly in FSN (Filter Smoke Number) units. Soot measurements of FSN were transformed into specific emissions (g/kWh) by means of the factory AVL calibration.

2.2. In-Cylinder Pressure Signal Analysis

The combustion analysis was performed with an in-house developed one-zone model named CALMEC. This combustion diagnosis tool uses the in-cylinder pressure signal and the mean variables recorded during the experiments (engine speed, coolant, oil, inlet and exhaust temperatures, air, EGR, and fuel mass flow) as its main inputs. The full description of the model can be found in [29], and the main hypotheses are enumerated next:

- The pressure is supposed homogeneous in the combustion chamber. This hypothesis is generally accepted since the fluid and the flame propagation velocity are lower than the speed of sound.
- The fluid that evolves inside the combustion chamber is considered as a mixture of air, gaseous fuel, and burned products, which are considered to evaluate the thermodynamic conditions of the mass trapped in the cylinder.
- An ideal gas behavior is assumed for the mixture that evolves in the combustion chamber. It is reasonable to accept this assumption for the air and burned products, however, it could seem inadequate for the gaseous fuel. In this sense, Lapuerta [30] compared the results from the combustion analysis model using different state equations for the gaseous fuel. The results confirmed that the differences in the mean temperature and Rate of Heat Release (RoHR) are small enough to accept the hypothesis.

During the experiments, the in-cylinder pressure was measured with a resolution of 0.2 CAD using a Kistler 61215C pressure transducer coupled with a Kistler 5011B10 charge amplifier. The pressure traces from 150 consecutive engine cycles were recorded in order to compensate the cycle-to-cycle variation during engine operation. Then, the individual pressure data of each engine cycle was smoothed using a Fourier series low-pass filter. Once filtered, the collected cycles were ensemble averaged to yield a representative cylinder pressure trace, which was used to perform the analysis. The first law of thermodynamics was applied between intake valve closing (IVC) and exhaust valve opening (EVO), considering the combustion chamber as an open system because of the blow-by and fuel injection.

The main result of the model used in this work was the Rate of Heat Release (RoHR), which is calculated as stated in Equation (1).

$$RoHR = m_{cyl} \cdot \Delta u_{cyl} + \Delta Q_W + p \cdot \Delta V - \left(\overline{h}_{f,iny} - u_{f,g} \right) \cdot \Delta m_{f,evap} + R_{cyl} \cdot T_{cyl} \cdot \Delta m_{bb} \tag{1}$$

The different terms found in the equation are explained below:

- *RoHR*: This term corresponds to the thermal energy released by the fuel assuming a constant heat power along the combustion event.
- $m_{cyl} \cdot \Delta u_{cyl}$: This is the sensible internal energy variation of the gas trapped in the control volume. As detailed in Lapuerta [31], this term is calculated by means of a specific correlation for each specie. For each temporal step, these correlations are solved as a function of the mean temperature in the control volume while pondering by the mass fraction of each specie.
- ΔQ_w: This terms accounts the heat transfer from the gas trapped in the control volume to the surrounding surfaces of the piston, liner, cylinder-head and valves. The model do not consider

the possibility of fuel impinged in the wall. The instantaneous heat transfer coefficient between the gas and the different surfaces is based on Woschni [32] with some improvements detailed in Payri et al. [33]. For the calculation of the different wall temperatures, a nodal heat transfer model was implemented [34].

- $p \cdot \Delta V$: This term represents the total work made by the gas trapped in the control volume during the calculation period. For the instantaneous calculation of the combustion chamber volume, a mechanical deformations model is considered. This submodel takes into account both the pressure made by the gas on the piston head and the inertial forces generated by the alternative movement of the masses.
- $\left(\overline{h}_{f,iny} - u_{f,g}\right) \cdot \Delta m_{f,evap}$: This term includes all the energetic considerations associated to the fuel injection process, i.e., the flow work, the heat needed to reach the evaporation temperature, and the heating-up of the vapor fuel until reaching the combustion chamber temperature.
- $R_{cyl} \cdot T_{cyl} \cdot \Delta m_{bb}$: Finally, the energy lost due to the blow-by through the piston rings is also considered. The blow-by mass is calculated using an isentropic nozzle model to simulate the gas evolution from the combustion chamber to the oil sump.

Once the RoHR is obtained, the start of combustion (SOC) is defined as the crank angle position in which the cumulated heat release reached a value of 5% (CA5) and the combustion phasing is defined as the crank angle position of 50% fuel mass fraction burned (CA50). Combustion duration was calculated as the difference between CA90 and SOC.

The ideal gas equation of state was used to calculate the mean gas temperature in the chamber. In addition, the in-cylinder pressure signal allowed obtaining the gas thermodynamic conditions in the chamber to feed the convective and radiative heat transfer models, as well as the filling and emptying model that provided the fluid-dynamic conditions in the ports, and the heat transfer flows in these elements. The convective and radiative models are linked to a lumped conductance model to calculate the wall temperatures [35].

2.3. Results Processing

A merit function [36] was used to select the best operating conditions in each one of the different studies. The merit function (MF) is defined as shown in Equation (2):

$$\text{MF} = \sum_i \max\left(0, \frac{x_i}{x_i^*} - 1\right) \tag{2}$$

where x_i is the value of the ith constrained parameter at the given conditions, x_i^* is the constraint of the ith parameter and i is the index over all the constraints.

The values of the constraints used to calculate the merit function were NOx = 0.4 g/kWh, soot = 0.01 g/kWh, and maximum pressure rise rate (PRR) of 15 bar/CAD. These limitations were aimed to fulfill ultra-low emissions while preserving the engine mechanical integrity. Thus, the contribution to the merit function from a given variable will be zero if only the measured value is less than or equal to the specified limit. When the merit function is non-zero, the contribution from each constrained parameter can be examined separately to quantify the severity of its non-compliance.

If various operating conditions fulfilled all the constraints for the same specific study (which results in a merit function value of zero), the best condition among these was considered the one that minimized the fuel consumption.

3. Paths for Increasing RCCI Efficiency

During RCCI combustion, the fuel energy is typically apportioned in the ranges shown in Figure 2, where the exact values obtained depend on the specific operating conditions and engine characteristics. The present work investigates the effectiveness of different ways to modify the energy outgoing flow path for minimizing the energy losses, thus providing the guidelines to maximize the efficiency of this combustion concept. The different paths are studied in a progressive way, so that the best solution coming from a previous path was maintained for the following steps.

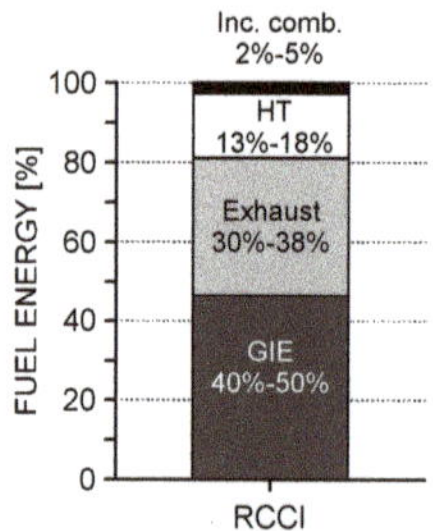

Figure 2. Typical apportionment of the fuel energy among the different outgoing energy paths during RCCI combustion.

The first path studied to increase the RCCI efficiency is the reduction of the losses associated to incomplete combustion. As seen in Figure 2, this source of energy loss is not very relevant if compared to the others, as it only represents the 5% of the input energy in the worst case. However, the incomplete combustion results in very high levels of HC and CO emissions, which can compromise the effectiveness of the aftertreatment systems and other subsystems. As the literature demonstrates, higher combustion losses occur at low loads, where lower in-cylinder pressure and temperature occur [22]. The solution explored in this work to minimize combustion losses is the optimization of the engine settings, because it is the most straightforward solution to manage the in-cylinder reactivity.

The second way proposed for increasing the gross indicated efficiency is the reduction of the heat transfer losses. In this sense, it is expected that only a portion of the recovered energy will be extracted as additional gross work, while the rest will be merely rejected as thermal exhaust loss due to the higher exhaust temperature. In any case, rising the exhaust temperature is more desirable than rejecting the heat to the coolant, because it contributes to increasing the conversion efficiency of the aftertreatment systems.

The third path to maximize RCCI efficiency relies on the fuel properties modification to look for a proper in-cylinder fuel reactivity that enhances the combustion propagation. In this sense, several studies confirm that, in order to achieve high efficiency while reducing NOx and soot emissions, the higher portion of the energy should come from the low reactivity fuel [22,26]. This fact suggests that the low reactivity fuel characteristics and its amount in the blend must have a key role on the in-cylinder reactivity, so this will be the fuel source varied during the investigation.

3.1. Engine Settings Combination

The first path for increasing the RCCI efficiency relies on the engine settings combination. To do this, it was decided to control three variables that govern the in-cylinder reactivity. Two of them were the EGR rate and intake gas temperature (T), which define the gas charge thermodynamic properties. The third variable studied was the gasoline fraction (GF), defined as the mass ratio of gasoline versus the total fuel injected, because it defines the in-cylinder fuel reactivity.

The tests were done at 1200 rpm and 7.5 bar IMEP, which corresponds to 25% load in this engine platform. This load was selected as representative as it is the minimum load considered in the World Harmonized Stationary Cycle (WHSC), proposed by the EURO VI regulation. The combustion phasing (CA50) was kept constant at +5 CAD ATDC, because this combustion phasing was found to offer a good compromise between engine-out emissions and performance. The injection strategy for diesel fuel was fixed at $-60/-50$ CAD ATDC, and the pair of values EGR + GF and T + GF were adjusted to keep the combustion phasing at the desired value.

tab:applsci-07-00036-t004 summarizes the engine settings tested with both strategies as well as those for the baseline operating condition.

Table 4. Summary of all the tests performed at 7.5 bar to evaluate the two optimization strategies.

	EGR [%]	Intake T [°C]	Intake P [bar]	Gasoline Fraction [%]	Diesel SOI [CAD ATDC]
Baseline	45			75	
	50			50	
Strategy 1 (EGR + GF)	46.5	40		61	
	43			69	
	38		1.35	78	$-60/-50$
		30		50	
Strategy 2 (T + GF)	45	40		61	
		50		71	
		60		79	

Figure 3 shows the RoHR traces for the two strategies, EGR + GF (left) and T + GF (right). Comparing both graphs, it is confirmed that EGR + GF strategy leads to higher maximum RoHR peaks than T + GF. In the case of EGR + GF, the start of the low temperature heat release (LTHR) and the LTHR peaks are equal for the four conditions tested. By contrast, the tests of T + GF denote a clear dependence of the LTHR onset on the intake temperature, showing earlier onset as temperature rises. In addition, it can be seen that the magnitude of the LTHR increases as diesel fuel mass increases (GF is reduced).

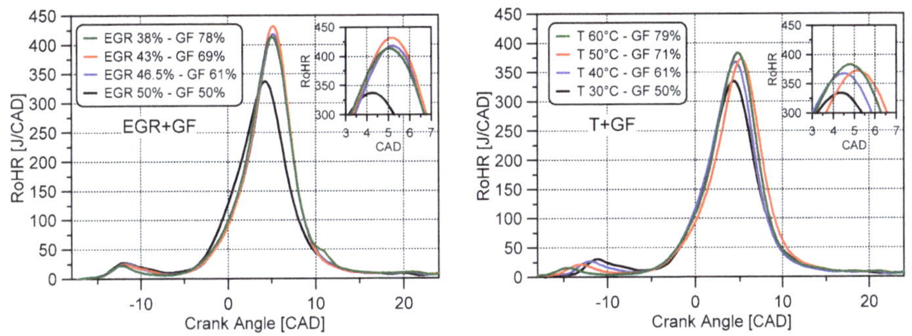

Figure 3. RoHR traces for the experiments with both strategies: EGR + GF (left) and T + GF (right) at 7.5 bar IMEP and 1200 rpm.

In order to select the best tests for each strategy, the merit function defined in Section 2.3 was applied to the complete batch of tests. Figure 4 represents the engine-out emissions and efficiency for the best tests of the two strategies as well as for the baseline condition. It is worth noting that the tests selected for both strategies are those that have near 70% of GF (EGR = 43% and T = 50 °C). The results show that both methods allow the decrease of unburned products and improve the GIE without penalizing NOx and soot emissions as compared to the baseline condition. Note that soot emissions are not depicted in the graph because all the tests provided soot levels under the detection limit of the smoke meter. Moreover, it is clear that EGR+GF provides higher gross indicated efficiency (GIE) than the T + GF strategy.

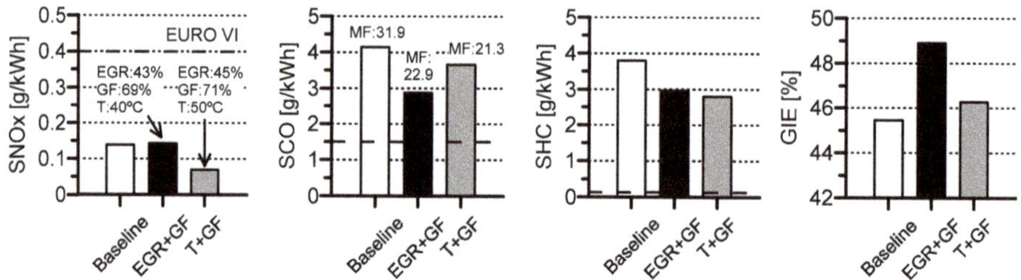

Figure 4. Comparison of the emissions and efficiency obtained with the two strategies versus the baseline condition at 7.5 bar IMEP and 1200 rpm. The merit function values for each case are shown in the second subplot.

The results of Figure 4 can be explained looking at Figure 5, where the RoHR traces and bulk gas temperature for the three cases are represented. As seen from Figure 5, the RoHR profiles of baseline and T + GF conditions are very similar, with only slightly earlier LTHR and HTHR onset in the case of T + GF due to the higher intake temperature (+10 °C). The RoHR peak of the EGR + GF strategy is 50 J/CAD greater than the other two cases, which results in higher bulk gas temperature, even using an intake temperature 10 °C lower than T + GF.

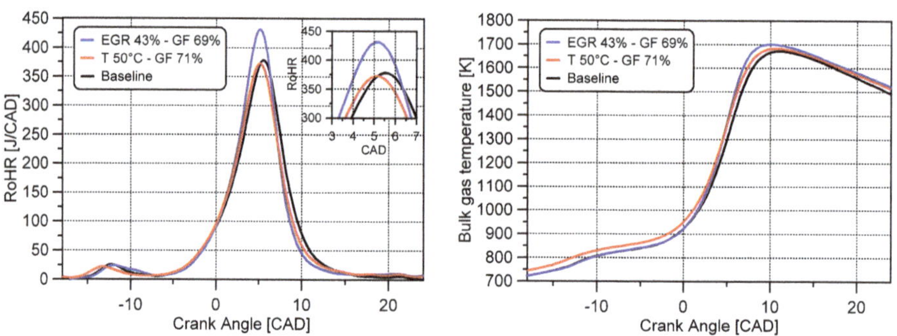

Figure 5. RoHR traces and bulk gas temperature for the best tests of both strategies and that for the baseline condition at 7.5 bar IMEP and 1200 rpm.

The reduction of the unburned products found in Figure 4 with the two strategies, should be related with the faster expansion period as compared to the baseline case. On the other hand, CO emissions are greatly influenced by the in-cylinder temperature. As it can be seen, the bulk gas temperate exceeds 1400 K in all cases, which is the threshold temperature to accelerate the CO oxidation. Since the bulk gas temperature of EGR+GF and T + GF is higher than that of the baseline operating condition, it is expected that a greater part of the in-cylinder charge experiences temperatures greater than 1400 K, explaining the CO reduction observed with the two strategies.

In the light of the results, it is possible to conclude that to achieve low emissions and high efficiency under the operating conditions tested, the EGR rate should be in the range of 43%–45%, GF 69%–71%, and intake temperature 40–50 °C. Moreover, the most efficient strategy seems to be achieved with slightly lower GF and higher oxygen concentration than the others. This knowledge will be used to define the boundary conditions in the next studies.

3.2. Piston Bowl Geometry Optimization

As a second path to increase RCCI efficiency, it was considered to modify the piston geometry for reducing the in-cylinder heat transfer losses. For this purpose, two new piston bowl geometries were defined. The volumes of the designed combustion bowl geometries were matched to keep the same geometric compression ratio as with the stock piston, which is 14.4:1. The three geometries are illustrated in the cross-sectional views presented in Figure 6.

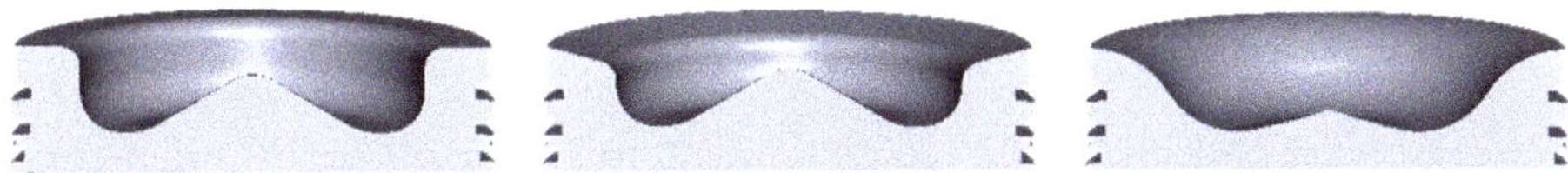

Figure 6. Cross-sectional views of the stock (left), tapered (middle), and bathtub (right) piston bowl geometries.

The first new piston was called 'tapered'. This bowl shape maintains the same central geometry than the stock piston, with slightly higher height necessary to keep the same compression ratio. The major change versus the stock piston is the tapered shape of the piston crown, which has two main purposes. The first one is to limit the heat transfer in this region through the heat transfer coefficient reduction due to the lower squish flow velocities [37]. Second, the tapered shape is intended to improve the penetration of high temperature gas into the squish and near-liner regions, where a great amount of gasoline gets trapped [22]. Moreover, this geometry resulted in near 6% less piston surface area than the stock piston which also contributes to the heat transfer reduction.

The second geometry was called 'bathtub' as it follows some of the design guidelines provided by Splitter et al. [38,39], which suggested that the efficiency of RCCI improves as the piston bowl radius increases and the bowl depth decreases. The application of these findings to the piston blanks, resulted in a piston bowl geometry with near 16% less piston surface area than the stock piston. This large reduction in surface area, in combination with the more quiescent combustion chamber created by the resulting flatter bowl geometry, should significantly reduce the heat transfer losses.

To evaluate the influence of the piston bowl geometry on RCCI combustion, a batch of parametric studies of the key variables governing the fuel reactivity stratification (diesel injection timing and GF) were performed from low to high load at 1200 rpm. tab:applsci-07-00036-t005 summarizes all the engine settings tested. Note that the effective compression ratio (CR_{eff}) was reduced down to 11:1 in the case of 18 bar IMEP to avoid the excessive knocking levels provoked by the sudden ignition of the high amount of homogeneously mixed gasoline. The CR reduction was done by shortening the intake event duration (early Miller cycle), through the VVA system.

Table 5. Summary of all the tests performed to evaluate the three piston geometries.

7.7 bar IMEP			
GF [%]	Diesel SOI_{pilot}	Diesel SOI_{main}	CR_{eff} [-]
60 to 85	−60 to −40	−40 to −15	14.4:1
13.5 bar IMEP			
60 to 70	−60 to −40	−40 to −9	14.4:1
18 bar IMEP			
50 to 75	−60 to −40	−16 to −4	11:1

The bar chart shown in Figure 7, in which all bars have the same baseline value, summarizes the best merit function results for each piston and load. The tests at 7.7 and 13.5 bar correspond to double injection, while single injection was found to be more suitable at 18 bar. As it can be seen EURO VI NOx emissions levels are reached with all the pistons. However, only the stock and tapered geometries allow working in the region under 0.01 g/kWh of soot emissions. From the second subfigure, it can be inferred that the bathtub piston has a higher sooting tendency than the other two geometries. This is thought to be related to the less prominent bowl, with reduces the turbulence near the top dead center (TDC) and worsens the air-fuel mixing process. This hypothesis is in line with the results, since the sooting tendency is more evident at high loads, where near TDC single injections are used and therefore the bowl shape plays a key role on the mixing process. In terms of CO and HC emissions, all the pistons are far from EURO VI levels. As it can be seen, the stock piston leads to less unburned products than the new geometries at low and medium loads. The inversion of this trend at high load is thought to be related with the change from double to single injection strategy. In this sense, the more prominent bowl of the stock piston confines the diesel injection and avoids increasing the reactivity in the crevice zone, which worsens the burning of the gasoline trapped in this region.

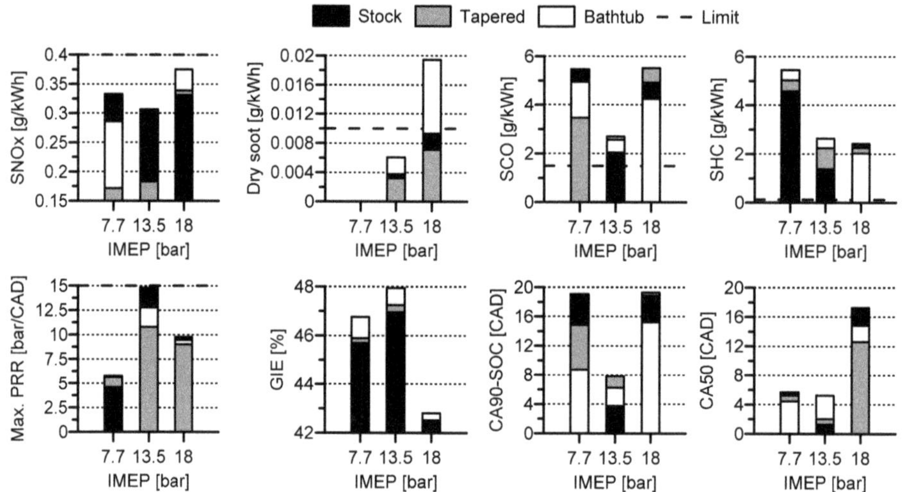

Figure 7. Best merit function results for each piston geometry at the different engine loads at 1200 rpm. Note that all bars have the same baseline value.

Maximum PRR is reduced with the two new geometries, providing a great margin to the limit at medium load conditions. The combustion duration (CA90-SOC) decreases when moving from 7.7 to 13.5 bar, and later increases. This occurs because the more reactive in-cylinder conditions at 13.5 than 7.7 bar allow introducing greater amount of gasoline, which becomes homogeneously-mixed and promotes much faster heat release, even having similar combustion phasing (CA50). At 18 bar IMEP, the injection strategy follows a single pattern to reduce the knocking levels. This leads to some diffusion combustion period, which provokes an increase of both the combustion duration and combustion phasing. Finally, as it can be inferred from the figure, the GIE has an inverse trend with the piston surface area, i.e., higher efficiency as bowl surface area reduced. This fact suggests that heat transfer reduction is contributing to the efficiency gain with the two new geometries. However, considering the excessive sooting tendency of the bathtub piston at high load and that the tapered piston does not provide a notable GIE increase versus the stock geometry, it was decided to keep the stock piston mounted in the engine for the next studies.

3.3. Fuel Autoignition Qualities Modification

The third path to maximize RCCI efficiency is based on modifying the fuel properties in order to look for a suitable combination of high reactivity fuel (HRF) and low reactivity fuel (LRF) that improves RCCI combustion.

Considering the mandatory presence of biofuels in the future context of road transport [40], the ability of ethanol to be blended with gasoline [41], and the main conclusions extracted from RCCI literature regarding ethanol [42], the low reactivity fuels selected to perform this study are E10-95, E10-98, and E20-95. In addition, a diesel fuel containing the maximum biodiesel percentage currently allowed to be distributed as a regular fuel grade in Europe, 7% by volume, has been used as high reactivity fuel during all the study. This will be referred to as diesel B7. The main characteristics of the four fuels are listed in tab:applsci-07-00036-t006. All the properties were obtained following the American Society for Testing and Materials (ASTM) standards.

Table 6. Physical and chemical properties of the fuels used along the study.

	Fuels			
Properties	Diesel B7	E10-95	E20-95	E10-98
Density [kg/m^3] (T = 15 °C)	837.9	739	745	755
Viscosity [mm^2/s] (T = 40 °C)	2.67	-	-	-
RON [-]	-	98.8	99.1	103
MON [-]	-	85.2	85.6	90
Ethanol content by volume [%]	-	9.7	19.7	9.7
Cetane number [-]	54	-	-	-
Lower heating value [MJ/kg]	42.61	41.32	40.05	41.29

As can be seen in tab:applsci-07-00036-t007, the lower heating value (LHV) of E20-95 is lower than the other two LRF. This is because of the greater ethanol content in the blend. To take into account this fact during the comparison, the premixed energy ratio (PER) is presented in Equation (3). The PER is defined as the energy ratio of the LRF versus the total delivered energy, so that it ensures that during the tests all the LRF are compared at equal conditions in terms of energy delivered to the engine.

Table 7. Summary of all the tests performed to evaluate the three combinations of HRF + LRF.

7.7 bar IMEP			
PER [%]	Diesel SOI$_{pilot}$	Diesel SOI$_{main}$	Energy/cycle [J]
65		−20 to −35	
70	−60	−20 to −35	2890
75		−20 to −30	
80		−20	

13.5 bar IMEP			
60			
65	−60	−9 to −21	4900
70			
75			

18 bar IMEP			
50			
55	-	−9 to +3	7200
60			
65			

$$PER[\%] = \frac{m_{LRF} \cdot LHV_{LRF}}{m_{HRF} \cdot LHV_{HRF} + m_{LRF} \cdot LHV_{LRF}} \tag{3}$$

As done in the previous section, to evaluate the influence of the LRF properties on RCCI combustion, a batch of parametric studies varying the diesel injection timing and PER were proposed at low, medium, and high loads at 1200 rpm. The different settings studied are depicted in tab:applsci-07-00036-t007.

Figure 8 synthetizes the results of the best tests extracted from the merit function. As it can be seen, all the fuels allow operating under the NOx and soot emissions limits. It is interesting to remark that, at 7.7 bar gross IMEP, soot emissions were under the detection limit of the smoke meter for all the fuels. This is due to the high PER used (low diesel amount injected) and the large advance of the diesel injection timing [43], which provides enough mixing time to avoid soot formation [44]. Focusing on CO and HC emissions, it is possible to state that, as a general trend, the operation with biofuels results in higher levels of unburned products than with regular diesel and gasoline fuels. Moreover, it is seen that the higher levels of both emissions are produced with E20-95 fuel, which has a large ethanol content. Considering this, the higher CO and HC levels are thought to be related with the greater enthalpy of vaporization of ethanol, which results in a cooling effect in the intake manifold and therefore lower temperature peak at TDC.

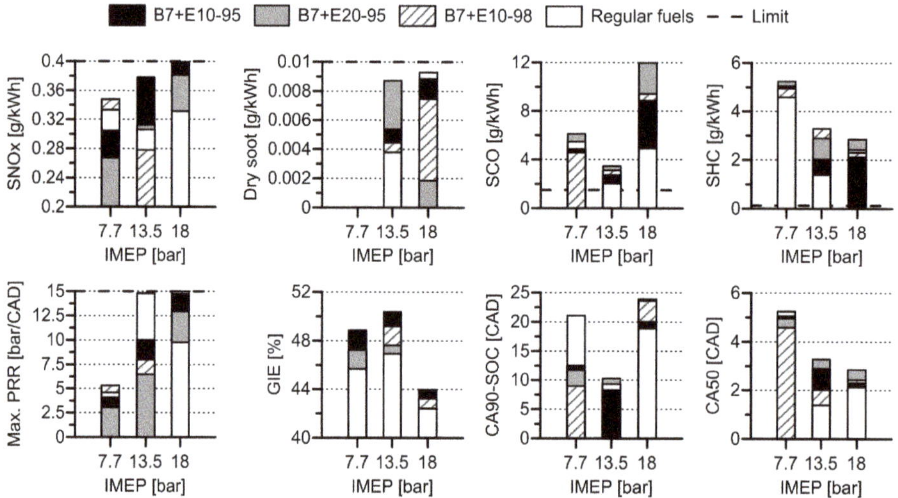

Figure 8. Best merit function results for each fuel combination at the different engine loads at 1200 rpm. Note that all bars have the same baseline value.

The maximum PRR increases as load increases. In this case, even using a single injection pattern at 18 bar, values near the 15 bar/CAD limit are observed. This is because the combustion phasing (CA50) has been reduced substantially as compared to the results shown in the previous section. On the other hand, following the same reasoning as the previous section, the combustion duration (CA90-CA10) reduces first, and later increases. Regarding engine performance, the figure shows that all the biofuels investigated allow more efficient operation than regular fuels (note that the GIE values for E20-95 and E10-98 at low load are equal). In addition, it is clear that B7 + E10-95 performs better than the rest of the fuels in all the load range, promoting the highest increase of GIE at low load (3% higher GIE than regular fuels).

4. Conclusions

The present work has investigated the effectiveness of different strategies to maximize the efficiency of RCCI combustion in a single-cylinder heavy-duty diesel engine. The solutions explored consisted of the optimization of the engine settings, modification of the piston bowl geometry, and variation of the low reactivity fuel properties. The major findings from this study are summarized as follows:

- The study regarding the engine settings combination revealed that both strategies (EGR + GF and T + GF) improve as unburned products (CO and HC) as the GIE without increasing NOx and soot emissions with respect to baseline condition. In addition, the results suggested that the EGR + GF strategy provides more efficient operation than the T + GF strategy, allowing GIE peaks near 49%.

- Maintaining equal compression ratio (14.4:1), different piston geometries were tested. The results showed that the bowl shape is a key parameter on the mixing process when delayed diesel injection timings are used, i.e., when moving towards high loads. Indeed, only two of these geometries, tapered and stock, were able to work below 0.01 g/kWh of soot levels at 18 bar IMEP, whereas the bathtub geometry produces excessively high levels of soot due to the absence of a prominent bowl shape, which reduces the turbulence at TDC. The stock piston produced lower unburned HC than the other geometries. However, the two new geometries showed improved thermal efficiency versus the stock piston, findings that are in line with the heat transfer reduction due to the lower bowl surface area.

- Following the path to improve the RCCI concept, fuels have been studied in order to establish criteria about the types of HRF and LRF which enhance the RCCI combustion. The study has been performed with ethanol-added gasoline and B7 biodiesel fuel. Results demonstrate that this type of fuel can promote more efficient operation than regular diesel and gasoline, achieving up to 3% greater GIE with the B7 + E10-95 fuel combination. On the other hand, excessively higher unburned HC and CO emissions were obtained. This fact should be further studied in the future, since it can compromise the efficiency of the aftertreatment systems.

Acknowledgments: The authors acknowledge VOLVO Group Trucks Technology for supporting this research and express their gratitude to the Spanish economy and competitiveness ministry for partially funding this investigation under the project HiReCo (TRA2014-58870-R). The author J. Monsalve-Serrano thanks the Universitat Politècnica de València for his predoctoral contract (FPI-S2-2015-1531), which is included within the framework of Programa de Apoyo para la Investigación y Desarrollo (PAID).

Author Contributions: J.B. and A.G. conceived and designed the experiments; J.M.-S. performed the experiments; J.M.-S. and V.B. analyzed the data and wrote the paper. All authors discussed the results and commented on the manuscript at all stages.

Conflicts of Interest: The authors declare no conflict of interest. The founding sponsors had no role in the design of the study; in the collection, analyses, or interpretation of data; in the writing of the manuscript, and in the decision to publish the results.

References

1. Posada, F.; Chambliss, S.; Blumberg, K. *Costs of Emission Reduction Technologies for Heavy-Duty Diesel Vehicles*; ICCT White Paper; International Council on Clean Transportation: San Francisco, CA, USA, 2016.
2. Zheng, M.; Asad, U.; Reader, G.T.; Tan, Y.; Wang, M. Energy efficiency improvement strategies for a diesel engine in low-temperature combustion. *Int. J. Energy Res.* **2009**, *33*, 8–28. [CrossRef]
3. Yanagihara, H.; Sato, Y.; Minuta, J. A simultaneous reduction in NOx and soot in diesel engines under a new combustion system (Uniform Bulky Combustion System e UNIBUS). In Proceedings of the 17th International Vienna Motor Symposium, Vienna, Austria, 1996; pp. 303–314.
4. Jacobs, T.J.; Assanis, D.N. The attainment of premixed compression ignition low-temperature combustion in a compression ignition direct injection engine. *Proc. Combust. Inst.* **2007**, *31*, 2913–2920. [CrossRef]

5.   Zhu, L.; Cheung, C.S.; Zhang, W.G.; Huang, Z. Effect of charge dilution on gaseous and particulate emissions from a diesel engine fueled with biodiesel and biodiesel blended with methanol and ethanol. *Appl. Therm. Eng.* **2011**, *31*, 2271–2278. [CrossRef]
6.   Wang, Y.; Zhao, Y.; Xiao, F.; Li, D. Combustion and emission characteristics of a diesel engine with DME as port premixing fuel under different injection timing. *Energy Convers. Manag.* **2014**, *77*, 52–60. [CrossRef]
7.   Payri, F.; Desantes, J.M.; Pastor, J.V. LDV measurements of the flow inside the combustion chamber of a 4-valve engine with axisymmetric piston-bowls. *Exp. Fluids* **1996**, *22*, 118–128. [CrossRef]
8.   Qiu, L.; Cheng, X.; Liu, B.; Dong, S.; Bao, Z. Partially premixed combustion based on different injection strategies in a light-duty diesel engine. *Energy* **2016**, *96*, 155–165. [CrossRef]
9.   Mathivanan, K.; Mallikarjuna, J.M.; Ramesh, A. Influence of multiple fuel injection strategies on performance and combustion characteristics of a diesel fuelled HCCI engine—An experimental investigation. *Exp. Therm. Fluid Sci.* **2016**, *77*, 337–346. [CrossRef]
10.   Singh, A.P.; Agarwal, A.K. Combustion characteristics of diesel HCCI engine: An experimental investigation using external mixture formation technique. *Appl. Energy* **2012**, *99*, 116–125. [CrossRef]
11.   Maurya, R.K.; Agarwal, A.K. Experimental investigation on the effect of intake air temperature and air–fuel ratio on cycle-to-cycle variations of HCCI combustion and performance parameters. *Appl. Energy* **2011**, *88*, 1153–1163. [CrossRef]
12.   Zhang, X.; Wang, H.; Zheng, Z.; Reitz, R.; Yao, M. Experimental investigations of gasoline partially premixed combustion with an exhaust rebreathing valve strategy at low loads. *Appl. Therm. Eng.* **2016**, *103*, 832–841. [CrossRef]
13.   Shen, M.; Tuner, M.; Johansson, B. Effects of EGR and Intake Pressure on PPC of Conventional Diesel, Gasoline and Ethanol in a Heavy Duty Diesel Engine. *SAE Tech. Pap.* **2013**. [CrossRef]
14.   Kalghatgi, G.T.; Gurubaran, R.K.; Davenport, A.; Harrison, A.J.; Hardalupas, Y.; Taylor, A.M.K.P. Some advantages and challenges of running a Euro IV, V6 diesel engine on a gasoline fuel. *Fuel* **2016**, *108*, 197–207. [CrossRef]
15.   Leermakers, C.A.J.; Bakker, P.C.; Nijssen, B.C.W.; Somers, L.M.T.; Johansson, B.H. Low octane fuel composition effects on the load range capability of partially premixed combustion. *Fuel* **2014**, *135*, 210–222. [CrossRef]
16.   Benajes, J.; Molina, S.; García, A.; Monsalve-Serrano, J.; Durrett, R. Conceptual model description of the double injection strategy applied to the gasoline partially premixed compression ignition combustion concept with spark assistance. *Appl. Energy* **2014**, *129*, 1–9. [CrossRef]
17.   Benajes, J.; Molina, S.; García, A.; Monsalve-Serrano, J.; Durrett, R. Performance and engine-out emissions evaluation of the double injection strategy applied to the gasoline partially premixed compression ignition spark assisted combustion concept. *Appl. Energy* **2014**, *134*, 90–101. [CrossRef]
18.   Park, S.H.; Youn, I.M.; Lim, Y.; Lee, C.S. Influence of the mixture of gasoline and diesel fuels on droplet atomization, combustion and exhaust emission characteristics in compression ignition engine. *Fuel Process. Technol.* **2013**, *106*, 392–401. [CrossRef]
19.   Bessonette, P.; Schleyer, C.; Duffy, K.; Hardy, W.; Liechty, M. Effects of fuel property changes on heavy-duty HCCI combustion. *SAE Tech. Pap.* **2007**. [CrossRef]
20.   Kazuhisa, I.; Takayuki, F.; Kazuaki, N.; Kiyomi, N.; Ichiro, S. Dual-Fuel PCI Combustion Controlled by In-Cylinder Stratification of Ignitability. *SAE Tech. Pap.* **2006**. [CrossRef]
21.   Kokjohn, L.; Hanson, M.; Splitter, A.; Reitz, D. Experimental Modeling of Dual-Fuel HCCI and PCCI Combustion Using In-Cylinder Fuel Blending. *SAE Tech. Pap.* **2010**, *2*, 24–39.
22.   Kokjohn, L.; Hanson, M.; Splitter, A.; Reitz, D. Fuel reactivity controlled compression ignition (RCCI): A pathway to controlled high-efficiency clean combustion. *Int. J. Engine Res.* **2011**, *12*, 209–226. [CrossRef]
23.   Dempsey, A.B.; Adhikary, B.D.; Viswanathan, S.; Reitz, R.D. Reactivity controlled compression ignition using premixed hydrated ethanol and direct injection diesel. *J. Eng. Gas Turbines Power* **2012**, *134*, 82806. [CrossRef]
24.   Benajes, J.; Molina, S.; Garcia, A.; Belarte, E.; Vanvolsem, M. An Investigation on RCCI combustion in a heavy duty diesel engine using in-cylinder blending of diesel and gasoline fuels. *Appl. Therm. Eng.* **2014**, *63*, 66–76. [CrossRef]
25.   Hanson, M.; Kokjohn, L.; Splitter, A.; Reitz, D. An experimental investigation of fuel reactivity controlled PCCI combustion in a heavy-duty engine. *SAE Int. J. Engines* **2010**, *3*, 700–716. [CrossRef]

26. Benajes, J.; Pastor, J.V.; Garcia, A.; Monsalve-Serrano, J. The potential of RCCI concept to meet Euro VI NOx limitation and ultra-low soot emissions in a heave-duty engine over the whole engine map. *Fuel* **2015**, *159*, 952–961. [CrossRef]

27. Ma, S.; Zheng, Z.; Liu, H.; Zhang, Q.; Yao, M. Experimental investigation of the effects of diesel injection strategy on gasoline/diesel dual-fuel combustion. *Appl. Energy* **2013**, *109*, 202–212. [CrossRef]

28. Reitz, R.; Duraisamy, G. Review of high efficiency and clean reactivity controlled compression ignition (RCCI) combustion in internal combustion engines. *Prog. Energy Combust. Sci.* **2015**, *46*, 12–71. [CrossRef]

29. Payri, F.; Olmeda, P.; Martín, J.; García, A. A complete 0D thermodynamic predictive model for direct injection diesel engines. *Appl. Energy* **2011**, *88*, 4632–4641. [CrossRef]

30. Lapuerta, M.; Ballesteros, R.; Agudelo, J.R. Effect of the gas state equation on the thermodynamic diagnostic of diesel combustion. *Appl. Therm. Eng.* **2006**, *26*, 1492–1499. [CrossRef]

31. Lapuerta, M.; Armas, O.; Hernandez, J.J. Diagnosis of DI Diesel combustion from in-cylinder pressure signal by estimation of mean thermodynamic properties of the gas. *Appl. Therm. Eng.* **1999**, *19*, 513–529. [CrossRef]

32. Woschni, G. A universally applicable equation for the instantaneous heat transfer coefficient in the internal combustion engines. *SAE Tech. Pap.* **1967**. [CrossRef]

33. Payri, F.; Margot, X.; Gil, A.; Martin, J. Computational study of heat transfer to the walls of a DI diesel engine. *SAE Tech. Pap.* **2005**. [CrossRef]

34. Torregrosa, A.J.; Olmeda, P.; Degraeuwe, B.; Reyes, M. A concise wall temperature model for DI Diesel engines. *Appl. Therm. Eng.* **2006**, *26*, 1320–1327. [CrossRef]

35. Payri, F.; Olmeda, P.; Martin, J.; Carreño, R. A New Tool to Perform Global Energy Balances in DI Diesel Engines. *SAE Int. J. Engines* **2014**, *7*, 43–59. [CrossRef]

36. Cheng, A.; Upatnieks, A.; Mueller, C. Investigation of Fuel Effects on Dilute, Mixing-Controlled Combustion in an Optical Direct-Injection Diesel Engine. *Energy Fuels* **2007**, *21*, 1989–2002. [CrossRef]

37. Kono, M.; Basaki, M.; Ito, M.; Hashizume, T.; Ishiyama, S.; Inagaki, K. Cooling Loss Reduction of Highly Dispersed Spray Combustion with Restricted In-Cylinder Swirl and Squish Flow in Diesel Engine. *SAE Int. J. Engines* **2012**, *5*, 504–515. [CrossRef]

38. Splitter, D.; Wissink, M.; Kokjohn, S.; Reitz, D. Effect of Compression Ratio and Piston Geometry on RCCI Load Limits and Efficiency. *SAE Tech. Pap.* **2012**. [CrossRef]

39. Hanson, R.; Curran, S.; Wagner, R.; Kokjohn, S.; Splitter, D.; Reitz, D. Piston Bowl Optimization for RCCI Combustion in a Light-Duty Multi-Cylinder Engine. *SAE Int. J. Engines* **2012**, *5*, 286–299. [CrossRef]

40. European Parliament, Council of the European. Directive 2009/28/EC of the European parliament and of the council of 23 April 2009 on the promotion of the use of energy from renewable sources. *Off. J. Eur. Union* **2009**, *140*, 16–62.

41. Owen, K.; Coley, T. *Automotive Fuels Reference Book*, 2nd ed.; Society of Automotive Engineers: Warrendale, PA, USA, 1995.

42. Tutak, W. Bioethanol E85 as a fuel for dual fuel diesel engine. *Energy Convers. Manag.* **2014**, *86*, 39–48. [CrossRef]

43. Desantes, J.M.; Lopez, J.J.; García, J.M. Evaporative diesel spray modeling. *Atomization Sprays* **2007**, *17*, 193–231. [CrossRef]

44. Desantes, J.M.; Arregle, J.M.; Lopez, J.J. Scaling laws for free turbulent gas jets and diesel-like sprays. *Atomization Sprays* **2006**, *16*, 443–473. [CrossRef]

Article

Developing Computational Fluid Dynamics (CFD) Models to Evaluate Available Energy in Exhaust Systems of Diesel Light-Duty Vehicles

Pablo Fernández-Yáñez [1], Octavio Armas [1,*], Arántzazu Gómez [1] and Antonio Gil [2]

[1] Universidad de Castilla La Mancha, Campus de Excelencia Internacional en Energía y Medioambiente, Escuela de Ingeniería Industrial, Real Fábrica de Armas, Edificio Sabatini, Av. Carlos III, s/n 45071 Toledo, Spain; Pablo.FernandezYanez@uclm.es (P.F.-Y.); aranzazu.gomez@uclm.es (A.G.)

[2] Universitat Politècnica de València, Instituto CMT-Motores Térmicos, Camino de Vera, s/n 46022 Valencia, Spain; angime@mot.upv.es

* Correspondence: octavio.armas@uclm.es; Tel.: +34-925-268-800 (ext. 3825)

Academic Editor: Kuang-Chao Fan

Received: 3 April 2017; Accepted: 1 June 2017; Published: 8 June 2017

Abstract: Around a third of the energy input in an automotive engine is wasted through the exhaust system. Since numerous technologies to harvest energy from exhaust gases are accessible, it is of great interest to find time- and cost-efficient methods to evaluate available thermal energy under different engine conditions. Computational fluid dynamics (CFD) is becoming a very valuable tool for numerical predictions of exhaust flows. In this work, a methodology to build a simple three-dimensional (3D) model of the exhaust system of automotive internal combustion engines (ICE) was developed. Experimental data of exhaust gas in the most used part of the engine map in passenger diesel vehicles were employed as input for calculations. Sensitivity analyses of different numeric schemes have been conducted in order to attain accurate results. The model built allows for obtaining details on temperature and pressure fields along the exhaust system, and for complementing the experimental results for a better understanding of the flow phenomena and heat transfer through the system for further energy recovery devices.

Keywords: CFD (computational fluid dynamics); model; exhaust; diesel; engine; energy recovery

---

1. Introduction

Around a third of the fuel energy in a light-duty diesel engine is wasted through the exhaust system [1]. Rising awareness of environmental issues together with the fuel economy have encouraged research on energy recovery [1,2] from exhaust gas. An obvious first step for energy recovery is evaluating the heat source. The nature of exhaust flow changes during engine operation. Hence, there is a need for models that allow evaluating exhaust gas in a cost- and time-efficient manner under different engine conditions.

Computational fluid dynamics (CFD) is becoming a very popular tool for numerical predictions in the automotive industry. CFD models are employed in five major areas: vehicle aerodynamics, thermal management (cooling and climate control), cylinder combustion, engine lubrication and exhaust system performance. The main applications of CFD in exhaust systems are: flow distribution in the front of the monolith, pressure loss through exhaust system, skin temperature prediction, heat loss analyses and emission-related studies [3].

High levels of energy lost through engine exhaust have brought attention to heat transfer in exhaust systems and how to model associated phenomena.

In Konstantinidis et al. [4], the status of knowledge regarding heat transfer phenomena in automotive exhaust systems was summarized, and a transient model covering diverse exhaust piping configurations was presented. Using this model, Kandylas et al. [5] studied heat transfer in automotive exhaust pipes for steady and transient conditions. It is stated that the code is suitable to support a complete and efficient methodology for design optimization.

Not only are exhaust heat transfer and temperature models important for energy recovery, but also for on-board control and diagnosis of the after-treatment system. In the work of Guardiola et al. [6], diesel oxidation catalyst (DOC) inlet temperature was modelled depending on fuel mass flow and engine speed using an engine map look-up table.

Traditionally, one-dimensional (1D) simulations using in-house or commercial codes have been used to study exhaust flow. In Shayler et al. [7], a 1D model of system thermal behavior was developed. The exhaust system is modelled as connected pipe and junction elements with lumped thermal capacities. Heat transfer correlations for quasi-steady and transient conditions were investigated. The computational model supports studies of exhaust and after-treatment system design, the investigation of thermal conditions, and performance characteristics. Model predictions and experimental data were in good agreement.

In Kapparos et al. [8], a 1D heat transfer model in an automotive diesel exhaust is presented. The external natural convective heat transfer coefficient was given by Churchill and Chu's correlation. A sensitivity analysis of the main variables (heat transfer coefficient, external pipe emissivity and others) affecting heat transfer in exhaust pipes was carried out. Fu et al. [9] developed a 1D model of an exhaust pipe (without the catalytic converter). Apart from typical variables studied, effects of pipe geometry and flow regime were investigated.

One-dimensional tools have the advantage of shorter computational time, but the accuracy of the results is limited by their inability to simulate 3D flow effects in regions such as the inlet and outlet cone. Usually strong turbulent flow exists when angled and asymmetrical cones are presented [3].

Three different approaches were used in Fortunato et al. [10] to simulate the exhaust and underbody of the car: 1D, 3D finite elements and 3D complete thermo-fluid-dynamics. Both 3D models allowed to detect the most critical points in terms of higher temperatures

In Chuchnowski et al. [11], temperature distribution and heat flow analyses on the exhaust system of a mining diesel engine to determine technical parameters of a heat recuperation system are presented. Three-dimensional numerical simulations show it is possible to develop a suitable design of an energy recovery device.

Simulations in a coupled three-dimensional thermal model including the underhood, underbody and exhaust systems of a vehicle were conducted by Guoquan et al. [12]. Pulsating flow and steady flow effects were compared. Pulsating exhaust showed an enhancement of more than a 10% in heat transfer. Influence of this effect was higher at the final components of the exhaust, such as the muffler.

Research has also been focused on finding optimal position for exhaust components. Durat et al. [13] compared experimental and CFD heat transfer results in an exhaust system of a spark-ignition engine. An optimal position of a close-coupled catalytic converter in terms of light-off time was found using the CFD model. Liu [14] developed a 3D exhaust model to study the optimal position of a heat recovery device. Three different simulations were carried out varying the position of a thermoelectric generator along the system. Higher surface temperature and lower backpressure were obtained when the recovery device was placed between catalytic converter and muffler. The influence of materials and insulation on heat transfer in catalytic converters has also been discussed employing a 3D model [15].

In the same way preceding authors have employed CFD simulations successfully to find an optimal position for exhaust components or to assist in their design, CFD can also be used to gain knowledge of the relevant characteristics of the flow in terms of its usage in energy recovery. Previous works [16] state that knowledge of heat transfer within the gas is critical to enhance harvested energy.

In exhaust systems, not only should heat transfer to surroundings be reproduced but also chemical reactions in catalytic processes, since after-treatment devices can modify gas temperature. Most mathematical kinetics models rely on the Langmuir-Hinshelwood expressions derived on pellet-type

Pt catalysts for the CO and HC oxidation with modified parameters (activation energies and activity factors), best suited to the case modeled. This approach is widely employed because it provides acceptable accuracy in the most common automobile range of operation.

Measurements on pellet-type platinum-alumina catalysts have been processed [17] to derive kinetic rate expressions for the oxidation reactions of CO and C_3H_6 under oxygen-rich conditions for a better understanding of platinum oxidation catalysts in automotive emission control systems. Terms accounting for CO, C_3H_6 and NO inhibition were included. The isothermal reactor approach in a numerical integration-optimization method was used to fit the kinetic parameters. These expressions, especially as written by Oh et al. [18], have been widely used [19–22]. This approach has also been used to build simpler kinetic models over a range of interest [23].

Detailed kinetics for a Pt/Rh three-way catalyst were described in Chatterjee et al. [24]. The model consists of 60 elementary reaction steps and one global reaction step, involving 8 gas species and 23 site species. It considers the steps of adsorption of the reactants on the surface, reaction of the adsorbed species, and desorption of the reaction products. This approach includes a mechanism of C_3H_6 oxidation on Pt/Al_2O_3, a mechanism of NO reduction on Pt, and a mechanism for NO reduction and CO oxidation on Rh. This model has been used when more accuracy is required, as in Kumar et al. [25].

Nevertheless, due to the variety of washcoat materials and loadings of the monoliths and to ageing effects of after-treatment devices, the kinetic parameters published have difficulties in fitting to conditions or catalytic converters different from the study cases. To overcome the difficulty of finding kinetic parameters fitted to their application, it would be of interest to have a simple method for researchers to develop their own kinetic constants for CFD exhaust models.

The real physical space of a full-size monolith converter contains thousands of tiny channels and could present an extremely time-consuming problem to solve. There are several approaches to modeling the monolith of a catalytic converter. For simulations of the converter performance, most models use a volume-averaging method and treat the monolith as a continuous porous medium [23–26]. With the porous medium, approach channels are not simulated but a macroscopic understanding of the flow is achieved while keeping the computational requirements affordable.

The scope of this work is to develop a combined theoretical and experimental methodology to build an exhaust model to obtain information about temperature and heat losses along the system. Results were used to provide an insight into those flow characteristics that must be taken into account when harvesting energy, especially but not exclusively for thermoelectric modules. The model does not only intend to reproduce the behavior of the exhaust system at a single operation point but in the whole most-used part of the engine map. This information is needed to evaluate the exhaust gas energy available for recovery under real operating conditions.

2. Materials and Methods

2.1. Materials

2.1.1. Engine

Tests were carried out in a test-bench with a Nissan YD22 four-stroke, turbocharged, four-cylinder diesel engine. The engine bore and stroke are, respectively, 86 mm and 130 mm, and the compression ratio is 16.7:1. The exhaust system (Figure 1) is equipped with a diesel oxidation catalyst (DOC) and a muffler.

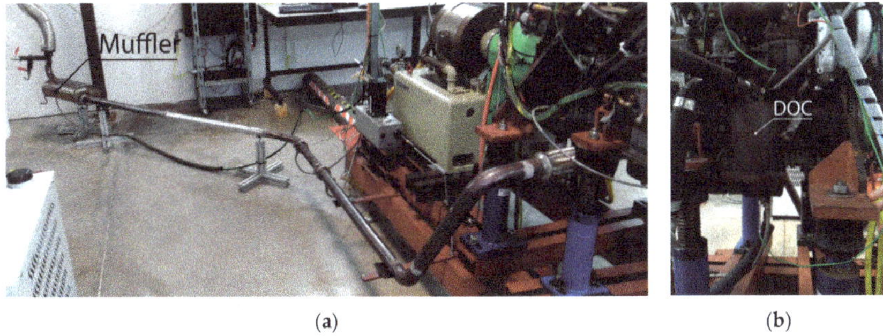

Figure 1. View of the engine test bench including (**a**) the exhaust system pipe and (**b**) the diesel oxidation catalyst (DOC).

2.1.2. Measurement Devices

Piezo resistive pressure sensors and K-type thermocouples were used for measuring pressure and temperature of the gas along the exhaust system (see Figure 2). The exhaust mass flow rate was calculated from the addition of fuel and air mass flow rates. Skin temperatures used in calculations were provided by the IR camera Gobi-384 GigE.

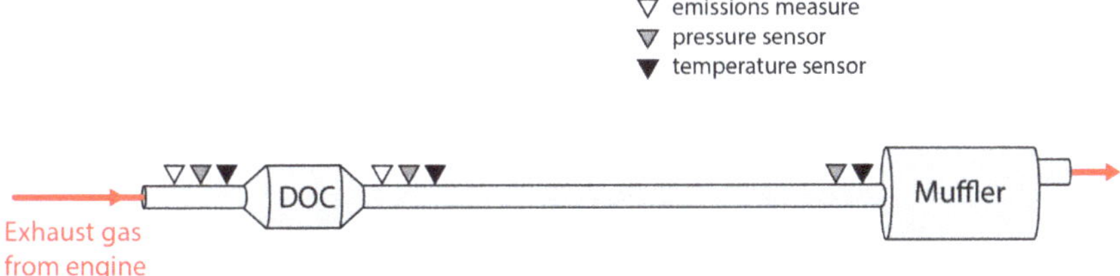

Figure 2. Sketch of the exhaust with measurement points involved in the development of the model. The exhaust system is 4 m long and pipe has a diameter of 5 cm (figure not to scale).

Pollutants are measured using an ENVIRONNEMENT manufacturer equipment. A TOPAZE 32M model was used as NO_x analyser, based on the chemiluminescence effect from NO oxidation by ozone (O_3). The GRAPHITE 52M gas analyser measures total hydrocarbons (THCs) by flame ionization detection while a MIR 2R gas analyser measures CO and CO_2 species, detecting the molecules absorption in the infrared spectrum. Other species compositions are estimated via chemical balances from fuel and air consumption.

2.2. Methods

2.2.1. Test Plan

The velocity profile imposed by the New European Driving Cycle (NEDC) used for light-duty vehicle certification was translated into engine operating conditions (torque and engine speed), as shown in Figure 3 (black squares), employing longitudinal dynamics equations [27,28]. Then, a matrix of nine steady-state modes (see Figure 3, circles) covering the most-used quarter of the engine map under both urban and extra-urban NEDC conditions was selected for testing.

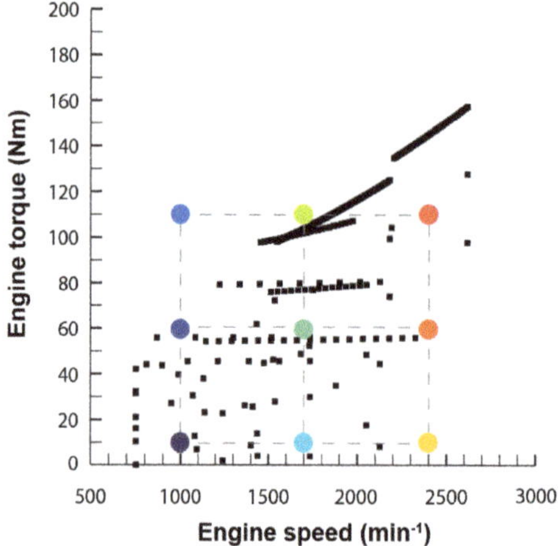

Figure 3. Test matrix. Circles represent selected engine operating points.

2.2.2. Experimental Characterization of Model Input Parameters

(a)   Monolith pressure loss characterization.

The classical approach that links pressure drop and velocity in flows through porous media is the Forchheimer equation (Equation (1)), that can be derived from the Navier–Stokes equation for one-dimensional, incompressible and steady laminar flow of a Newtonian fluid in a rigid porous medium [29]:

$$-\frac{dp}{dx} = \frac{1}{\kappa}(\mu \cdot v) + \beta\left(\rho \cdot v^2\right) \tag{1}$$

The second term in the right can be interpreted as a second-order correction to account for the contribution of inertial forces, but, at sufficiently low velocities, this effect is negligible and Equation (1) can be reduced to Darcy's law (Equation (2)) [30]:

$$-\frac{dp}{dx} = \frac{1}{\kappa}(\mu \cdot v) \tag{2}$$

Darcy's law can be rewritten in the form of Equation (3), relating the average fluid velocity v through the pores with pressure drop Δp along a segment of length L.

$$-\frac{\Delta p}{\mu L} = \frac{1}{\kappa} v \tag{3}$$

If the equation from the linear fit of the scatter plot of $-\frac{\Delta p}{\mu L}$ vs. v is (Equation (4)):

$$Y = BX \tag{4}$$

then Darcy's constant κ can be derived as follows (Equation (5)):

$$\kappa = \frac{1}{B} \tag{5}$$

(b)   Heat transfer to surroundings.

Natural convection to pipe surroundings in the test-bench needs to be included. Wall temperatures along the exhaust pipe were measured using thermography. Figure 4 shows, as an example, views of the external surface of the pipe and the DOC.

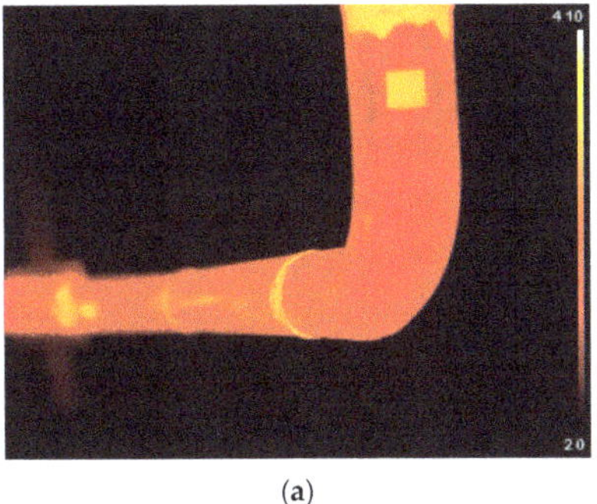

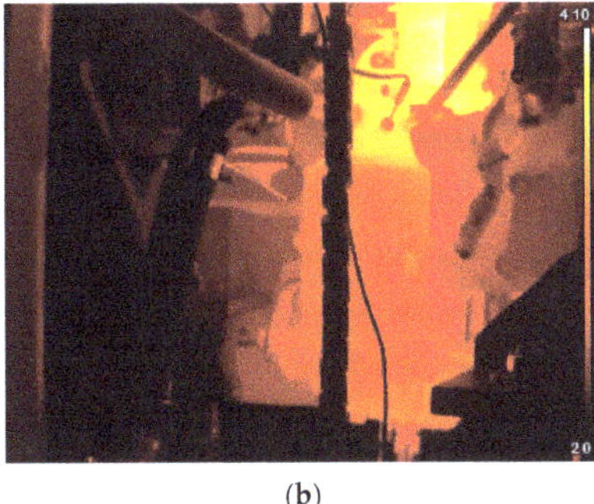

(a) (b)

Figure 4. Example of infrared images of the exhaust line in (**a**) the exhaust pipe and (**b**) the DOC. Temperatures in the scale are in °C. Squares seen in (**a**) are wall temperature measurement points.

An average skin temperature was used to calculate an effective convection heat transfer coefficient. This skin temperature simplification is considered as reasonable, since calculation of heat transfer rates is not very sensitive in the values of pipe wall temperatures [5].

Total transferred heat from pipe wall to surroundings is $\dot{Q}$ (W) (Equation (6)):

$$\dot{Q} = \bar{h}_\infty \pi d (T_{wall} - T_\infty) + \varepsilon \pi dL\sigma\left(T_{wall}^4 - T_{amb}^4\right) \qquad (6)$$

Here, all terms in the right side of Equation (6) are grouped in one single equivalent convection term (Equation (7)):

$$\dot{Q} = \bar{h}_{eff} \pi dL (T_{wall} - T_\infty) \qquad (7)$$

The effective convection coefficient accounts for convection and radiation heat losses, although radiation heat transfer to the environment is expected to be low, since wall temperature is below 400 °C [4]. A similar approach is explained in Kapparos et al. [8].

Based on the measured exhaust gas and pipe wall temperatures, an energy balance for the exhaust gas is employed for the calculation of the heat flux. The resulting heat fluxes are employed in the estimation of a mean gas-to-wall convection coefficient.

Since temperature at both ends of the exhaust pipe are measured, total heat losses $\dot{Q}$ (W) can be obtained as Equation (8):

$$\dot{Q} = \dot{m}_g c_{p,g} \left(T_{g,in} - T_{g,\,out}\right) \qquad (8)$$

Equaling Equations (7) and (8), $\bar{h}_{eff}$ can be derived as Equation (9):

$$\bar{h}_{eff} = \frac{\dot{m}_g c_{p,g} \left(T_{g,in} - T_{g,\,out}\right)}{\pi d (T_{wall} - T_\infty)} \qquad (9)$$

(c) Estimation of kinetic parameters.

Under light-off temperatures (about 200 °C) catalytic processes remain basically inactive. Until activation temperature is reached, reactions are chemically controlled. On the other hand, post-light-off reaction rates are limited mainly by mass transfer and, consequently, conversion efficiency depends on the residence time within the monolith, the surface to volume ratio of the monolith and the mass transfer [31]. In operating conditions from test matrix, catalytic reactions are normally within the light-off band. Thus, estimation of kinetic parameters is needed.

Two main reactions occurring in DOCs, as in Voltz et al. [17], are modeled (Equations (10) and (11)):

$$C_3H_6 + \frac{3}{2}\,O_2 \rightarrow 3\,CO_2 + 3\,H_2O \qquad (10)$$

$$CO + \frac{1}{2}O_2 \rightarrow CO_2 \tag{11}$$

Propylene is representative of the easily oxidized hydrocarbons, which constitute about 80% of the total hydrocarbons found in a typical exhaust gas. Other saturated hydrocarbons (typically represented as methane or propane) that are resistant to oxidation usually make up the remaining 20% [17]. Since only total hydrocarbon (THC) data is available and fast oxidation hydrocarbons represent the majority of the THC, propylene alone was used as representative of the HC (as is common [32]).

Usually adopted forms of the rates of CO and HC oxidations are as follows (Equations (12) and (13)):

$$R_{CO} = \frac{k_{CO}c_{CO}c_{O_2}}{G} \tag{12}$$

$$R_{C_3H_6} = \frac{k_{C_3H_6}c_{C_3H_6}c_{O_2}}{G} \tag{13}$$

where G is a term accounting for NO, CO and HC inhibition effects on oxidation (Equation (14)):

$$G = T_s\left(1 + K_1 c_{CO} + K_2 c_{C_3H_6}\right)^2 \left(1 + K_1 c_{CO} + K_2 c_{s,C_3H_6}\right)^2 \left(1 + K_3 c_{CO}^2 c_{C_3H_6}^2\right)\left(1 + K_4 c_{NO}^{0.7}\right) \tag{14}$$

Since the intended target is an empirical model to fit experimental data and power-law reactions showed good performance, an inhibition term was not included (Equations (15) and (16)):

$$R_{CO} = k_{CO}c_{CO}c_{O_2} \tag{15}$$

$$R_{C_3H_6} = k_{C_3H_6}c_{C_3H_6}c_{O_2} \tag{16}$$

Arrhenius kinetic constants k_m are defined as (Equation (17)):

$$k_m = A_m e^{-\frac{E_{a,m}}{RT_s}} \tag{17}$$

The problem is reduced to obtain the pre-exponential factors A_m and activation energies $E_{a,m}$. For both species, an iterative process varying the pre-exponential factor in CFD simulations is conducted until the deviation of outlet concentration value from the experimental is as much as 0.1%. Once this is achieved, values of k_m are obtained. This tuning process is done for the different operating conditions, and then k_m of each reaction is obtained as a function of monolith temperature (provided by CFD results). This iterative process needs to be followed just once (to obtain the above-mentioned constants and implement them in the model) and not for every prospective 3D simulation.

The simulations were performed on a 2D axisymmetric DOC model. The 2D model maintained the same characteristic lengths but with a simplified geometry (see Figure 5), allowing fast iterative simulations. Other aspects about set-up of these 2D simulations are the same than those presented below for the 3D calculations.

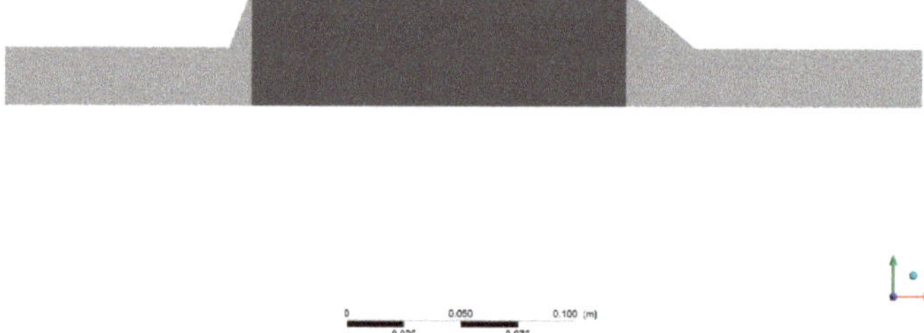

Figure 5. Geometry of the 2D axisymmetric DOC model for the tuning process of kinetic parameters. Darker shade of grey represents the monolith.

From results, the kinetic exponential law (Equation (17)) can be obtained, since (Equation (18)):

$$\ln k_m = C_m - \frac{D_m}{T_s} \tag{18}$$

where C_i and D_i are constants from the linear fit. Sought kinetic constants are (Equations (19) and (20)):

$$A_m = e^{C_m} \tag{19}$$

$$E_{a,m} = D_m R \tag{20}$$

2.2.3. Three-Dimensional CFD Model

(a)   Simulations setup

The numerical solution of the continuity, momentum, energy and species equations was computed using a CFD proprietary code (ANSYS Fluent 16), based on the finite volume method. In this work, a steady state, three-dimensional, viscous, turbulent and incompressible (since the maximum Mach number is below 0.3) flow was assumed. Pressure–velocity coupling was taken care of by the segregated, pressure-based solver, the Semi-Implicit Method for Pressure-Linked Equations (SIMPLE) algorithm [33]. Summary details of spatial discretization are presented in Table 1. The convergence criteria were set to 10^{-6} for the thermal energy and chemical residuals and 10^{-4} for residuals from mass, momentum, turbulence kinetic energy and turbulence energy dissipation rate. Relevant physical and chemical quantities were monitored to assure convergence.

Table 1. Summary of the spatial discretization.

Parameter	Value
Gradient	Least-squares cell-based
Pressure	Standard
Momentum Species conservation Turbulent kinetic energy	Second order upwind

The governing equations solved are the continuity equation (Equation (21)), momentum equation (Equation (22)), energy equation (Equation (23)) and conservation equation for chemical species (Equation (24)):

$$\frac{\partial \rho}{\partial t} + \frac{\partial}{\partial x_i}(\rho u_i) = 0 \tag{21}$$

$$\frac{\partial}{\partial t}(\rho u_i) + \frac{\partial}{\partial x_i}(\rho u_i u_j) = -\frac{\partial p}{\partial x_i} + \frac{\partial}{\partial x_i}\left[\mu\left(\frac{\partial u_j}{\partial x_i} + \frac{\partial u_i}{\partial x_j}\right)\right] + \frac{\partial}{\partial x_i}\left(-\overline{\rho u_i' u_j'}\right) + \rho g_i + F_i \tag{22}$$

$$\frac{\partial}{\partial t}(\rho c_p T) + \frac{\partial}{\partial x_j}(\rho u_j c_p T) - \frac{\partial}{\partial x_j}\left(K\frac{\partial T}{\partial x_j}\right) = S_T \tag{23}$$

$$\frac{\partial}{\partial t}(\rho Y_m) + \frac{\partial}{\partial x_j}(\rho u_j Y_m) = -\frac{\partial}{\partial x_j} J_{m,j} + R_m + S_m \tag{24}$$

Notice that for steady simulations, time-dependent terms become zero.

(b)   Turbulence

A laminar regime was forced in the monolith because of small Reynolds numbers due to low velocity and very small hydraulic diameter inside channels.

In the other parts of the domain, Reynolds-averaged Navier–Stokes (RANS) approach was employed with the realizable $k - \varepsilon$ model [34] selected. This model has the same turbulent kinetic energy equation as the standard $k - \varepsilon$ model but holds an improved equation for ε. Compared to standard $k - \varepsilon$, it shows better performance for flows involving: planar and round jets (predicts round jet spreading correctly), boundary layers under strong adverse pressure gradients or separation, rotation, recirculation or/and strong streamline curvature [34].

(c)   Computational domain

Simulation domain comprehends the area of interest for energy recovery. Great energy is lost in the muffler [35] and energy harvesting must be performed before it. After-treatment processes, such as chemical reactions within the DOC, might be affected by a temperature change caused by some sort of energy recovery device [36]. Consequently, energy recovery should take place between after-treatment devices and the muffler, as pointed out in [14]. The simulation domain includes the DOC and the exhaust pipe (see Figure 6). DOC must be included, since exothermal chemical reactions taking place in it can modify gas temperature. In order to obtain a properly-developed velocity profile at the inlet, an extruded entrance zone is added to the physical domain.

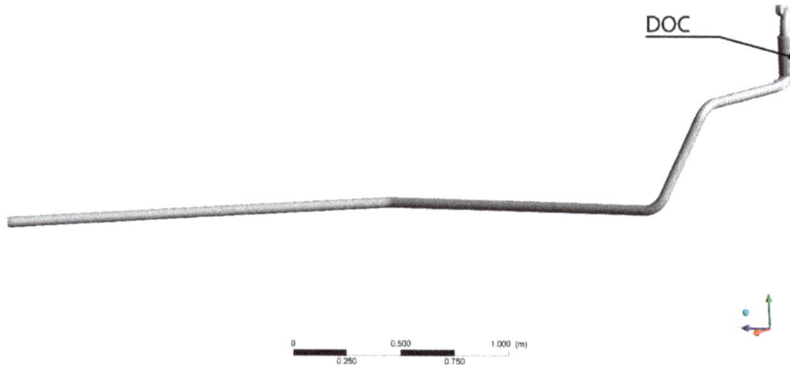

Figure 6. Three-dimensional geometry model of the exhaust system.

(d)   Computational mesh

Given the nature of the geometry, a hybrid Tet/Hex mesh (see Figure 7) has been used to define the computational domain. Complex 3D features such as inlet and outlet DOC cones have been meshed by means of tetrahedral elements, and hexahedral meshes have been used for the monolith and the duct. In order to ensure the accuracy of the calculations, a grid independence study was conducted and mesh independency was achieved with a 6.5×10^5 elements grid with an average mesh size of 3×10^{-3} m.

Figure 7. Mesh detail. Tetrahedral (DOC inlet and outlet cones) and hexahedral zones (monolith and pipe) can be distinguished.

(e) Boundary conditions

Table 2 shows boundary condition types selected for the model. A commonly employed porous medium approach was used for the monolith, with a one-directional pressure gradient. Boundary conditions values are presented in Table 3.

Table 2. Boundary conditions included in the model.

Boundary	Type
Inlet	Mass-flow inlet
Outlet	Pressure-outlet
Walls	Wall with convection heat transfer
Monolith	Anisotropic porous media Porosity: 0.74 Viscous pressure loss coefficient: 5.9×10^7 m^{-2}

Table 3. Measured temperatures, mass flow values and outlet pressure. These values have been employed as boundary conditions for the model.

Engine Speed (min^{-1})	1000			1700			2400		
Engine Torque (Nm)	10	60	110	10	60	110	10	60	110
Exhaust mass flow (kg/s)	0.017	0.019	0.020	0.023	0.027	0.041	0.031	0.045	0.066
$T_{gas,in}$ (°C)	131.0	244.3	385.9	164.0	302.0	363.7	199.1	305.4	348.9
Outlet relative pressure (Pa)	70	220	370	330	560	1690	720	1940	4540

A total amount of seven species are considered in gas composition (see Table 4). Oxidations of carbon monoxide and propylene, as explained in the previous section, are accounted for.

Table 4. Modeled species and volumetric fraction range in gas inlet composition within the test matrix.

Engine Speed (min^{-1})	1000			1700			2400		
Engine Torque (Nm)	10	60	110	10	60	110	10	60	110
N_2	7.8×10^{-1}	7.6×10^{-1}	7.4×10^{-1}	7.8×10^{-1}	7.6×10^{-1}	7.6×10^{-1}	7.7×10^{-1}	7.7×10^{-1}	7.6×10^{-1}
O_2	1.7×10^{-1}	1.3×10^{-1}	7.0×10^{-2}	1.6×10^{-1}	1.1×10^{-1}	9.6×10^{-2}	1.7×10^{-1}	1.2×10^{-1}	1.1×10^{-1}
H_2O	2.2×10^{-2}	5.1×10^{-2}	8.5×10^{-2}	3.2×10^{-2}	6.4×10^{-2}	7.0×10^{-2}	2.4×10^{-2}	5.5×10^{-2}	6.0×10^{-2}
CO_2	2.2×10^{-2}	5.7×10^{-2}	9.9×10^{-2}	3.1×10^{-2}	6.7×10^{-2}	7.0×10^{-2}	3.5×10^{-2}	5.6×10^{-2}	6.6×10^{-2}
NO	1.6×10^{-4}	7.2×10^{-4}	1.2×10^{-3}	1.3×10^{-4}	3.0×10^{-4}	7.4×10^{-4}	1.3×10^{-4}	2.5×10^{-4}	4.3×10^{-4}
CO	2.2×10^{-4}	1.1×10^{-4}	1.6×10^{-4}	4.1×10^{-4}	2.0×10^{-4}	1.9×10^{-4}	5.9×10^{-4}	3.8×10^{-4}	1.7×10^{-4}
C_3H_6	8.8×10^{-5}	5.4×10^{-5}	9.4×10^{-5}	1.3×10^{-4}	6.7×10^{-5}	5.0×10^{-5}	1.5×10^{-4}	9.0×10^{-5}	5.4×10^{-5}

(f) Boundary layer sensitivity analysis.

A mesh with a special refinement (three layers) near the walls and other without it were tested. Only a part of the domain was used (DOC and a section of the exhaust pipe) in order to reduce computational requirements. As can be seen in Table 5, refinement along the boundary layer adds a considerable number of extra cells and does not provide significant variations.

Table 5. Boundary layer sensitivity analysis results.

Case	Transferred Heat (W)	Number of Mesh Cells
Without boundary layer refinement	455.7	4×10^5
With boundary layer refinement	459.5	2.1×10^6

Boundary layer refinement results differed from the base case by less than 1% and required five times more cells. It can be stated that the model is not boundary layer-sensitive and, hence, boundary layer refinement is omitted.

(g)   Wall treatment sensitivity analysis.

Three different wall approaches were tested to study their influence in the results of the model: standard, non-equilibrium and enhancement wall functions (Figure 8).

The standard wall functions proposed by Launder and Spalding [37] have been widely used, but the assumption of logarithmic velocity distribution treatment may not be adequate for complex non-equilibrium flows. To overcome this, non-equilibrium wall functions are based on pressure-gradient sensitized Launder and Spalding's [37] log-law for mean velocity.

Enhanced wall treatment is a near-wall modelling method that combines a two-layer model with enhanced wall functions. A one-equation relationship is used to evaluate the laminar sub-layer with fine mesh and transition to log-low function for the turbulent part of the boundary layer. The restriction that the near-wall mesh must be suitably fine might impose large computational requirements.

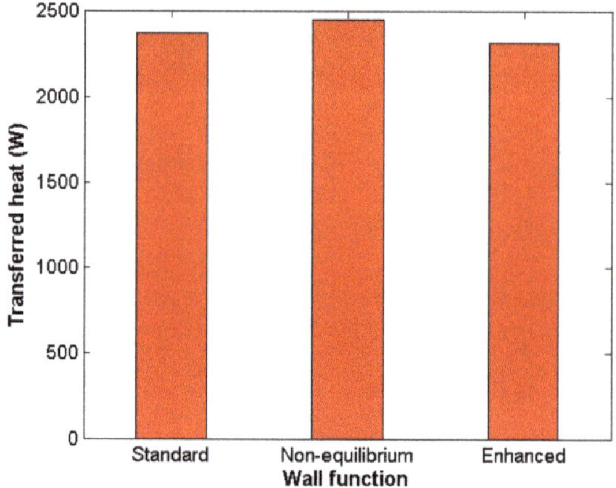

Figure 8. Results of test runs with different wall functions. No great disagreement between the three approaches was obtained.

More accurate wall functions such as non-equilibrium or enhanced showed no more than a 3% discrepancy with the standard approach, while requiring eight times more mesh elements. Consequently, the standard wall functions were selected.

3. Results

3.1. Pressure Loss Coefficient

Different exhaust mass flows and their corresponding pressure drops were measured to obtain accurate predictions. Empirical data (Figure 9) show a clear linear correlation between velocity and pressure drop (being v the horizontal axis and $-\frac{\Delta p}{\mu L}$ the vertical axis) and Darcy's law can be employed. The resulting Darcy's constant is $\kappa = 1.695 \times 10^{-8}$ m^2.

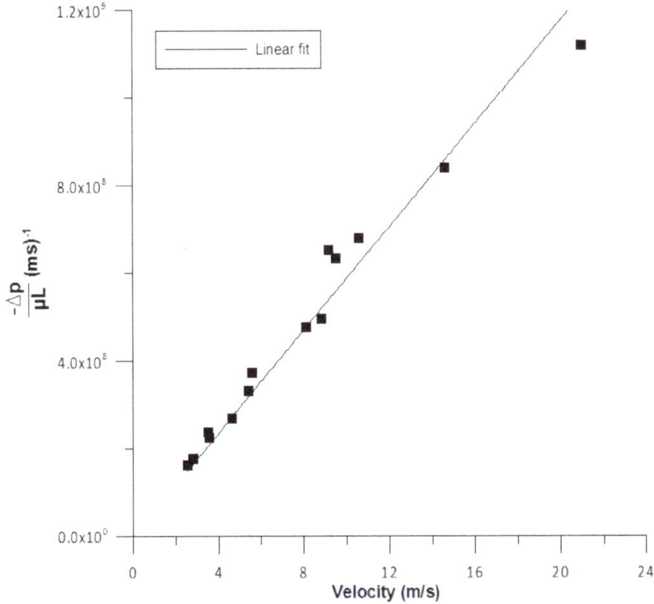

Figure 9. Plot for viscous resistance derivation. Linear fit constant: $B = 5.9 \times 10^7 \ \mathrm{m}^{-2}$.

3.2. Convection Heat Transfer Coefficient

Effective convection coefficients derived in each operation mode are presented in Table 6.

Table 6. Experimentally-derived effective convection heat transfer coefficients for each operating mode in test matrix.

Engine Speed (min^{-1})	Torque (Nm)	$\bar{h}_{eff}$ (W/m^{2}K)
1000	10	11.1
1000	60	14.6
1000	110	17.9
1700	10	13.3
1700	60	15.8
1700	110	17.9
2400	10	14.1
2400	60	16.7
2400	110	18.6

3.3. Kinetic Parameters

Complete and zero species consumption at DOC are not useful for estimating kinetic parameters. Therefore, data close to 0 or 100% conversion were excluded from calculations. Given that light-off bands seemed to be narrow, particularly for CO, two sets of experiments needed to be used. Due to intended repeatability of the experiments, similar tests would result in similar results, adding no new information. It was decided to repeat the test matrix under very different external ambient conditions, so that different exhaust temperature results in the range of interest are obtained (see Figure 10).

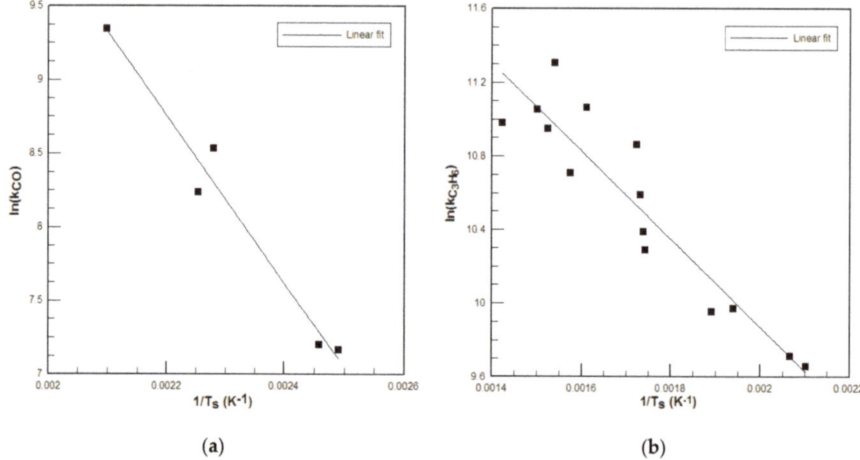

Figure 10. Plot for (**a**) CO kinetic parameters derivation. Linear fit: C_{CO} = 21.261, D_{CO} = 5685.10 K (**b**) C_3H_6 kinetic parameters derivation Linear fit: C_{C3H6} = 14.647, D_{C3H6} = 2388.55 K.

Resulting kinetic parameters are shown in Table 7:

Table 7. Obtained kinetic parameters for CO and C_3H_6 oxidations.

$A_{CO}(\text{m}^3/(\text{mol·s}))$	3.42×10^6
$E_{CO}(\text{m}^3/(\text{mol·s}))$	4.73×10^4
$A_{C_3H_6}(\text{m}^3/(\text{mol·s}))$	2.30×10^3
$E_{C_3H_6}(\text{J}/\text{mol})$	1.98×10^4

3.4. Chemical Results

Kinetic model results are shown in Table 8. Measured and modelled conversions of CO and propylene are included.

Table 8. Experimental versus chemical model conversion results.

Engine Speed (min^{-1})	Torque (Nm)	Measured CO Conversion (%)	Modelled CO Conversion (%)	Error in CO Conversion (%)	C_3H_6 Conversion (%)	Modelled C_3H_6 Conversion (%)	Error in C_3H_6 Conversion (%)
1000	10	27.3	25.8	−1.4	1.4	0.0	−1.4
1000	60	90.9	99.2	8.3	83.3	88.4	5.1
1000	110	93.7	100.0	6.2	79.3	85.5	6.2
1700	10	51.2	39.6	−11.6	3.6	0.0	−3.6
1700	60	95.0	99.9	4.9	69.0	80.8	11.8
1700	110	99.5	100.0	0.5	58.0	69.8	11.8
2400	10	66.10	64.7	−1.4	72.0	76.3	4.3
2400	60	97.37	99.1	1.8	73.6	67.4	−6.2
2400	110	100.0	98.9	−1.1	58.8	55.4	−3.4

3.5. Temperature and Heat Results

To assess the complete model, experimental tests were conducted in five operating points from the test matrix. Modelled outlet temperature $T_{g,\,out}$ and heat loss through the exhaust pipe $\dot{Q}$ are compared with experimental results in the five calibrating points (see Table 9).

Table 9. Experimental versus model thermal results in calibration engine modes.

Engine Speed (min^{-1})	Torque (Nm)	Measured $T_{g,\,out}$ (°C)	Modelled $T_{g,\,out}$ (°C)	Error in $T_{g,\,out}$ (%)	Experimental $\dot{Q}$ (W)	Modelled $\dot{Q}$ (W)	Error in $\dot{Q}$ (%)
1000	60	176	169	4.0	1384	1407	1.6
1700	10	125	123	1.6	915	876	−4.3
1700	60	235	226	3.7	2016	2063	−2.3
1700	110	299	288	3.7	2977	3107	−4.4
2400	60	263	256	2.7	2423	2508	3.5

The remaining four operating points of the test matrix are used to evaluate the model out of training points. Results are shown in Table 10.

Table 10. Experimental versus model thermal results out of calibration engine modes.

Engine Speed (min^{-1})	Torque (Nm)	Measured $T_{g,\,out}$ (°C)	Modelled $T_{g,\,out}$ (°C)	Error in $T_{g,\,out}$ (%)	Experimental $\dot{Q}$ (W)	Modelled $\dot{Q}$ (W)	Error in $\dot{Q}$ (%)
1000	10	94	96	1.4	590	562	−4.9
1000	110	269	253	5.9	2587	2639	2.0
2400	10	167	164	1.8	1287	1357	5.4
2400	110	310	300	3.0	3286	3364	2.4

3.6. Flow Temperature Distribution at the DOC Outlet

As commented in Section 2.2.3, it is suggested that energy recovery processes should take place downstream of after-treatment devices in order not to interfere with their operation. However, they should also occur sufficiently close in order to minimize further thermal energy dissipation to the surroundings. Therefore, the analysis of the flow abandoning the DOC is essential to know the working conditions of energy-harvesting devices.

Low uniformity in inlet flow occurring in most catalysts leads to different flow paths with different residence times. Flow inside the DOC is subject to cooling because external convection and heating because of chemical reactions in the DOC. These different paths (see Figure 11) caused by the inlet cone cause a gradient of temperature in the outlet section of the catalytic converter, as can be seen in Figure 12.

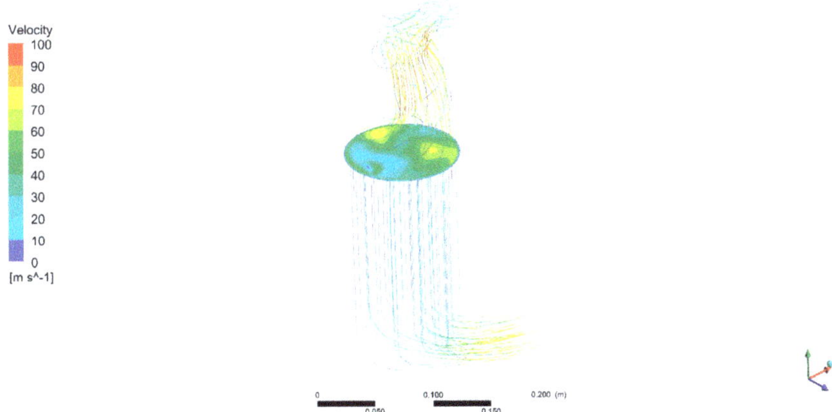

Figure 11. Streamlines and inlet velocity distribution in the monolith. An example of usual bad flow uniformity caused by inlet cones in automotive catalysts can be seen.

Coefficient of variation (standard deviation divided by mean value) of temperature distribution across the DOC outlet area (distribution shown in Figure 12b) was calculated from results of the CFD model. This statistical coefficient accounts for the dispersion from the average temperature flow. As can be seen in Figure 13, variation in temperature is enhanced in engine conditions with low mass flows, since flow thermal inertia is also lower. Engine modes in which catalytic reactions are active (see black dots in Figure 13) present more uneven distribution than those in which they are inactive (see white dots in Figure 13).

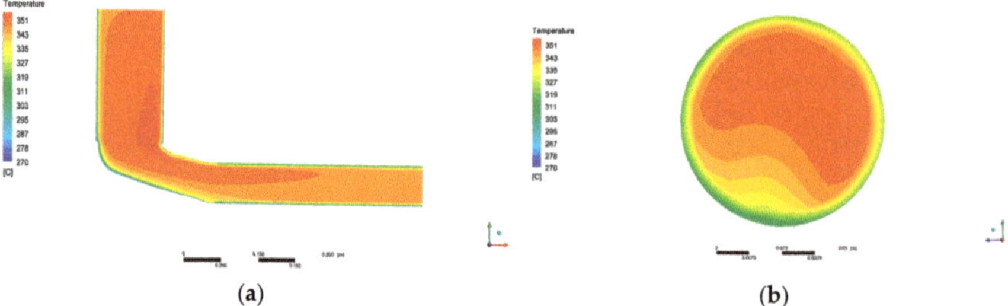

(a)   (b)

Figure 12. Temperature distribution for the 2400 rpm–110 Nm mode in (**a**) monolith, outlet cone and exhaust pipe and (**b**) cross-section of DOC outlet (before exhaust pipe).

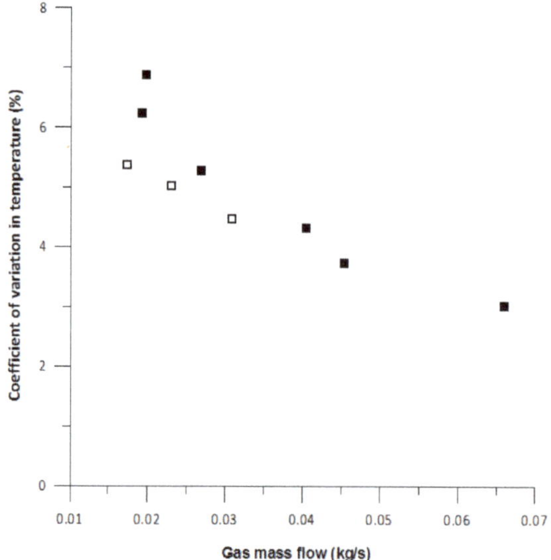

Figure 13. Plot for coefficient of variation in temperature distribution at DOC outlet. White dots represent engine modes in which chemical processes were not active.

3.7. Temperature Loss along the Exhaust Pipe

Sometimes, because of limitations in available space or ease of installation, the selected place for energy recovery devices moves away from the location with the maximum temperature available (immediately before the after-treatment systems). Average temperature of the exhaust gas in the first 50 cm downstream of the DOC was obtained from the validated model to quantify the loss in temperature when moving the energy recovery device away from the DOC.

Test-bench results with natural convection were compared with external forced convection (as in a moving vehicle with velocity v_∞). Forced convection was simulated as boundary condition in pipe walls. Two different modes (lowest and highest engine power within the test matrix) were studied for each external condition. Forced convection coefficients were calculated with external parallel flow correlations [38] taking into account velocities of the vehicle at those engine conditions. Notice that natural convection conditions can also be relevant in moving vehicles since the part of the exhaust pipe adjacent to the DOC could not be in contact with car underbody air.

Average temperature decreases ($\Delta T/L$) for external natural convection (test-bench) range from 0.9 °C/dm in the lower power mode to 1.2 °C/dm in the higher power mode. For forced convection, decrease in temperature ranges from 1.2 °C/dm to 2.6 °C/dm (see Table 11).

Table 11. Cooling in the exhaust gas for the part of the pipe adjacent to the DOC (first 50 cm).

Engine Mode	1000 rpm–10 Nm		2400 rpm–110 Nm	
Convection	Natural	Forced (v_∞ = 20 km/h)	Natural	Forced (v_∞ = 120 km/h)
$\Delta T/L$ (°C/dm)	−0.9	−1.2	−1.2	−2.6

4. Discussion

4.1. Accuracy of the Model

As can be seen, maximum error in chemical conversion was 11.58% and 11.81% for the CO model and the C_3H_6 model, respectively. Notice also that at relative low temperature modes with very low conversion rates, the model shows a 0% conversion. This is due to exclusion of near-to-zero points while developing the kinetic model, in order to better predict the whole operating range. Similarly, the CO kinetic model tends to show near total conversion for all high conversion modes, since these points were excluded when developing it for the same reason as above.

As can be seen, the maximum error at calibrating points was 4% in outlet temperature and 4.4% in heat losses. Out of training conditions, the maximum error in outlet was 5.9% in outlet temperature and 5.4% in heat losses.

It is difficult to find published kinetic parameters that fit the performance of a particular DOC, due mainly to the different washcoat materials and loadings of the monoliths and also to ageing effects of the DOC. A method to develop tailored kinetic parameters in the area of interest using numerical simulations was presented. The results show that although error in chemical conversion predictions reaches almost 12%, the error in the overall performance of the thermal model is lower. This is due to the low concentration of reactant species in the exhaust gas (see Table 4): small errors in conversion do not involve big changes in total generated enthalpy of chemical reactions in the DOC. It can be concluded that the model shows good agreement with empirical data.

Regarding pressure drop, the linear correlation fits the data appropriately but for high velocities, the deviation can be larger (as seen in Figure 9).

Complex numerical schemes and boundary layer refinement were proven not necessary to obtain reliable results. Standard wall-functions showed good results in comparison with other, more demanding approaches.

4.2. Energy Recovery Considerations for Exhaust Systems

DOC outlet flow analysis showed that modules in thermoelectric generator devices placed after catalytic converters may have different performances not only because of the internal design of the devices but also because of the nature of the exhaust flow. It was found that the smaller the flow, the more uneven the flow temperature distribution is.

This can affect recovery devices being placed in the exhaust system. For instance, modules in a thermoelectric generator expected to have the same behaviour (for instance, upper and lower part thermoelectric modules in a flat-shaped, symmetrical design) may show a different performance due to exhaust flow structures.

Awareness of this effect be used to allow for electrical connections between modules depending on differences in voltage output from modules caused by variations in temperature. Even various types of modules can be selected in each zone to optimize the power output result for the different ranges of temperature.

Loss in gas temperature with distance from DOC outlet was quantified. This will help to assess the trade-off between placing the recovery device near after-treatment systems (higher temperatures) or not (easiness in installation). Notice that temperature falls fast in a relatively short distance, reducing the amount of achievable harvested power.

5. Conclusions

Experimental exhaust data obtained for most common light-duty diesel engines was used to construct a 3D model. A complete methodology to build 3D CFD exhaust models is presented, including how to estimate main parameters needed and numerical approaches. A 2D axi-symmetrical complementary model was shown to be useful to overcome the necessity of chemical parameters fitted to a specific model of catalytic converter.

A computational model incorporating these findings will lead to accurate and faster numerical calculations of analysis of recoverable energy in exhaust systems. This will allow more designs to be explored in a given time.

It has been pointed out that the particular nature of the exhaust flow leaving catalytic converters has to be taken into account for energy-harvesting calculations, since the end of after-treatment systems is the indicated position to place energy recovery devices.

Furthermore, temperature losses caused by placing the recovery device distant from the DOC were evaluated. This information can be used in the evaluation process of the position of a recovery device in an automotive exhaust system. Since temperature falls promptly, it is advised to place the device as close as possible.

In works regarding simulations of recovery devices, if not enough experimental information is available, it is encouraged to include upstream elements of the exhaust that could alter the flow significantly in order to better predict their performance when installed in vehicles.

Acknowledgments: Authors wish to thank the financial support provided by the Spanish Ministry of Economy and Competitiveness to the project POWER Ref. ENE2014-57043-R and Universidad de Castilla—la Mancha for the pre-doctoral funding [2015/4062]. Authors also wish to thank the technical support provided by NISSAN Iberica Co. and CMT-Motores Térmicos Institute at Universitat Politécnica de Valencia.

Author Contributions: Pablo Fernández-Yáñez developed the model; Octavio Armas directed and supervised all the work done; Arántzazu Gómez assisted with the experimental work; Antonio Gil assisted with the development of the model.

Conflicts of Interest: There is no conflicts of interest.

Nomenclature

A	Pre-exponential factor ($m^3/(mol \cdot s)$)
B	Pressure drop linear fit constant (m^2)
C	Reaction rate linear fit constant
D	Reaction rate linear fit constant (K)
c_p	Specific heat at constant pressure ($J/(kgK)$)
d	Diameter (m)
F	External body forces (N)
g	Gravitational acceleration (m/s^2)
G	Inhibition factor (K)
h	Convection heat transfer coefficient (W/m^2K)
J	Diffusion term (kg/m^2s)

k	Arrhenius kinetic constant ($m^3/(mol{\cdot}s)$)
K	Thermal conductivity (W/mK)
L	Length (m)
$\dot{m}$	Mass flow (kg/s)
p	Pressure (Pa)
$\dot{Q}$	Heat loss (W)
R	Reaction rate ($mol/m^3 s$)
S	Source term
T	Temperature ($^\circ$C, K)
u	Vector component of velocity (m/s)
x	Cartesian coordinate (m)
X	Horizontal axis variable
Y	Vertical axis variable

Greek

Δ	Variation
β	Inertial forces coefficient (m)
ε	Surface material emissivity
κ	Darcy's constant (m^2)
μ	Dynamic viscosity
ρ	Density (kg/m^3)
σ	Stefan-Boltzmann constant ($W/(m^2 K^4)$)

Subscripts

∞	Ambient
eff	Effective
g	Exhaust gas
i	Cartesian coordinate
in	Inlet
out	Outlet
m	Species
S	Solid
T	Thermal
$wall$	Pipe wall

1. Hossain, S.N.; Bari, S. Waste heat recovery from exhaust of a diesel generator set using organic fluids. *Procedia Eng.* **2014**, *90*, 439–444. [CrossRef]
2. Galindo, J.; Dolz, V.; Royo-Pascual, L.; Haller, R.; Melis, J. Modeling and Experimental Validation of a Volumetric Expander Suitable for Waste Heat Recovery from an Automotive Internal Combustion Engine Using an Organic Rankine Cycle with Ethanol. *Energies* **2016**, *9*, 279. [CrossRef]
3. Zhang, X.; Romzek, M. Computational Fluid Dynamics (CFD) Applications in Vehicle Exhaust System. *SAE Tech. Pap.* **2008**. [CrossRef]
4. Konstantinidis, P.A.; Koltsakis, G.C.; Stamatelos, A.M. Transient heat transfer modelling in automotive exhaust systems. *Proc. Inst. Mech. Eng. C J. Mech. Eng. Sci.* **1997**, *211*, 1–15. [CrossRef]
5. Kandylas, I.P.; Stamatelos, A.M. Engine exhaust system design based on heat transfer computation. *Energy Convers. Manag.* **1999**, *49*, 1057–1072. [CrossRef]
6. Guardiola, C.; Dolz, V.; Pla, B.; Mora, J. Fast estimation of diesel oxidation catalysts inlet gas temperature. *Control Eng. Pract.* **2016**, *56*, 148–156. [CrossRef]
7. Shayler, P.J.; Hayden, D.J.; Ma, T. Exhaust System Heat Transfer and Catalytic Converter Performance. *SAE Tech. Pap.* **1999**. [CrossRef]
8. Kapparos, D.J.; Foster, D.E.; Rutland, C.J. Sensitivity Analysis of a Diesel Exhaust System Thermal Model. *SAE Tech. Pap.* **2004**. [CrossRef]
9. Fu, H.; Chen, X.; Shilling, I.; Richardson, S. A One-Dimensional Model for Heat Transfer in Engine Exhaust Systems. *SAE Tech. Pap.* **2005**. [CrossRef]

10. Fortunato, F.; Caprio, M.; Oliva, P.; D'Aniello, G.; Pantaleone, P.; Andreozzi, A.; Manca, O. Numerical and Experimental Investigation of the Thermal Behavior of a Complete Exhaust System. *SAE Tech. Pap.* **2007**. [CrossRef]

11. Chuchnowski, W.; Tokarczyk, J.; Stankiewicz, K.; Woszczynski, M. Method for modelling temperature distribution in exhaust system of diesel engine in the light of mine systems of heat recuperation. *J. KONES Powertrain Transp.* **2001**, *18*, 101–108.

12. Guoquan, X. Transient simulation of heat transfers for vehicle exhaust system. In Proceedings of the 7th International Conference on Fluid Mechanics, Qingdao, China, 24–27 May 2015.

13. Durat, M.; Parlak, Z.; Kapsiz, M.; Parlak, A.; Fiçici, F. CFD and experimental analysis on thermal performance of exhaust system of a spark ignition engine. *J. Therm. Sci. Technol.* **2013**, *33*, 89–99.

14. Liu, X. Experiments and simulations on heat exchangers in thermoelectric generator for automotive application. *Appl. Therm. Eng.* **2014**, *71*, 364–370. [CrossRef]

15. Hamedi, M.R.; Tsolakis, A.; Herreros, J.M. Thermal Performance of Diesel Aftertreatment: Material and Insulation CFD Analysis. *SAE Tech. Pap.* **2014**. [CrossRef]

16. Fernández-Yáñez, P.; García-Contreras, R.; Gutiérrez, J.; Agudelo, A.F. Potential for energy recovery from heat transfer rate through the exhaust pipe wall in a light duty diesel engine. In Proceedings of the Energy and Environment Knowledge Week, Paris, France, 28–29 October 2016.

17. Voltz, S.E.; Morgan, C.R.; Liederman, D.; Jacob, S.M. Kinetic Study of Carbon Monoxide and Propylene Oxidation on a Platinum Catalyst. *Ind. Eng. Chem. Prod. Res. Dev.* **1973**, *12*, 294–301. [CrossRef]

18. Oh, S.H.; Cavendish, J.C. Transients of Monolithic Catalytic Converters: Response to Step Changes in Feedstream Temperature as Related to Controlling Automobile Emissions. *Ind. Eng. Chem. Prod. Res. Dev.* **1982**, *21*, 29–37. [CrossRef]

19. Dubien, C.; Schweich, D.; Mabilon, G.; Martin, B.; Prigent, M. Three-way catalytic converter modelling: Fast- and slow-oxidizing hydrocarbons, inhibiting species, and steam-reforming reaction. *Chem. Eng. Sci.* **1998**, *53*, 471–481. [CrossRef]

20. Koltsakis, G.C.; Konstantinidis, P.A.; Stamatelos, A.M. Development and Application Range of Mathematical Models for 3-way Catalytic Converters. *Appl. Catal. B Environ.* **1997**, *12*, 161–191. [CrossRef]

21. Wurzenberger, J.; Wanker, R.; Schüßler, M. Simulation of Exhaust Gas Aftertreatment Systems—Thermal Behavior During Different Operating Conditions. *SAE Tech. Pap.* **2008**. [CrossRef]

22. Guojian, W.; Song, T. CFD Simulation of the effect of upstream flow distribution on the light-off performance of a catalytic converter. *Energy Convers. Manag.* **2005**, *46*, 2010–2031. [CrossRef]

23. Hayes, R.E.; Fadic, A.; Mmbaga, J.; Najaf, A. CFD modelling of the automotive catalytic converter. *Catal. Today* **2012**, *188*, 94–105. [CrossRef]

24. Chatterjee, D.; Deutschamnn, O.; Warnatz, J. Detailed surface reaction mechanism in a three-way catalyst. *Faraday Discuss.* **2001**, *119*, 371–384. [CrossRef]

25. Kumar, R.; Sonthalia, A.; Goel, R. Experimental study on waste heat recovery from an internal combustion engine using thermoelectric technology. *Therm. Sci.* **2011**, *15*, 1011–1022. [CrossRef]

26. Pong, H.; Wallace, J.; Sullivan, P.E. Modeling of Exhaust Gas Treatment for Stationary Applications. *SAE Tech. Pap.* **2012**. [CrossRef]

27. Cárdenas, M.D.; Armas, O.; Mata, C.; Soto, F. Performance and pollutant emissions from transient operation of a common rail diesel engine fueled with different biodiesel fuels. *Fuel* **2016**, *185*, 743–762.

28. Cárdenas, M.D. Estudio de las Emisiones de Motores Diésel de Automoción en Condiciones de Funcionamiento Transitorias al Usar Biodiesel. Ph.D. Thesis, Universidad de Castilla-La Mancha, Albacete, Spain, 2016. (In Spanish)

29. Dullien, F.A. *Porous Media–Fluid Transport and Pore Structure*; Academic: New York, NY, USA, 1979.

30. Andrade, J.S.; Costa, U.M.S.; Almeida, M.P.; Makse, H.A.; Stanley, H.E. Inertial Effects on Fluid Flow through Disordered Porous Media. *Phys. Rev. Lett.* **1999**, *82*, 5249–5252. [CrossRef]

31. Benjamin, S.F.; Liu, Z.; Roberts, C.A. Automotive catalyst design for uniform conversion efficiency. *Appl. Math. Model.* **2004**, *28*, 559–572. [CrossRef]

32. Mladenov, N.; Koop, J.; Tischer, S.; Deutschmann, O. Modeling of transport and chemistry in channel flows of automotive catalytic converters. *Chem. Eng. Sci.* **2010**, *65*, 812–826. [CrossRef]

33. Patankar, S.V.; Spalding, D.B. A calculation procedure for heat, mass and momentum transfer in three-dimensional parabolic flows. *Int. J. Heat Mass Transf.* **1972**, *15*, 1787–1806. [CrossRef]

34. Shih, T.; Liou, W.W.; Shabbir, A.; Yang, Z.; Zhu, J. A New k-e Eddy-Viscosity Model for High Reynolds Number Turbulent Flows—Model Development and Validation. *Comput. Fluids* **1995**, *24*, 227–238. [CrossRef]

35. Agudelo, A.F.; García-Contreras, R.; Agudelo, J.R.; Armas, O. Potential for exhaust gas energy recovery in a diesel passenger car under European driving cycle. *Appl. Energy* **2016**, *174*, 201–212. [CrossRef]

36. Vazquez, J.; Sanz-Bobi, M.A.; Palacios, R.; Arenas, A. State of the art of thermoelectric generators based on heat recovered from the exhaust gases of automobiles. In Proceedings of the 7th European Workshop on Thermoelectrics, Pamplona, Spain, 3–4 October 2002.

37. Launder, B.E.; Spalding, B.I. The numerical computation of turbulent flows. *Comput. Methods Appl. Mech. Eng.* **1974**, *3*, 269–289. [CrossRef]

38. Incropera, F.P.; DeWitt, D.P. *Introduction to Heat Transfer*; John Wiley & Sons: New York, NY, USA, 2011.

Article

Hybrid Electric Vehicle Performance with Organic Rankine Cycle Waste Heat Recovery System

Amin Mahmoudzadeh Andwari [1,2,*], Apostolos Pesiridis [1], Apostolos Karvountzis-Kontakiotis [1] and Vahid Esfahanian [2]

[1] Centre for Advanced Powertrain and Fuels Research (CAPF), Department of Mechanical, Aerospace and Civil Engineering, Brunel University London, London UB8 3PH, UK; apostolos.pesiridis@brunel.ac.uk (A.P.); a.karvountzis@brunel.ac.uk (A.K.-K.)

[2] Vehicle, Fuel and Environment Research Institute, School of Mechanical Engineering, College of Engineering, University of Tehran, Tehran 1439956191, Iran; evahid@ut.ac.ir

* Correspondence: amin.mahmoudzadehandwari@brunel.ac.uk; Tel.: +44-(0)-1895-267901

Academic Editor: Jose Ramon Serrano
Received: 6 February 2017; Accepted: 20 April 2017; Published: 26 April 2017

Abstract: This study examines the implementation of a waste heat recovery system on an electric hybrid vehicle. The selected waste heat recovery method operates on organic Rankine cycle principles to target the overall fuel consumption improvement of the internal combustion engine element of a hybrid powertrain. This study examines the operational principle of hybrid electric vehicles, in which the internal combustion engines operates with an electric powertrain layout (electric motors/generators and batteries) as an integral part of the powertrain architecture. A critical evaluation of the performance of the integrated powertrain is presented in this paper whereby vehicle performance is presented through three different driving cycle tests, offering a clear assessment of how this advanced powertrain configuration would benefit under several different, but relevant, driving scenarios. The driving cycles tested highlighted areas where the driver could exploit the full potential of the hybrid powertrain operational modes in order to further reduce fuel consumption.

Keywords: internal combustion engine; organic Rankine cycle; hybrid electric vehicle; waste heat recovery; brake specific fuel consumption; New European Driving Cycle

---

1. Introduction

The current trend of the automotive industry focuses not only on maximizing the vehicle's performance, but also in minimizing the emissions and fuel consumption of the vehicle [1,2]. Several technologies have been developed since the 1990s, when worldwide emissions standards started to impose boundaries on the levels of acceptable emissions of vehicles, which resulted in ever-reducing levels of emissions, as well as fuel consumption [3,4]. These technologies include the adoption of forced induction with the use of a supercharger and/or a turbocharger, usage of electric energy through high-capacity batteries, and combinations of both [5–7]. This study aims to examine vehicle technology that uses the combination of the above technologies, which are known as electric hybrid vehicles, and they use an internal combustion engine (ICE) combined with an electric powertrain as part of the propulsion architecture of the vehicle. As a significant amount of fuel energy used to move any vehicle equipped with an ICE is lost as heat, several systems are developed to exploit the energy of this wasted heat in order to offer it back to the engine or vehicle in the form of increased power or as an additional source of electrical energy [8–10]. The organic Rankine cycle (ORC), which is examined in this work, uses the exhaust gases of the engine to heat an organic fluid and pass it through a series of devices in order to produce mechanical or electrical power, which can be employed to enhance various aspects of engine or vehicle performance. The final stage of the current study includes the evaluation of the developed hybrid model through three driving cycles: the New European Driving Cycle (NEDC) and two EPA federal tests (FTP-75, US06) used in USA.

The motivation for this particular project arises from the fact that limited research has been conducted for the use of ORC waste heat recovery in combination with gasoline engine-equipped light hybrid electric vehicles. The ORC system is a new technology that is not yet implemented in lightweight vehicles, not only because the cost is fairly high, but also because there is lack of evidence proving that this subsystem could operate efficiently with smaller capacity engines [11–14]. This study intends to assess the merits of implementing an ORC waste heat recovery system as part of a hybrid vehicle's powertrain, in particular its fuel consumption reduction potential. For this purpose the implementation of the ORC system is tested through two widely-used current drive cycles to further the utility of this assessment, thus increasing its value.

2. Operational Modes of Hybrid Vehicles

When a vehicle is moving on the road surface three driving modes can be defined: traction mode, braking mode, and coasting mode. In traction mode the vehicle is accelerating and its force overcomes inertia, while in braking mode the vehicle is decelerating and the brakes dissipate the kinetic energy. The coasting mode refers to the phase where the vehicle is free-rolling without propulsive power from the engine or braking force from the brakes being supplied. Similarly, for hybrid electric vehicles the same operational modes apply, but are more complicated, as the electric motor extends the range of the available driving modes. Moreover, there is a distinction between the operational modes of a series hybrid vehicle and a parallel hybrid vehicle, as their power delivery operating principles differ. These operational modes include the engine drive, electric drive, hybrid, power split and braking mode, which engage different powertrain parts in each type of hybrid electric vehicle. Parallel hybrid electric vehicles use the engine drive mode as a propulsive mode when the electric motor is switched off. The electric drive mode, on the other hand, uses only the electric motor to provide vehicle motion while the engine is disengaged. In hybrid mode, the vehicle is driven by both the ICE and the electric motor, while in the power split mode the engine is not only providing motion, but also charging the battery pack [15–19]. Finally, the braking mode includes the regenerative brake, the energy of which is stored in the battery for further use. In a series hybrid electric vehicle these modes retain the same operating principles, while the differences are based on the components that are used in each mode. The engine is disengaged from the drivetrain and is only connected to a generator in a series hybrid electric vehicle and the battery alone, transfers power to the traction motors [20,21].

3. Waste Heat Recovery

A large proportion of fuel energy in an internal combustion engine is lost as exhaust gas heat, a fact that reduces the overall engine efficiency of conventional vehicles [22]. Typical state-of-the-art waste heat recovery (WHR) technologies include mechanical or electric turbocompounding, bottoming cycles, and thermoelectric generators [1,5]. A mechanical turbocompound system uses the exhaust gas energy to spin a turbine, which can then be linked to the crankshaft, offering more output power and as much as 5% better fuel economy. On the other hand, an electric turbocompound setup can exploit the exhaust gas energy to gain electrical power that can be saved in a battery or used to support several electric components of the vehicle, increasing the fuel economy up to 10%. Currently, approximately a 2% reduction in fuel consumption can be offered by thermoelectric generators, which use the exhaust gas heat to produce electricity directly via thermoelectric conversion means [2,17,20]. This study focuses on the bottoming cycle type of waste heat recovery, which employs thermodynamic cycles to gain energy out of the exhaust gas heat. These cycles operate with a working fluid and heat exchangers, which absorb the heat of exhaust gases to change the fluid's state, which is then channeled towards a turbine for power generation. The most widely used thermodynamic cycles for this application are Rankine and Brayton, but in this particular work only the Rankine cycle is taken into consideration as the most widely-accepted and employed means of waste heat recovery [23,24]. A typical parameter that can predict the efficiency of a waste heat recovery system is the temperature of the exhaust gases, and as this temperature is increased, the amount and quality of energy that can be exploited will increase. Specifically, in the automotive industry the main types of vehicles that can benefit from a waste heat recovery system are the heavy-duty vehicles. These vehicles not only consume fuel in an excessive manner, but also produce great amounts of hazardous emissions, including NOx and CO_2 gases. The implementation of a waste heat recovery system in such vehicles can offer energy recovery benefits in various forms that can further increase the efficiency of the powertrain by approximately 10–15% and also reduce harmful emissions [4,9,14,25].

4. Organic Rankine Cycle (ORC)

The ORC system uses a series of devices in a closed loop in order to recover energy, which can then be fed back to the engine or vehicle as required. It consists of a heat exchanger (the evaporator) that the hot gases pass through and increase the temperature of the working fluid to the extent that it changes states from liquid to superheated vapor. After the evaporator, the working fluid expands isentropically providing work in an expansion device, which is connected to an electric generator. Then the expanded vapor passes through a condenser to change its state back to liquid and reaches the pump which increases its pressure and pushes the working fluid back to the evaporator to repeat the process. Figure 1a represents a temperature-enthalpy (T-S) diagram of the Rankine cycle, while Figure 1b illustrates a schematic view of a typical Rankine cycle with all major components. Theoretically, the ORC systems can offer several benefits to the whole powertrain configuration; at best, power output of the powertrain can be increased by a maximum of approximately 15% depending on the application [26–28]. As a consequence of the increased efficiency of the engine, the fuel consumption can be decreased, making the ORC systems desirable for various applications [1,9,29].

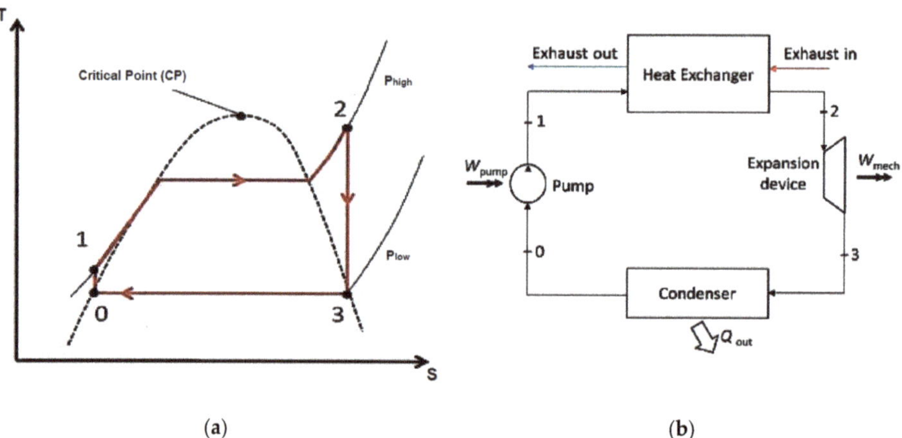

Figure 1. Working principle of typical Rankine cycle: (**a**) temperature-enthalpy diagram; and (**b**) schematic of Rankine cycle components.

Furthermore, a significant advantage of the ORC systems is that they do not depend on the pressure of the exhaust gases, as the only thing that affects the efficiency of the system are the exhaust gas temperature and mass flow rate. Comparing a conventional steam Rankine cycle to an ORC, the main difference arises from the fact that ORCs use an organic fluid (instead of water) and do not need a superheater in order for the fluid to reach the desired temperature for evaporating in order to spin the turbine [15,16,18].

5. Engine Model Calibration

In order to validate the engine model, a MATLAB (2015b, MathWorks, Natick, MA, USA, 2015) code was employed to produce a Brake Specific Fuel Consumption (BSFC) map which was identical to a chosen theoretical one. This theoretical BSFC map corresponded to the selected engine profile of a Ford 2000cc Zetec-SE DOHC engine (Ford Motors Company, London, UK), which is a four-cylinder engine with a 16-valve design and has a maximum Brake Mean Effective Pressure (BMEP) value of approximately 10 bar and a maximum engine speed at 5000 rpm. The BSFC map is shown in the figures below.

Following the selection of the theoretical engine configuration and BSFC map, the MATLAB code needed to be tested to determine the level of its accuracy. To accomplish this, the map was discretized at 29 different points that represent respective engine speed values and BMEP values. Particularly, the BSFC map was divided into five columns representing engine speeds from 1000 to 5000 rpm, and six rows representing BMEP values from 2 bar until reaching the maximum curve. This discretization is shown in Figure 2a with red dots.

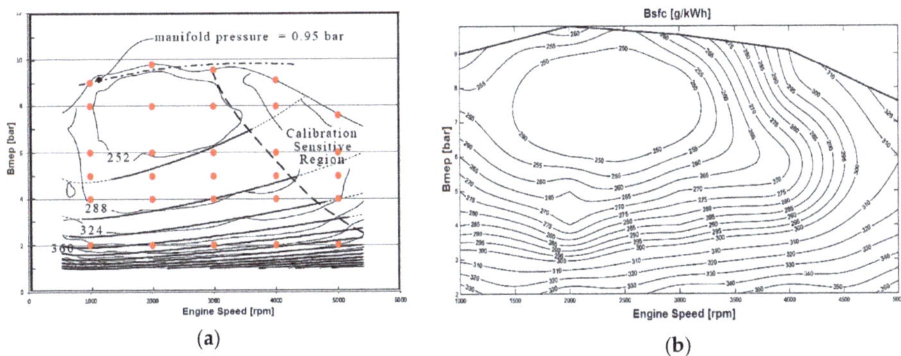

Figure 2. BSFC map of Ford 2.0 L Zetec-SE DOHC engine: (**a**) predicted and discretized BSFC map of 29 points; and (**b**) the BSFC map given from optimized MATLAB code.

All of these 29 points correspond to different BSFC values, which were estimated visually and gathered into a table that was later used as an input in the MATLAB code. The MATLAB code can produce different performance maps, depending on the input values given from the user and for this current simulation the BSFC values were used, along with the engine speed (*x*-axis) and BMEP values (*y*-axis). After the optimization a new BSFC map was obtained as shown in Figure 2b, pointing out to the satisfactory accuracy of the code. For validation of the code results the Ford engine was modelled in GT-Power and the results correlated against the MATLAB code data. The GT-Power models of both the engine and the dynamic HEV sub-models are shown in Figure 3a,b, respectively.

The complete model included a throttle controller that was adjusted accordingly in the simulations, while the values of the inlet and exhaust ports were estimated after benchmarking and suggestions of the software. The table that included these BSFC values was given as an input to the MATLAB code and the BSFC that was generated matched the BSFC values of the theoretical model accurately. The next step after validating that the engine model can produce accurate values was to modify the model so that the new engine displacement was implemented. When the changes were implemented, the simulations were run for thirty discrete engine/load cases. It is important to point out that this work did not aim to evaluate the effect of a dynamic ORC bottoming cycle on the overall driving cycle of the vehicle, but to provide a discrete-point assessment of a conventional ORC layout for implementation in a hybrid vehicle.

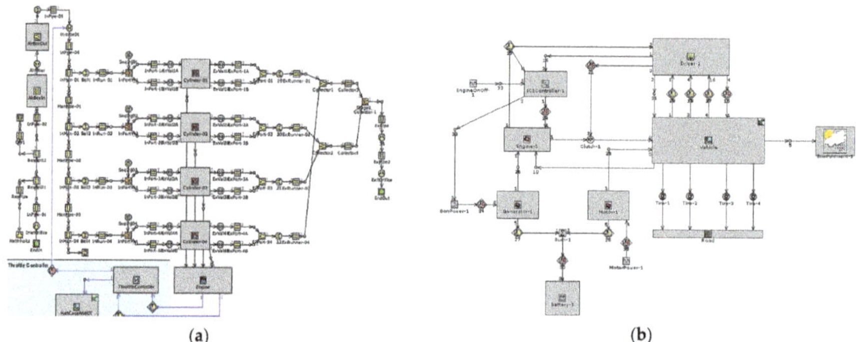

(a)   (b)

Figure 3. Ford 2.0 L Zetec-SE DOHC Engine modeled in GT-Power software (7.3b2, Gamma Technologies LLC, Westmont, IL, USA, 2015): (**a**) integrated conventional engine's components; (**b**) dynamic Hybrid Electric Vehicle (HEV) subsystem integrated with the base engine components.

6. Organic Rankine Cycle Fluid Selection and System Optimization

The selection of the appropriate organic fluid for each respective application is a demanding challenge for engineers since the performance and efficiency of the organic Rankine cycle system is seriously affected by the working fluid. However, the properties of the fluid are not the only criteria for selection, as the cost and the environmental impact of each fluid may limit the list of available fluids. As far as the cost is concerned, the engineer should decide which fluid would decrease the payback period and offer the maximum output and thermal efficiency at the same time. The properties of the organic fluids can be divided into four categories, each one of them being equally important for the efficient and safe operation of the ORC system. Thermodynamic properties of the organic fluids vary in several aspects, as do the density, viscosity, boiling point temperature, pressure, and the latent heat of vaporization. Each one of these parameters affect not only the thermal efficiency of the system, but also the design and construction of the respective internal combustion engine configuration. Critical and maximum operating conditions constitute the process-related properties and are linked with the efficiency of the organic Rankine cycle system. As far as the safety and environmental aspects are concerned, the toxicity and flammability of the fluid concerns engineers, while the global warming potential and ozone depletion danger are the major dangers for the environment. The working fluids should compromise among several criteria specified below [10,11,20,21,25,30–34]:

- Low condensation temperature
- Very low freezing point
- No need to superheat (dry fluid)
- Eco-friendly (low global warming potential and ozone depletion potential)
- Low flammability and toxicity

Considering all of aforementioned requirements, R245fa was selected as the working fluid in this study entirely considering prior experience and potential for widespread use. This organic fluid has no ozone impact (ODP), low global warming impact (GWP), it is non-flammable, and its thermodynamic properties fulfill the above criteria. After the selection of the working fluid, the final model was created. In GT-Power software some assumptions were considered and are shown below [2,12,13,18,22]:

- The evaporator exhaust gas initial pressure and temperature are equal with the exhaust gas outlet pressure and temperature, respectively.
- The condenser coolant initial pressure and temperature are equal with the coolant outlet pressure and temperature, respectively.
- No heat is lost to the surroundings.

The input values for the ORC model, such as exhaust temperature and exhaust mass flow rate, were obtained from a table regarding thirty different cases from 2000 cc engines at several engine speeds and BMEP ranges. Figure 4a illustrates the schematic view of the ORC system modeled in GT-Power. Data obtained included parameters such as the evaporator energy, turbine power, pump efficiency, turbine efficiency, and the pressure rise in the pump. The design parameters of the ORC system used in the simulation are presented in Table 1.

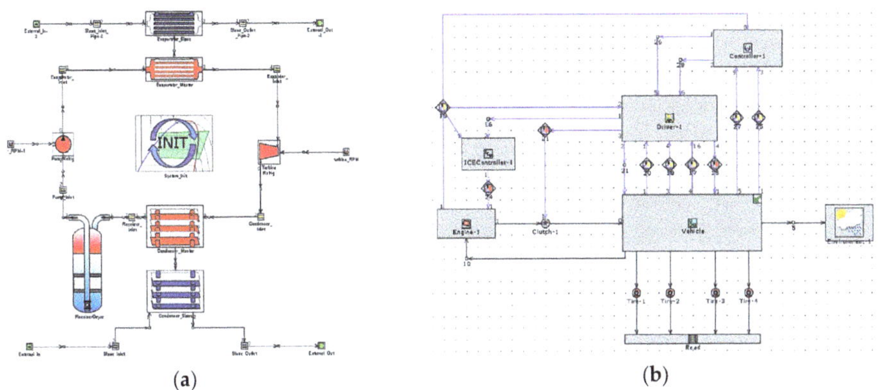

(a)   (b)

Figure 4. Schematic view of the engine system model in GT-Power: (**a**) organic Rankine cycle system sub-model; and (**b**) complete model used for driving cycle testing.

Table 1. Organic Rankine cycle component design parameters using R245fa refrigerant.

Design Parameters	ORC's Main Components					
	Evaporator (Slave)	Evaporator (Master)	Condenser (Slave)	Condenser (Master)	Turbine Expander	Pump
Average Inlet Pressure (bar)	1.00102	24.9	2.15	3.28	24.3	2.6
Average Outlet Pressure (bar)	1	24.3	2	2.6	3.28	24.9
Average Pressure Drop (bar)	0.0010197	0.631	0.148264	0.674932	-	-
Average Inlet Temperature (K)	973.1	315.8	296.1	405.1	445.2	314.1
Average Outlet Temperature (K)	450.7	445.2	302.6	314.1	405.263	315.8
Average Mass Flow Rate (g/s)	140	269.2	3394.6	269.3	0.269	0.269
Combined Energy Rate out of Fluid (kW)	78.7	−78.7	−73.2	73.2	-	-
Average Speed (rpm)	-	-	-	-	1350	2000
Average Map Pressure Ratio	-	-	-	-	7.37	-
Average Efficiency (%)	-	-	-	-	51.61	61.42
Average Power (kW)	-	-	-	-	5.3	0.75
Average Pressure Rise (bar)	-	-	-	-	-	22.3
Heat Exchanger Volume (L)	11.75	3.33	6.5	5.65	-	-
Heat Exchanger Reference Length (m)	0.007	0.007	0.003	0.003	-	-
Heat Exchanger Heat Transfer Area (m^2)	1.87	1.87	9.33	9.33	-	-
Heat Exchanger Flow Area (m^2)	0.0138	0.0038	0.02	0.02	-	-

7. Driving Cycles Testing

The complete development and simulation of the ORC system model provided useful data regarding the efficiency and fuel consumption of the new, improved powertrain. This fact enables a comparison to be made between the initial engine setup and the new, improved one. The best way to compare these two powertrains is to assign them to a conventional lightweight vehicle and test them for various driving cycles. In this way the percentage difference between these two powertrains can be obtained, showing how much difference the ORC system produced. The fuel consumption benefit was the main target of this assessment, but emissions were also computed, and the major concern of this study remains the minimization of the fuel consumption of a lightweight passenger vehicle with the use of an ORC WHR recovery system.

With the use of three typical driving cycles, the fuel consumption of the initial powertrain and the powertrain with the ORC system was measured to define the extent that the ORC system benefits the vehicle. The driving cycles include:

- NEDC (New European Driving Cycle, EU)
- FTP-75 (Federal Test Procedure, US)
- US06 (a more realistic, aggressive supplement to FTP-75)

All of these driving cycles represent different driving scenarios, different durations, and average speeds. In this way, the comparison between the two powertrains (the original standalone hybrid and the hybrid powertrain equipped with the ORC WHR system) is fair and also shows where space for improvement exists if the ORC system is to be implemented in a HEV vehicle. When the developed ORC system is assigned to a HEV vehicle, different hybrid driving modes can be selected for further improved fuel economy in each driving cycle, regardless of the configuration of the HEV vehicle (series, parallel, or complex). These tests were run in GT-Power, where a virtual vehicle was modeled with an average weight of 1500 kg and several parameters set to simulate the driving scenarios. The library of the software includes all of the selected driving cycles, which ensures the precision and consistency of the results. The complete model for this test includes several parameters that were calibrated to achieve a model that fulfilled the requirements of the simulation, which is shown in Figure 4b.

8. Results and Discussion

In this section the results from the ORC simulations are presented, analyzing how efficiently the integrated system operated and where room for improvement could be identified. Lastly, the exact figures of the vehicle's fuel consumption are shown, respectively, for each of the three different driving cycles and constitute the deciding factor for the selection of the optimum hybrid profile for the integrated powertrain. Several aspects that control the efficiency of the ORC system and how this system is combined with the IC engine to provide more power to the wheels are analyzed in this section. It is worth stating at this stage that the global assumptions considered for the ORC configuration of the HEV powertrain, namely, that even though the backpressure caused by the ORC's evaporator placed in the exhaust pipe, and also all of the extra weight imposed by ORC components to the vehicle system, deteriorate the overall performance and the fuel consumption of the vehicle to some extent. In this early evaluation study of ORC WHR systems for HEVs, it has been assumed that the effects of these are negligible, but should be borne in mind by the reader when considering the results.

The main target set for the results of the ORC analysis is that the maximum efficiency of the turbine and the pump does not have a percentage difference of higher than 15%. Moreover, the power output of the turbine is critical, as it is added to the power output of the IC engine to provide the total power of the powertrain, so it is required to be as high as possible. Another parameter that is tested in this analysis is the pressure rise in the pump which indicates the degree to which the pump ensures the successful provision of the required pressure level throughout the entire cycle. Finally, this section concludes with the presentation of the new BSFC map of the upgraded powertrain (with ORC). The distribution of the turbine and pump efficiency is illustrated in Figure 5a,b, respectively, over the tested engine load and speed range. In the figures it can be observed how the efficiencies of these two components fluctuate with engine speed and the BMEP, as well as how the contours of constant efficiency are formed inside the plot. The maximum value of turbine efficiency is 61% and it is obtained at part-load, while the maximum efficiency of the pump is 72% at higher BMEP values and engine speeds. Great effort was expended to reach the highest efficiencies of these two components and resulted in an efficiency improvement of almost 10% from the initial model.

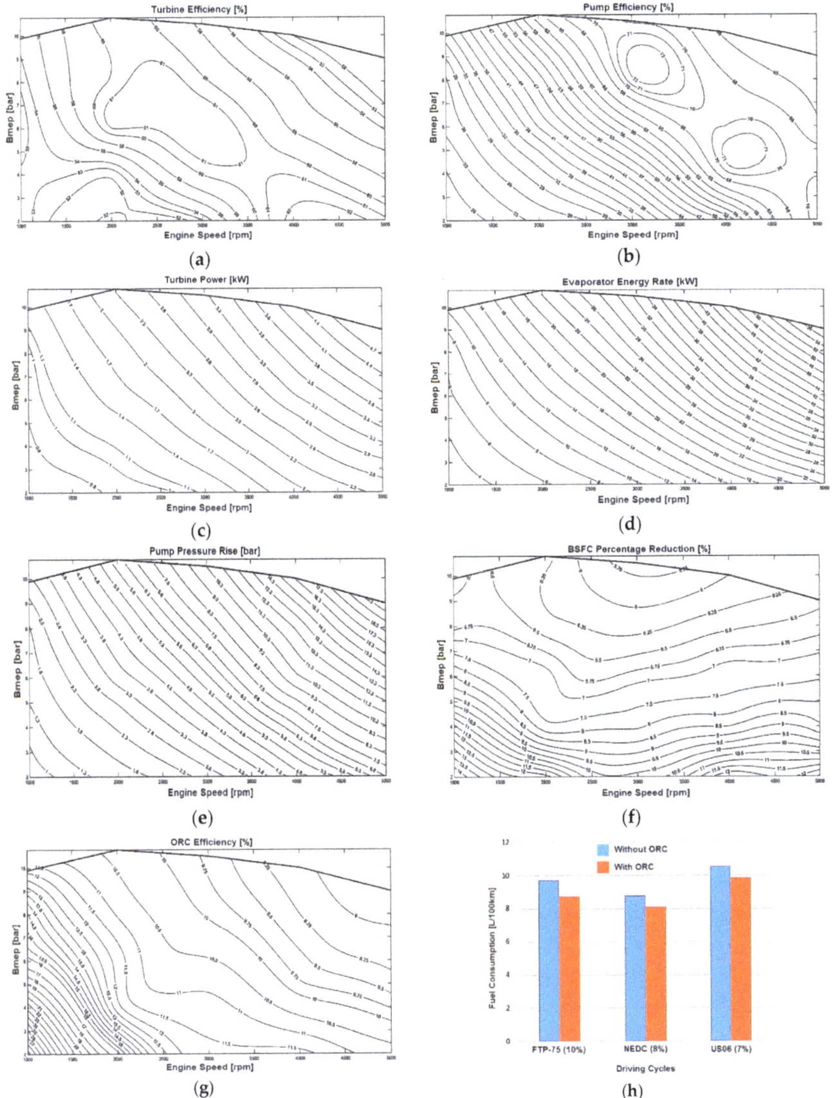

Figure 5. Schematic view of the engine system model in GT-Power code: (**a**) turbine efficiency distribution; (**b**) pump efficiency distribution; (**c**) turbine power distribution; (**d**) EVAPORATOR energy rate; (**e**) Pump pressure rise; (**f**) BSFC percentage reduction distribution; (**g**) ORC efficiency distribution; and (**h**) fuel consumption per driving cycle.

Figure 5c represents the allocation of the turbine power output over the tested range of engine loads and speeds. It can be observed that the maximum power is obtained at full load and high engine speeds because the exhaust gases have greater temperature and mass flow at these speeds. The plot shows that the maximum power output that the turbine produces is almost 5 kW at high engine speed and load. At part-load (around the median of 6 bar) and lower engine speeds the turbine can produce typical values of around 2.5 kW. These values may seem small at first sight, but they proved to be high enough to cause a decrease in BSFC of almost 7% under the same operating conditions.

Figure 5d,e provides information about the evaporator energy rate and the pump pressure rise, respectively. Figure 5f illustrates the BSFC reduction contour plot superimposed on the engine map. It can be interpreted from the plot that the BSFC percentage reduction differs for different engine loads and speeds with the maximum values of the reduction being obtained at the lower BMEP values, reaching a maximum 12% decrease. In addition, as the BMEP values increase at the mid-load range (6 bar), the average percentage reduction is approximately 7%, which is higher than the set target. This region is the one of highest interest as the vehicle spends most of the driving time in this region, so the 7% decrease in fuel consumption can highly benefit the hybrid vehicle. At higher engine loads and BMEP values of 9 bar, the BSFC reduction percentage is lower, but always above 5%.

One final parameter that can be examined regarding the ORC system is the overall efficiency of the system that produced the above reduction in BSFC. The efficiency of the ORC system is defined as the fraction of the turbine power output over the evaporator energy rate, showing how efficiently the available energy is used inside the ORC WHR system. However, as for this investigation a four-cylinder 2000 cc engine was selected, the available energy rate is limited and the constraints inside the system are greater when compared to a larger (commercial-type) diesel engine (which can exploit the full potential of the system and reach higher efficiencies). In order to mitigate this drawback, the ORC system has been modified to give the higher possible efficiency.

Figure 5g shows the distribution of the ORC efficiency superimposed on the engine map. It can be seen that, overall, the ORC efficiency fluctuates at different engine speeds and loads, as well as the efficiency values are not very high. At lower engine speeds and BMEP values the ORC efficiency reaches a maximum of 27%, which is expected because the energy rate is fairly low, while at least the simulated turbine power has already reached near-maximum efficiencies even from lower loads and speeds. As the engine speeds and loads rise, the evaporator energy rate increases, resulting in the decrease of the efficiency because the turbine power is not increasing accordingly due to restrictions in the ORC system design and in the exhaust mass flow rate and temperature. The ORC efficiency reaches an average value of 10.5% at the engine speed of 3000 rpm and BMEP of 6.5 bar, which is an important region of the powertrain operating points. A further increase in engine speed and BMEP values show an efficiency of 9%, which is the minimum percentage obtained in this study and for the entire engine operating range. In general, the efficiency of the ORC is significant, but it is not reaching its full potential due to several inherent limitation of the smaller capacity powertrain. However, with this efficiency distribution, the ORC system is able to produce an average decrease in BSFC values of 7%, which is significant compared to a conventional HEV configuration.

Hereinafter, the overall results of the project are further interpreted in order to obtain a clear ORC WHR system characterization of its performance and benefit that it brings to the conventional HEV architecture. In addition, the discussion indicates how the different hybrid modes of the vehicle should be chosen in order to further decrease the fuel consumption of the vehicle; specifically, the discussion identifies which hybrid mode should be chosen in the different driving scenarios, such that the vehicle can operate more efficiently, while consuming the minimum amount of fuel.

During the various driving cycle tests that were performed in the engine simulation software, the vehicle model was run with and without the ORC configuration, so that the absolute difference between these two setups could be distinguished. Moreover, the three different driving cycle tests can validate the efficiency of the whole powertrain in different driving conditions, providing useful information about the hybrid mode which emerges in each case. Finally, these tests revealed possible areas where the ORC system could be further optimized for greater HEV vehicle benefit.

For the FTP-75 driving cycle test the obtained data from the analysis showed that the fuel consumption of the vehicle without the ORC system was 9.74 L/100 km. When the ORC system was included in the powertrain, the new fuel consumption was calculated to be 8.76 L/100 km, which is almost 1 L less per 100 km.

On the other hand, on the NEDC, the results showed that the fuel consumption without the ORC system was 8.79 L/100 km, while when the ORC system was included in the analysis the new fuel consumption was 8.07 L/100 km. The difference between these two values is almost 0.8 L/100 km.

The final driving cycle test was the US06, giving a fuel consumption of 10.57 L/100 km without the ORC system. When the ORC system was added in the analysis, the improved fuel consumption was decreased to 9.84 L/100 km, which is almost 0.7 L/100 km less. This driving cycle is more aggressive than the previous two, which is why the powertrain reached higher fuel consumption values in comparative terms to NEDC and FTP-75.

Moreover, the results from the driving cycle analysis can assist the understanding of how the ORC system cooperates with the engine under different loads, accelerations, and decelerations. The three selected driving cycles constitute a representative example of driving regulations in use today and which were required to validate HEV performance with the novel component (ORC WHR system).

In Figure 5h the bar charts illustrate the differences in fuel consumption between the original HEV engine and the ORC-equipped equivalent in order to minimize fuel consumption. Furthermore, the percentage difference is calculated, so that the absolute difference of the improved powertrain can be differentiated. This bar chart includes information for all three of the driving cycle tests and presents the exact amount of burned fuel per 100 km of driving. It can be noticed that in all of the driving cycle scenarios, the fuel consumption was decreased after the implementation of the ORC system. The largest reduction was found to be in the FTP-75 driving cycle and the calculation of the absolute percentage difference showed a 10% lower fuel requirement for 100 km. The NEDC required the least amount of fuel, as it is not as aggressive as the other two cycles. Moreover, in this driving cycle the ORC system improved the fuel consumption by 8%, which is a significant reduction if the strict European regulations are taken into consideration. Finally, the US06 driving cycle required the most fuel overall, but the ORC system managed to lower the fuel consumption of the vehicle by 7%, which resulted in a drop of fuel consumption below 10 L/100 km.

A closer look into the three driving cycles was required to reveal the driving strategy needed in order to maximize the powertrain performance in the electric hybrid vehicle. Taking into account the three major hybrid operation modes, which are the traction mode, braking mode, and coasting mode, the driver can be guided in how to handle the accelerator pedal, the braking pedal, and the clutch pedal to maximize the efficiency of the overall vehicle.

As far as the NEDC is concerned, it can be identified that it consists of numerous accelerations and decelerations, and the average cruising speed is 60 km/h. This driving cycle profile can exploit the braking and coasting operation modes of the hybrid configuration to gain advantages regarding the fuel consumption of the vehicle. During the decelerations the brake should be lightly applied and this will lead to the battery being charged by dissipating the braking energy and, hence, more power can be delivered to the wheels during the acceleration zones. Moreover, the driver can use the coasting mode, while no brakes or throttle are applied, before reaching the deceleration zones and prevent, in that way, the engine from burning fuel in a non-beneficial manner.

The FTP-75 driving cycle has a complicated profile of accelerations and decelerations, which makes it difficult for the driver to adopt the optimum driving strategy. However, as there are multiple fluctuations in the speed profile of this cycle, the driver can exploit the braking mode to charge the battery of the hybrid setup. This gained power can be used during steep acceleration gradients to assist the work of the IC engine, leading to less fuel being burned during the numerous accelerations. As long as there are no flat speed regions in this driving cycle, the coasting mode can hardly be applied and cannot influence a further reduction of the fuel consumption.

Finally, the US06 driving cycle results exhibit lower fluctuations in the speed profile compared to the FTP-75, which means that the coasting mode can be applied, offering significant benefits in fuel consumption reduction. This driving cycle pushes the vehicle to reach high speeds of more than 120 km/h, which undoubtedly increases the consumption of fuel of the vehicle. In order to counteract this, judicious use of the accelerator pedal would be recommended, along with use of the power of the electric motor to assist with significant accelerations. On the other hand, the braking mode can be used to gather the essential electric power in the battery and the coasting mode can be used when the car needs to decrease its speed without the strict application of the brakes.

9. Conclusions

A simulation study has been conducted to prove that a waste heat recovery method can be employed in a hybrid electric vehicle and offer significant benefits regarding the overall performance of the vehicle, primarily in terms of fuel consumption reduction. An organic Rankine cycle-based waste heat recovery system was chosen as the method by which to extract further work from the conventional elements of this hybrid powertrain. After utilizing this system in the powertrain, the results showed the overall fuel consumption of the vehicle was decreased significantly. In the simulation the engine was run for three different driving cycle tests, including FTP-75, NEDC, and US06, in order to validate the real improved performance of the developed powertrain in different driving conditions, and ultimately led to an efficient driving strategy for the driver of the electric hybrid vehicle that could lower fuel consumption even further.

Acknowledgments: This work was not financially supported or funded by any organization/company and the authors would like to acknowledge Marios Kalamakis who performed the simulation procedure.

Author Contributions: Amin Mahmoudzadeh Andwari and Vahid Esfahanian have written the paper context and performed results presentation. Apostolos Pesiridis and Apostolos Karvountzis-Kontakiotis have carried out the design of experiment in the simulation and have extracted the data analysis.

Conflicts of Interest: The authors declare no conflict of interest.

1. Pesiridis, A. *Automotive Exhaust Emissions and Energy Recovery*; Nova Science Publishers: New York, NY, USA, 2014.
2. Oyewunmi, O.A.; Markides, C.N. Thermo-economic and heat transfer optimization of working-fluid mixtures in a low-temperature organic Rankine cycle system. *Energies* **2016**, *9*, 448. [CrossRef]
3. Andwari, M.A.; Aziz, A.A.; Farid, M.; Said, M.; Zulkarnain, A.L. Controlled auto-ignition combustion in a two-stroke cycle engine using hot burned gases. *Appl. Mech. Mater.* **2013**, *388*, 201–205. [CrossRef]
4. Teng, H.; Regner, G.; Cowland, C. Waste heat recovery of heavy-duty diesel engines by organic Rankine cycle Part I: Hybrid energy system of diesel and Rankine engines. *SAE Tech. Pap.* **2007**. [CrossRef]
5. Feneley, A.J.; Pesiridis, A.; Andwari, A.M. Variable geometry turbocharger technologies for exhaust energy recovery and boosting—A review. *Renew. Sustain. Energy Rev.* **2017**, *71*, 959–971. [CrossRef]
6. Kulkarni, K.; Sood, A. Performance analysis of organic Rankine cycle (ORC) for recovering waste heat from a heavy duty diesel engine. *SAE Tech. Pap.* **2015**. [CrossRef]
7. Stanzel, N.; Streule, T.; Preißinger, M.; Brüggemann, D. Comparison of cooling system designs for an exhaust heat recovery system using an organic Rankine cycle on a heavy duty truck. *Energies* **2016**, *9*, 928. [CrossRef]
8. Amin, M.A.; Azhar, A.A. Homogenous Charge Compression Ignition (HCCI) technique: A review for application in two-stroke gasoline engines. *Appl. Mech. Mater.* **2012**, *165*, 53–57. [CrossRef]
9. Zhang, X.; Zeng, K.; Bai, S.; Zhang, Y.; He, M. Exhaust recovery of vehicle gasoline engine based on organic Rankine cycle. *SAE Tech. Pap.* **2011**. [CrossRef]
10. Zhou, L.; Tan, G.; Guo, X.; Chen, M.; Ji, K.; Li, Z.; Yang, Z. Study of energy recovery system based on organic Rankine cycle for hydraulic retarder. *SAE Tech. Pap.* **2016**. [CrossRef]
11. Ringler, J.; Seifert, M.; Guyotot, V.; Hübner, W. Rankine cycle for waste heat recovery of IC engines. *SAE Int. J. Engines* **2009**, *2*, 67–76. [CrossRef]
12. Shu, G.; Zhao, J.; Tian, H.; Wei, H.; Liang, X.; Yu, G.; Liu, L. Theoretical analysis of engine waste heat recovery by the combined thermo-generator and organic Rankine cycle system. *SAE Tech. Pap.* **2012**. [CrossRef]
13. Sprouse, C., III; Depcik, C. Organic Rankine cycles with dry fluids for small engine exhaust waste heat recovery. *SAE Int. J. Altern. Powertrains* **2013**, *2*, 96–104. [CrossRef]
14. Zhang, X.; Mi, C. *Vehicle Power Management; Modeling, Control and Optimization*; Springer: London, UK, 2011.

15. Arsie, I.; Cricchio, A.; Pianese, C.; Ricciardi, V.; De Cesare, M. Modeling and optimization of organic Rankine cycle for waste heat recovery in automotive engines. *SAE Tech. Pap.* **2016**. [CrossRef]

16. Boretti, A. Improving the efficiency of turbocharged spark ignition engines for passenger cars through waste heat recovery. *SAE Tech. Pap.* **2012**. [CrossRef]

17. Kirmse, C.J.W.; Oyewunmi, O.A.; Haslam, A.J.; Markides, C.N. Comparison of a novel organic-fluid thermofluidic heat converter and an organic Rankine cycle heat engine. *Energies* **2016**, *9*, 479. [CrossRef]

18. Cipollone, R.; Di Battista, D.; Perosino, A.; Bettoja, F. Waste heat recovery by an organic Rankine cycle for heavy duty vehicles. *SAE Tech. Pap.* **2016**. [CrossRef]

19. Cochran, D.L. Working fluids for high temperature, Rankine cycle, space power plants. *SAE Tech. Pap.* **1961**. [CrossRef]

20. El Chammas, R.; Clodic, D. Combined cycle for hybrid vehicles. *SAE Tech. Pap.* **2005**. [CrossRef]

21. Lodwig, E. Performance of a 35 HP organic Rankine cycle exhaust gas powered system. *SAE Tech. Pap.* **1970**. [CrossRef]

22. Heywood, J.B. *Internal Combustion Engine Fundamentals*; McGraw-Hill: New York, NY, USA, 1988.

23. Chen, T.; Zhuge, W.; Zhang, Y.; Zhang, L. A novel cascade organic Rankine cycle (ORC) system for waste heat recovery of truck diesel engines. *Energy Convers. Manag.* **2017**, *138*, 210–223. [CrossRef]

24. Hsieh, J.-C.; Fu, B.-R.; Wang, T.-W.; Cheng, Y.; Lee, Y.-R.; Chang, J.-C. Design and preliminary results of a 20-kW transcritical organic Rankine cycle with a screw expander for low-grade waste heat recovery. *Appl. Therm. Eng.* **2017**, *110*, 1120–1127. [CrossRef]

25. Read, M.; Smith, I.; Stosic, N.; Kovacevic, A. Comparison of organic Rankine cycle systems under varying conditions using turbine and twin-screw expanders. *Energies* **2016**, *9*, 614. [CrossRef]

26. Shu, G.; Wang, X.; Tian, H. Theoretical analysis and comparison of rankine cycle and different organic rankine cycles as waste heat recovery system for a large gaseous fuel internal combustion engine. *Appl. Therm. Eng.* **2016**, *108*, 525–537. [CrossRef]

27. Wang, E.; Yu, Z.; Zhang, H.; Yang, F. A regenerative supercritical-subcritical dual-loop organic Rankine cycle system for energy recovery from the waste heat of internal combustion engines. *Appl. Energy* **2017**, *190*, 574–590. [CrossRef]

28. Yang, F.; Zhang, H.; Yu, Z.; Wang, E.; Meng, F.; Liu, H.; Wang, J. Parametric optimization and heat transfer analysis of a dual loop ORC (organic Rankine cycle) system for CNG engine waste heat recovery. *Energy* **2017**, *118*, 753–775. [CrossRef]

29. Kolasiński, P.; Błasiak, P.L.; Jozef, R. Experimental and numerical analyses on the rotary vane expander operating conditions in a micro organic Rankine cycle system. *Energies* **2016**, *9*, 606. [CrossRef]

30. Reck, M.; Randolf, D. An organic Rankine cycle engine for a 25-passenger bus. *SAE Tech. Pap.* **1973**. [CrossRef]

31. Thaddaeus, J.; Pesiridis, A.; Karvountzis-Kontakiotis, A. Design of variable geometry waste heat recovery turbine for high efficiency internal combustion engine. *Int. J. Sci. Eng. Res.* **2016**, *7*, 1001–1017.

32. Karvountzis-Kontakiotis, A.; Pesiridis, A.; Zhao, H.; Franchetti, B.; Pesmazoglou, I.; Alshammari, F.; Tocci, L. Effect of an ORC waste heat recovery system on diesel engine fuel economy for off-highway vehicles. In Proceedings of the SAE World Congress, Detroit, MI, USA, 4–6 April 2017.

33. Franchetti, B.; Pesiridis, A.; Pesmazoglou, I.; Sciubba, E.; Tocci, L. Thermodynamic and technical criteria for the optimal selection of the working fluid in a mini-ORC. In Proceedings of the 29th International Conference on Efficiency, Cost, Optimization, Simulation and Environmental Impact of Energy Systems (ECOS 2016), Portorož, Slovenia, 19–23 June 2016.

34. Karvountzis-Kontakiotis, A.; Alshammari, F.; Pesiridis, A.; Franchetti, B.; Pesmazoglou, I.; Tocci, L. Variable geometry turbine design for off-highway vehicle organic Rankine cycle waste heat recovery. In Proceedings of the THIESEL 2016, Valencia, Spain, 13–16 September 2016.

MDPI
St. Alban-Anlage 66
4052 Basel
Switzerland
Tel. +41 61 683 77 34
Fax +41 61 302 89 18
www.mdpi.com

MDPI Books Editorial Office
E-mail: books@mdpi.com
www.mdpi.com/applsci